Muon Spin Rotation Spectroscopy
Principles and Applications in Solid State Physics

A Schenck

Institut für Mittelenergiephysik,
ETH, Zürich

Adam Hilger Ltd, Bristol and Boston

© Adam Hilger Ltd 1985

All rights reserved. No part of this publication may be produced, stored in a retrieval system or transmitted in any form or by any means, electronic, mechanical, photocopying, recording or otherwise, without the prior permission of the publisher.

British Library Cataloguing in Publication Data
Schenck, A.
  Muon spin rotation spectroscopy
  Principles and applications in solid state physics.
  1. Muon spin rotation   2. Spectrum analysis
  I. Title
  539.7'2114   QC 454.M/

ISBN 0-85274-551-6

Consultant Editor: **Professor E W J Mitchell**, Clarendon Laboratory, Oxford

Published by Adam Hilger Ltd
Techno House, Redcliffe Way, Bristol BS1 6NX, England
PO Box 230, Accord, MA 02018, USA

Typeset by Mid-County Press, London
and printed in Great Britain by J W Arrowsmith Ltd, Bristol.

Contents

Preface		ix
1	**Introduction**	1
2	**The Muon Spin Rotation Technique**	
	2.1 The basic principles	7
	2.1.1 Muon properties and muon decay	7
	2.1.2 Production of spin polarised muon beams	10
	2.1.3 Range and stopping distribution in matter	14
	2.1.4 μ_{SR} signal	15
	2.2 Experimental techniques	20
	2.2.1 Time differential spectroscopy in zero, longitudinal and transverse fields	20
	2.2.2 Integral techniques: stroboscopic and magnetic resonance spectroscopy	27
	2.3 Dipolar relaxation	43
	2.3.1 Transverse fields	43
	2.3.2 Inclusion of quadrupole effects	47
	2.3.3 Zero and longitudinal fields	48
	2.4 Total magnetic field at the μ^+	56
3	**Muon Spin Rotation in Metals: Localisation, Diffusion, Trapping**	
	3.1 Small polaron state of μ^+ in metals—a variation of the hydrogen in metals problem	60
	3.2 Classification of possible motional modes of μ^+ in metals	69
	3.3 Modification of static dipolar relaxation function in presence of diffusion	78
	3.3.1 Motional narrowing in transverse fields	78
	3.3.2 Effect of diffusion on zero field relaxation function	83

	3.3.3	Effect of trapping on transverse field relaxation function	87
	3.3.4	Effect of trapping on zero field ($Kubo$–$Toyabe$) relaxation function	93
3.4	Diffusion and relaxation in presence of dilute magnetic impurities		94
3.5	Coherent diffusion and spin relaxation in BCC ferromagnets		97
3.6	Longitudinal relaxation in magnetically ordered systems		100
3.7	Change of precession frequency and amplitude in the presence of trapping		104
3.8	Short review of μ^+ diffusion results in metals and a comparison with hydrogen data		105
	3.8.1	FCC metals: Cu, Al, Au, Ag	106
	3.8.2	BCC metals: Nb, Fe	118

4 Hyperfine Fields at the μ^+ in Metals

4.1	Hyperfine fields, Knight shift and their relation to local electronic structure		128
	4.1.1	Introduction	128
	4.1.2	Isotropic Knight shift	129
	4.1.3	Anisotropic Knight shift	133
	4.1.4	Diamagnetic screening	135
	4.1.5	Total Knight shift	135
4.2	Short survey on theoretical calculations of spin densities at the μ^+		136
4.3	Experimental results of μ^+ Knight shift investigations		141
	4.3.1	Systematics in simple metals	141
	4.3.2	Temperature dependence in Cd and $\underline{Cd}Mg$ and $\underline{Cd}Hg$ alloys	148
	4.3.3	Axial Knight shift	153
4.4	μ^+ Knight shift in transition metals and alloys/compounds		154
	4.4.1	Group VB metals: V, Nb, Ta	155
	4.4.2	Ni, Pd, Pt	157
	4.4.3	Ce and $Ce_{1-x}Th_x$—valence transition study	157
4.5	μ^+ Knight shift in semi-metals		161
4.6	Spontaneous hyperfine and dipole fields in ferromagnetic and antiferromagnetic materials		162
	4.6.1	Contributions to the local field at a μ^+: the dipole field	162
	4.6.2	Metals and compounds	166

5 Study of Dynamic Effects in Magnetism
5.1 Dynamic effects and critical phenomena   182
    5.1.1 Spin fluctuations of itinerant electrons in MnSi   182
5.2 Spin glasses   184
    5.2.1 μ^+ relaxation function models in zero, longitudinal and transverse fields   184
    5.2.2 Short review of μSR results in $\underline{Cu}Mn$, $\underline{Au}Fe$ and $\underline{Ag}Mn$ 191

6 Applications in Superconductors
6.1 Magnetic effects   203
6.2 μ^+ precession in mixed phase of type II superconductors   207

7 Muonium in Matter
7.1 Magnetic properties in the 1S ground state   215
    7.1.1 Hyperfine structure and Zeeman splitting   215
    7.1.2 Evolution of μ^+ polarisation in muonium ground state   225
    7.1.3 Inclusion of electron relaxation and chemical lifetime   230
    7.1.4 Transitions between muonium states with different hyperfine Hamiltonians   237
    7.1.5 Superhyperfine interaction   240
7.2 Muonium formation   245
7.3 Muonium in quartz and other insulators   251
    7.3.1 Longitudinal field studies   252
    7.3.2 Transverse and zero field studies: hyperfine properties   253
    7.3.3 Zero field studies: relaxation behaviour   260
    7.3.4 Double electron muon magnetic resonance   261
7.4 Muonium in Ge, Si and diamond   264
    7.4.1 Hyperfine properties of muonium states   266
    7.4.2 Temperature and pressure dependence of hyperfine parameters   271
    7.4.3 Determination of electronic g factor   275
    7.4.4 Relaxation   276
    7.4.5 Formation probabilities and transitions among different states   281
    7.4.6 Nature of muonium states   286
7.5 Surface interactions in oxide powders   291

8 Muon Spin Rotation in Insulators
- 8.1 Introduction — 300
- 8.2 αFe_2O_3 and Cr_2O_3 — 301
- 8.3 Orthoferrites — 306
- 8.4 MnO and V_2O_3 — 310
- 8.5 MnF_2 and CoF_2 — 311
- 8.6 $CoCl_2 \cdot 2H_2O$ — 312

Index — 317

Preface

Modern μSR spectroscopy was born in the western world at the Lawrence Radiation Laboratory (LBL) in Berkeley during 1969 and 1970. It emerged as a by-product of a magnetic moment determination of the positive muon with which I had the privilege to be involved. Perhaps it was the spirited intellectual atmosphere in Berkeley, still alive in those days of the people's park and anti-Vietnam war demonstrations, by which one was stimulated to venture beyond the immediate objectives of proposed experiments and to investigate new and sometimes crazy ideas. This was triggered to a great extent by R W Williams' (Seattle) vivid curiosity about the μ^+'s chemical fate in liquids and was further nurtured by K M Crowe's (Berkeley) enthusiasm and willingness to divert beam-time at the venerable 184" cyclotron to these early games with muons. It is with gratitude to both that I remember those days of lively discussion and wild speculation whilst the first experimental results were emerging.

Around the same time, although we were not really aware of it, the groups of I I Gurevich and V G Firsov at Dubna (USSR) had already started a number of ground-breaking experiments. It was not until somewhat later however, that I started to benefit profoundly from their and Ivanter and Smilga's publications. The reader will certainly notice traces of all of their work in this book. μSR started to spread to many more places a few years later, in particular to the so-called meson factories that were just coming into being. There is not enough space here to mention all those who have been instrumental since then in the further development of the field, nor those whose works have contributed so much to this book. I would, however, like to acknowledge my close associates, both past and present: M Camani, F N Gygax, A Hintermann, W Rüegg, H Schilling, W Studer and A J van der Wal, without whose help, encouragement and enthusiasm I would not have been in a position to write such a book.

Finally, I have to thank Mrs Rosa Bächli for diligently typing the manuscript without complaints and the publisher for his patience with my progress in writing the book.

Alexander Schenck
Villigen

March 1985

1
Introduction

The roots of Muon Spin Rotation (μSR) spectroscopy date back to the year 1957, when Garwin, Ledermann, and Weinrich detected that parity conservation was violated in the weak decay chain $\pi^\pm \rightarrow \mu^\pm \rightarrow e^\pm$. This result was published in *Physical Review*, next to the famous work of Wu *et al* (1957) which demonstrated for the first time that parity was not conserved in the weak interactions. The muon emerges from the pion decay with a 100% spin polarisation along its momentum in the rest frame of the pion because of this non-conservation. When these muons are stopped in a target, the spin polarisation manifests itself by an asymmetric distribution of the decay electrons with respect to the spin polarisation vector. In the experiment of Garwin *et al* the spin polarisation vector was precessing in a transversely applied field and detected by the concomitant change of the decay electron rate in a given direction. Knowing the applied field and the average angle of precession during the muon lifetime, a rough determination of the muon's magnetic moment could be achieved simultaneously. In addition, it was found that some of the initial polarisation carried by the incoming positive muons was lost in a nuclear emulsion target but not in graphite, calcium and polyethylene. In contrast, negative muons appeared always to be strongly depolarised. These observations showed that the $\mu^\pm$ must be subject to depolarising processes inside the stopping target which depend sensitively, first on the sign of the muon charge, and secondly on the target material used. It was immediately obvious that a new probe for the investigation of solid state properties had been found, and this led Garwin *et al* to make the statement: '...it seems possible that polarised positive and negative muons will become a *powerful tool* for exploring *magnetic fields* in *solids* and nuclei'.

Shortly after, Friedman and Telegdi (1957) proposed that the partial loss of polarisation of positive muons could be due to the formation of the atom muonium (μ^+e^-). This would be a completely analogous system to atomic hydrogen, in which the μ^+ would be subject to a strong hyperfine interaction with the electron. So, besides the bare μ^+, there emerged the prospect of another probe, namely the atom muonium with possibilities in chemistry as well. However, it took more than a decade before a systematic and well aimed use of the positive muon in solid state physics and chemistry started (this formative stage is well accounted for in Schenck and Crowe (1974) and Brewer *et al* (1975)) and roughly another decade before the application of the positive muon reached a mature and beneficial state. The use of negative muons, however, is still in its infancy for reasons which will be mentioned later. This slow development of the muon technique (now carrying the abbreviation μSR for Muon Spin Rotation, Relaxation, Resonance, Research, etc) may be contrasted with the development of some other well known microscopic techniques with which it can be compared. Nuclear magnetic resonance (NMR) was invented by Purcell *et al* and Bloch *et al* independently in 1946. By the time of the Garwin *et al* experiment it was already a flourishing and widely accepted method. The Mössbauer effect was detected in 1958 and immediately gained wide recognition among researchers. The first application of perturbed angular correlation measurements to solid state investigations took place in the early sixties, and soon after became an established technique in the field of hyperfine interactions. This is illustrated by the series of international conferences on hyperfine interactions, the first one of which was held at Asilomar in 1967 (Matthias and Shirley 1968). Neutron diffraction and scattering were first applied to condensed matter physics around 1950 and ever since have played a growing and indispensable role. Finally positron annihilation in condensed matter physics dates back to the mid fifties. The first international conference on this subject was held in 1965 at Wayne State University, Detroit (Stuart and Roellig 1967). In contrast, the first international conference on μSR was not held until 1978 in Rorschach (Gygax *et al* 1979) but has been followed by further conferences in Vancouver 1980 (Brewer and Percival 1981) and Shimoda, Japan, 1983 (Yamazaki and Nagamine 1984).

The reason for the relatively slow development of μSR can be traced back to several circumstances. First, compared with all the other probes mentioned, the muon is probably the most exotic particle and the idea to use it as a probe in condensed matter physics may have appeared alien to potentially interested people. Second, obviously one needs to have access to an accelerator laboratory at which polarised muon beams can be produced. Apart from there not being many such laboratories around, the environment there may not encourage a visiting solid state physicist to think of using muons. The experimental hall at a high energy accelerator laboratory has much more in common with a dirty, dusty factory hall than the clean laboratory a solid state physicist or chemist is used to. The situation at a nuclear reactor is, in this respect, much more appealing. Thirdly, the sophistication and scale of the equipment surpassed by far the instrumentation in a solid state or chemistry laboratory. This has all changed now with the introduction of microelectronics and on-line computing in all branches of physics. Finally, the attempts of particle physicists to apply their methods to other fields were mainly concentrated on pionic and muonic x-ray work in the so-called field of mesic chemistry (see for example Schneuvely 1977). These attempts proved to be largely inconclusive† and were never able to attract the permanent attention of chemists. Depolarisation studies with negative muons (e.g. Evseev 1975) have suffered a similar fate.

The advent of the so-called meson factories plus the improvement of existing accelerators and other new facilities, has meant that the availability of polarised muon beams has increased considerably in recent years. In addition, μSR experiments in the recent past have proved themselves able to provide relevant, new, unique, and sometimes novel information in the realm of condensed matter physics and physical chemistry. Consequently, the majority of μSR experiments are nowadays carried out by solid state physicists and only a minority of researchers have started their career in elementary particle or nuclear physics. As a result of these more recent developments, muon spin rotation spectroscopy is now an accepted technique and can be collected under the title 'nuclear methods in solid state physics', or in short, 'nuclear solid state physics' with the other methods: perturbed angular correlation and nuclear orientation, the Mössbauer effect, neutron diffraction and positron annihilation.

It seems, therefore, the appropriate time to present the method of muon spin rotation spectroscopy to a wider readership in the physics community and, at the same time, provide the experts in the field with a somewhat detailed collection of concepts, applications, models and key results. Hence the aim of this book is twofold: providing a general introduction into the field which is useful to potential users and graduate students, and meeting the need on the side of the researchers in the field for a kind of handbook on μSR spectroscopy. The present treatment is therefore designed to have the features of a textbook and not so much to provide a review of achievements and results. Excellent review articles on specific aspects of μSR have appeared in the literature, and mentioned here are some of the more recent ones

Advances in Muon Spin Rotation   Brewer and Crowe (1978)
Positive Muons as Probes in Ferromagnetic Metals   Denison et al (1979)
Muon Diffusion and Trapping in Solids   Stoneham (1980)
μSR in Semiconductors   Meier (1980)
The Use of Positive Muons in Metal Physics   Karlsson (1982)
The Electronic Structure of Hydrogen in Elemental Metals in the Light of Muon Spin Rotation Investigations   Schenck (1982)
Transport Mechanisms of Light Interstitials in Metals   Richter (1983).

Muon and muonium chemistry has recently found its own broad coverage in a book by Walker (1983). For this reason the widespread and increasingly specialised field of chemical applications of the positive muon is not contained

---

† An exception in this respect is the field of mesomolecular processes (e.g. Gershtein and Ponomarev 1975).

in the present text. Nevertheless, chapter 7 on muonium can serve as a basic introduction to the field of muonium chemistry.

We have also not included μ^-SR in the present text, since its application has been of a limited scope so far. This is related to the fact that negative muons are captured into atomic orbitals of the target atoms. Following a series of radiationless and radiative transitions, the μ^- will end up in a very short time in the atomic 1s ground state. It reaches this with a much reduced polarisation, because of spin–orbit coupling effects during the cascade. At best the polarisation is 'only' reduced to 1/6 of its initial value, but it may well be less than that. Once in the ground state its lifetime will be shortened by nuclear capture. The capture probability rises with the charge Z of the nucleus, approximately as Z^4 (Primakoff 1959). In high Z elements, an effective lifetime of about 100 ns is observed, but already in Na ($Z=11$) the effective lifetime is reduced to half the free lifetime (2.2 μs) (e.g. Weissenberg 1967). Both the reduced polarisation and the shortened lifetime (and the fraction of free decays accordingly reduced) render the observation of the μ^-SR signal much more difficult than the μ^+SR signal. Moreover, in all the elements with nuclear spin (except for the lighter ones), hyperfine effects will further reduce the polarisation and this will in practice inhibit the application of the μSR technique. In view of all these difficulties, it can be expected that μ^-SR spectroscopy will also be of limited interest in the future. However this does not imply that μ^-SR spectroscopy, in certain selected cases, will not be a unique and beneficial tool. For the current status of μ^-SR spectroscopy, see the contributions to the Yamada conference (Yamazaki and Nagamine 1984). Chapter 2 discusses the principles of μSR for both negative and positive muons.

So far the positive muon has found its widest application in various areas of metal physics. Its particular usefulness is related to the fact that it can be considered as a light isotope of the proton, allowing us to study phenomena in the field of hydrogen in metals, extended to a considerably smaller isotopic mass ($m_p/m_\mu \simeq 8.6$). In this respect, the diffusion and trapping of positive muons in metals has attracted particular attention and the discovery of anomalous features in the μ^+ diffusion at low temperatures has led to considerable theoretical efforts in the field of quantum diffusion of light interstitials. This will be studied in chapter 3. Electronic properties associated with hydrogen in metals are studied by hyperfine field and Knight shift measurements with positive muons. These relate to studies in ferromagnetic and antiferromagnetic systems as well as in simple non-transition and transition metals. Apart from information on the local electronic and magnetic properties in the vicinity of the μ^+, bulk properties also come into focus as do the problems currently of interest, such as valence transitions and fluctuations which have become the subject of μ^+SR investigations very recently. All this is the subject of chapter 4. Chapter 5 deals with the application of μ^+ to dynamical spin phenomena. This includes the study of critical phenomena near magnetic phase transitions, and in particular the investigation of spin glasses. The latter is probably the area to which μ^+SR has contributed most importantly. The nature of the transition into and the dynamics in the spin glass phase have in particular been the theme of μ^+SR experiments. The application of μ^+SR to superconducting systems is treated in chapter 6. Although only a few μ^+SR experiments have addressed phenomena in superconductivity, this area may be of considerable importance in the future with the anticipated extension of studies into the 10 mK temperature range.

Quite a different area of μ^+SR spectroscopy is looked at in chapter 7. It concerns the observation of muonium and muonium-like states in insulators, semiconductors and even on the surface of fine powder grains. A considerable fraction of all μ^+SR research activity takes place in this area, which has much in common with the field of point defects and colour centres as studied by electron spin resonance (ESR) and related methods. Since muonium is like atomic hydrogen, its chemistry in the solid state is of broader interest. Finally in chapter 8 the investigation of ferro- and antiferromagnetic insulators by μ^+SR is dealt with, allowing the study of a variety of interesting phenomena.

The introduction to the μSR method, its basic principles and the variety of techniques possible is given in chapter 2. We discuss there the conventional time differential, as well as integral and magnetic resonance techniques. A separate section is devoted to dipolar relaxation in zero, transverse and longitudinal applied fields.

Looking back over the records of μ^+SR spectroscopy, it must be said that μ^+SR has not only contributed to the solution of well recognised problems in condensed matter physics, but—the μ^+ being a new probe—has very often brought forward new phenomena, and hence new problems, calling for new solutions. This feature has given rise to criticism and it has been said that μSR spectroscopy deals mainly with self-generated problems. Such a view is certainly short sighted. By implanting a positive muon, say in a metal for example, a perturbation is introduced and what is really studied is the *response* of the many body host system to this perturbation. In life, as always, the real nature of things is often best revealed by exposing them to some stress, or quite literally giving them a kick. Our understanding of solid state properties would be incomplete if we did not understand the response to these perturbations. μ^+SR spectroscopy, therefore, provides the challenging opportunity to widen the horizon of our awareness and thus deepen our understanding of nature. In this sense the author hopes that the reader will enjoy this book as much as he himself has enjoyed practising μSR and seeing the development of this field.

References

Bloch F, Hansen W W and Packard 1946 *Phys. Rev.* **69** 127
Brewer J H and Crowe K M 1978 *Ann. Rev. Nucl. Part. Sci.* **28** 239

Brewer J H, Crowe K M, Gygax F N and Schenck A 1975 *Muon Physics* vol. III, ed. V. W Hughes and C S Wu (New York: Academic) pp 3—139

Brewer J H and Percival P W (ed.) 1981 *Proc. Second Int. Topical Meeting μSR Vancouver 1980* in *Hyperfine Interactions* **8** 307–853

Denison A B, Graf H, Kündig W and Meier P F 1979 *Helv. Phys. Acta* **52** 460

Evseev V S 1975 *Muon Physics* vol. III, ed. V W Hughes and C S Wu (New York: Academic) pp 235–98

Friedman J I and Telegdi V L 1957 *Phys. Rev.* **106** 1290

Garwin R L, Ledermann L M and Weinrich M 1957 *Phys. Rev.* **105** 1415

Gershtein S S and Ponomarev L I 1975 *Muon Physics* vol. III, ed V W Hughes and C S Wu (New York: Academic) pp 191–233

Gygax F N, Kündig W and Meier P F (ed.) 1979 *Proc. First Int. Topical Meeting μSR Rorschach, Switzerland 1978* in *Hyperfine Interactions* **6** 1–450

Karlsson E 1982 *Phys. Rep.* **82** 271

Matthias E and Shirley D (ed.) 1968 *Hyperfine Structure and Nuclear Radiation* (Amsterdam: North Holland)

Meier P F 1980 *Exotic Atoms '79* ed. K M Crowe, J Duclos, G Fiorentini and G Torelli (New York: Plenum) pp 331–78

Mössbauer R L 1958 *Z. Phys.* **151** 124

Primakoff H 1959 *Rev. Mod. Phys.* **31** 802

Purcell E M, Torrey H C and Pound R V 1946 *Phys. Rev.* **69** 37

Richter D 1983 *Neutron Scattering and Muon Spin Rotation, Springer Tracts in Modern Physics*, vol. 101 (Berlin: Springer)

Schenck A 1982 *Helv. Phys. Acta* **54** 471

Schenck A and Crowe K M 1974 *Proc. Topical Meeting Intermediate Energy Physics, Zuoz, Switzerland 1973* CERN rep. 74–8

Schneuvely H 1977 *Proc. First Course Int. School of Physics of Exotic Atoms, Erice, Italy* ed. G Fiorentini and G Torelli pp 254–354

Stoneham A M *Exotic Atoms '79* ed. K M Crowe, J. Duclos, G. Fiorentini and G Torelli (New York: Plenum)

Stuart T and Roellig L O (ed.) 1967 *Positron Annihilation* (New York: Academic)

Walker D C 1983 *Muon and Muonium Chemistry* (Cambridge: Cambridge University Press)

Weissenberg A O 1967 *Muons* (Amsterdam: North Holland)

Wu C S, Ambler E, Hayward R W, Hoppes D D and Hudson R P 1957 *Phys. Rev.* **105** 1413

Yamazaki T and Nagamine K (ed.) 1984 *Proc. Yamada Conf. Muon Spin Rotation Shimoda, Japan, 1983* in *Hyperfine Interactions* **17–19** 1–1034

2

The Muon Spin Rotation Technique

2.1 The Basic Principles

2.1.1 Muon properties and muon decay

Muons or μ mesons are weakly and electromagnetically interacting particles which may be viewed as heavy electrons. Like electrons or positrons, muons exist as positively or negatively charged particles with a point-like structure. Positive and negative muons are each others' antiparticles. Together with the muonic neutrinos and the strange and charmed quarks they form the second generation of the basic building blocks of nature. Electrons (e^-, e^+), muons (μ^-, μ^+), and their accompanying neutrinos (v_e, $\bar{v}_e$, v_μ, $\bar{v}_\mu$) and heavier relatives make up the family of leptons. The leptons are characterised by a lepton number which is strictly conserved in reactions and decays as far as we know.

It has been experimentally found (Goldhaber *et al* 1957, 1958) that neutrinos are left-handed particles only while antineutrinos are right-handed only (i.e. their spin is parallel or antiparallel, respectively, to their momentum; helicity $H = \pm 1$). These facts correspond with the V–A theory of the weak interaction and provide another illustration of parity violation in such interactions. It is also this property which makes μSR possible as we shall see.

Some other properties of muons are listed in table 2.1. Note that magnetic moment and spin are parallel for the μ^+ and antiparallel for the μ^- (as for positrons and electrons). The mass of the muon is about 1/9 the mass of the proton, while it is about 200 times heavier than the electron. In this respect the μ^+ may be better compared with protons than with positrons.

The muon of either sign is an unstable particle and decays with a mean lifetime of about 2 μs as follows

$$\mu^+ \to e^+ + v_e + \bar{v}_\mu$$
$$\mu^- \to e^- + \bar{v}_e + v_\mu. \tag{2.1}$$

We see that muon decay is a three-body decay. The kinetic energy of the emerging positron or electron may therefore vary continuously between zero

Table 2.1 Some properties of the muon where m_e is the electron mass, m_p the proton mass and μ_p the proton magnetic moment.

Property	Values
Mass (m_μ)	$206.76835(11) \times m_e$
	$= 0.1126096 \times m_p$
	$= 105.6595$ MeV c^{-2}
Charge	$+e, -e$
Spin (I)	$\frac{1}{2}\hbar$
Magnetic moment (μ_μ) (in units of μ_p)	3.1833452(10)
Gyromagnetic ratio ($\gamma_\mu/2\pi$)	13.55342 (± 0.51 ppm) kHz G^{-1}
g factor (g_μ)	2.002331848(17)
Direction of μ_μ	$\pm |\gamma_\mu| I$ ($+:\mu^+, -:\mu^-$)
Lifetime (τ_μ)	2.19714(7) μs

(the antiparallel travelling neutrinos carry away all the available kinetic energy) and

$$E_{max} = \tfrac{1}{2}m_\mu c^2[1+(m_e/m_\mu)^2] - m_e c^2 = 52.3 \text{ keV}$$

(neutrinos travel together and antiparallel to the electron/positron).

The quantitative treatment of the μ decay is based on the weak interaction Hamiltonian which is of the form of a current–current interaction

$$\mathcal{H} = (G/2^{1/2}) j_W \cdot j_W^* \qquad (G/2^{1/2}) = 1.01 \times 10^{-5} m_p^{-2}. \qquad (2.2)$$

The current j_W in the pure V–A picture is given by

$$j_W = \bar{u}_\mu \gamma_\mu (1+\gamma_5) u_{\nu_\mu} + \bar{u}_e \gamma_\mu (1+\gamma_5) u_{\nu_e} \qquad (2.3)$$

where u are spinor wavefunctions and γ_μ are the Dirac matrices. The decay probability of the muon follows the golden rule

$$dW = 2\pi^4 \frac{|\mathcal{H}^2|}{2E_\mu} \frac{d\mathbf{k}}{2(2\pi)^3 E} \frac{d\mathbf{q}_1}{2(2\pi)^3 W_1} \frac{d\mathbf{q}_2}{2(2\pi)^3 W_2} \delta(k+q_1+q_2-p) \qquad (2.4)$$

where $k = (E, \mathbf{k})$ is the momentum of the electron/positron, $q_{1,2} = (W_{1,2}, \mathbf{q}_{1,2})$ are the momenta of the neutrinos and $p = (E_\mu, \mathbf{p})$ is the momentum of the muon (all in relativistic notation).

We are here only interested in the probability with which the electron or positron is emitted in a certain direction with a certain energy. The neutrinos are not observed, and we have to integrate over their momenta. Expressing the electron/positron energy in units of its possible maximum energy $E_{max} \simeq \tfrac{1}{2} m_\mu c^2$ (i.e. $\varepsilon = E/E_{max}$), evaluation of equation (2.4) yields (see Okun 1965)

$$dW^\pm = W^\pm(\theta, \varepsilon)\, d\varepsilon\, d\cos\theta$$
$$= \{G^2 m_\mu^5 (3-2\varepsilon)[1 \mp (1-2\varepsilon)/(3-2\varepsilon)\cos\theta]\varepsilon^2/192\pi^3\}\, d\varepsilon\, d\cos\theta. \quad (2.5)$$

The upper and the lower sign refer to μ^+ and μ^- decay, respectively. The angle θ is the angle between the muon spin $\mathbf{I}$ and the momentum $\mathbf{k}$ of the emerging positron or electron. The asymmetry in the angular distribution can be traced back to the parity violating terms in the interaction Hamiltonian, equation (2.2). By integrating equation (2.5) over both ε and $\cos\theta$, the total decay probability per unit time is derived which is equal to the inverse of the muon lifetime τ_μ

$$W = 1/\tau_\mu = G^2 m_\mu^5/192\pi^3. \quad (2.6)$$

The energy spectrum of the positrons or electrons follows from equation (2.5) by integrating over $\cos\theta$ alone

$$dW(\varepsilon) = W(\varepsilon)\, d\varepsilon = [2(3-2\varepsilon)\varepsilon^2/\tau_\mu]\, d\varepsilon. \quad (2.7)$$

The energy spectrum is shown in figure 2.1. The asymmetry factor in front of the cosine in equation (2.5) depends on the electron/positron energy

$$A^\pm = \pm(2\varepsilon - 1)/(3 - 2\varepsilon). \quad (2.8)$$

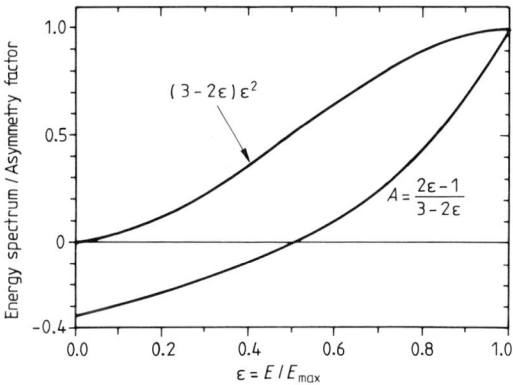

Figure 2.1 The energy spectrum of positrons/electrons from $\mu^\pm$ decay. Also shown is the energy dependence of the asymmetry factor, A.

The asymmetry factor becomes unity for $\varepsilon = 1$ which implies that no positrons with maximum energy are emitted antiparallel to the μ^+ spin (no electrons parallel to the μ^- spin). The asymmetry as a function of ε is also displayed in figure 2.1. Figure 2.2 displays the angular dependence of the decay probability for various asymmetries. It should be noted that the decay probability is

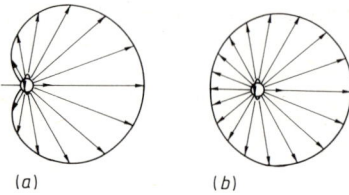

Figure 2.2 Angular distribution of positrons from the μ^+ decay, (a) with maximum energy $\frac{1}{2}m_\mu c^2$, and (b) integrated over all energies.

axially symmetric with respect to the muon spin. The energy average of the asymmetry factor is calculated from equation (2.5) to be $\bar{A}^\pm = \pm\frac{1}{3}$.

2.1.2 Production of spin polarised muon beams

The asymmetry in the angular distribution of positrons/electrons from the muon decay allows, in principle, the muon spin direction, or even its evolution in time to be determined after implantation into a suitable target. For that purpose one needs spin polarised muon beams, so that all muons enter the target with the same initial spin orientation.

Muons are obtained from pions which decay weakly according to

$$\pi^+ \to \mu^+ + \nu_\mu$$
$$\pi^- \to \mu^- + \bar{\nu}_\mu \qquad (2.9)$$

with an average lifetime of $\tau_\pi = 26$ ns. Pions can be produced by protons via the elementary reactions

$$p + p \to \pi^+ + p + n$$
$$p + n \to \pi^+ + n + n \qquad (2.10)$$
$$\to \pi^- + p + p$$

by directing a proton beam of sufficient energy (500–800 MeV) on a production target of suitable material (e.g. C, Be). The threshold energy for producing pions is about 180 MeV. At higher energies in the GeV region, multiple pion production also becomes possible. Proton beams of the desired properties for producing large intensities of pions are obtained from accelerators, a list of which is presented in table 2.2. In addition, π^+, π^- can also be produced with electron accelerators.

There are now two ways to produce spin polarised muon beams. In the first instance one makes use of very low energy pions which have already stopped near the surface of the primary production target. In the case of negative pions, these would be captured almost immediately by the target nuclei and the free decay into muons and neutrinos would not be observed. However positive

Table 2.2 List of proton accelerators with μSR facilities (in existence or planned) 1983†.

Name	Country	Maximum p energy (MeV)	Average p flux (μA)	Mode	Pulse structure burst width (ns) × rep. rate	Typical μ⁺ fluxes in x cm² area (s⁻¹)
SIN	Switzerland	580	100–140	DC	$1 \times (5 \times 10^7)$	10^5–10^8 in 30
LAMPF	USA	800	500–600	pulsed	$(7 \times 10^5) \times 120$	10^5–10^8 in 30
TRIUMF	Canada	520	30–100	DC	$5 \times (2.3 \times 10^7)$	$\sim 10^6$ in 100
CERN	Switzerland	600		DC		10^4 in 10
KEK/ BOOM	Japan	500	1–2	pulsed	50×20	2×10^5 in 50–100
BNL	USA	28 000	1	pulsed	1 s every 2.5 s or 5 ns every 1.5–2.5 s	$\sim 10^4$
Rutherford	UK	800	200	pulsed	two 100 ns 250 ns apart × 50	$\sim 10^7$

† Other μSR facilities exist at Dubna (USSR), Gatchina (Leningrad), NIKHEF (Amsterdam).

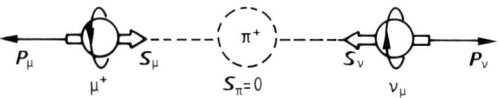

Figure 2.3 Illustration of the collinear decay of the π^+ into a μ^+ and a ν_μ in the rest frame of the pion.

pions now at rest remain somewhere between the target constituents and decay according to equation (2.9) (see figure 2.3). Positive muons emerging from the π^+ decay with the pions at rest are 100% spin polarised with respect to their momentum or flight path. This is easily seen as follows. First, the pion decay is a two-body decay which implies that μ^+ and ν_μ are collinearly emitted (see figure 2.3) with the muon carrying a kinetic energy of 4.1 MeV or a momentum of 29 MeV/c. Secondly, the spin of the pion is zero. The conservation of angular momentum requires that μ^+ and ν_μ together are also in a state of zero angular momentum. Since the helicity of the ν_μ is fixed at $H = -1$, i.e. the spin of the ν_μ is antiparallel to its momentum, the spin of the emerging μ^+ must also be antiparallel to its momentum, as is evident from figure 2.3. If muons of a certain momentum direction are collected into a beam, this beam will show an almost 100% spin polarisation. This type of beam is called a surface μ^+ beam and was first proposed and put into operation at LBL, Berkeley by Pfifer et al (1976) from the University of Arizona. Hence this type of beam is sometimes also called an Arizona beam. Besides possessing a high spin polarisation, this beam has the other property of being a very low momentum beam which implies a very short range in matter of approximately 170 mg cm^{-2} (see also §2.2.3).

The second type of muon beam is formed from pions decaying in flight. A modern layout of such a beam line is shown in figure 2.4. The pions emerging from the production target are collected over a certain solid angle by a quadrupole magnet. They are directed on to an 8 m decay section which consists of a long superconducting solenoid producing a field of 5 T. The primary purpose of this field is to confine the emerging muons now moving in helical trajectories towards the end of the decay section. If the momentum of the pions is sufficiently small (selected by the bending magnet before the solenoid) a large fraction of the pions will have decayed before they reach the end of the solenoid. The exit of the solenoid may be viewed as a diffuse source of muons. Further quadrupole magnets and a bending magnet for momentum selection collect and transport the muons on to a target where they may be stopped and used for further studies.

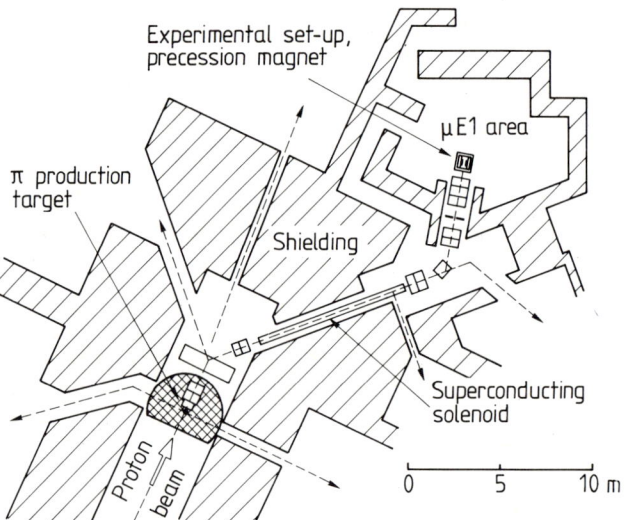

Figure 2.4 Pion–muon beam line μE1 at SIN.

As indicated, the first bending magnet serves to select the pion momentum p_π, and the second bending magnet the muon momentum p_μ. Given a certain pion momentum p_π, the distribution of muon momenta p_μ in the laboratory frame follows from relativistic kinematics (see e.g. Hagedorn 1963). Firstly the energy spectrum of the muons in the laboratory frame, $S(E)$, is constant and extends between a minimum energy E_{min} and a maximum energy E_{max} given by

$$E_{\substack{max\\min}} = \gamma E'_\mu(1 \pm v_\pi v'_\mu c^2). \tag{2.11}$$

E'_μ is the total energy of the muon in the rest frame of the pion (i.e. $E'_\mu = 109.8$ MeV), v'_μ is the velocity of the muon in the rest frame (i.e. $v'_\mu/c = 0.27$) and v_π is the velocity of the pion. Figure 2.5 displays the relation between p_π and the backward and forward momenta $p_{\mu,\text{min}}$ and $p_{\mu,\text{max}}$.

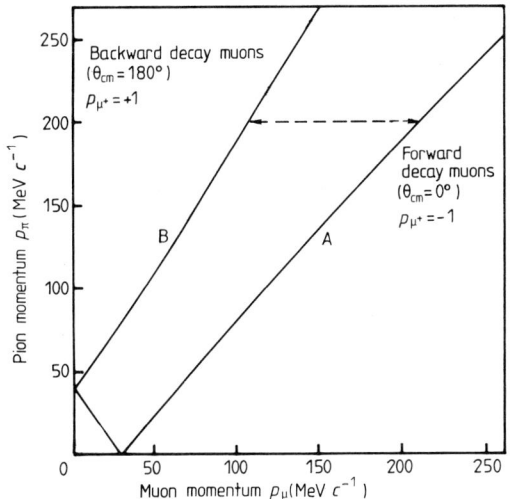

Figure 2.5 Relation between pion momentum p_π and the muon momenta p_μ for muons emitted: parallel (forward) to p_π (A); antiparallel (backward) to p_π (B).

The angular distribution of the muons with respect to p_π in the laboratory frame follows from

$$\tan\theta = \frac{\sin\theta'}{\gamma[(v_\pi/v'_\mu) + \cos\theta']}. \tag{2.12}$$

Here θ is the angle between p_π and p_μ in the laboratory frame and θ' is the angle between p_π and p'_μ in the rest frame of the pion. Since the pion decay is isotropic, θ' varies between 0 and 360° with equal probability. For $v_\pi > v'_\mu$, which is usually valid for a conventional decay beam line with $p_\pi \simeq 200$ MeV/c, equation (2.12) shows that the muons are emitted into a forward cone along p_π with a maximum opening angle of

$$\cot\tfrac{1}{2}\theta_{\text{max}} = \gamma[(v'_\mu/v_\pi)^2 - 1]^{1/2}. \tag{2.13}$$

Of prime interest is now the spin polarisation P_μ of muons with various p_μ in the laboratory frame

$$P_{\mu\pm} = S \cdot p_\mu/|S| |p_\mu| \qquad \text{with } S = \pm p'_\mu. \qquad (2.14)$$

Two situations can immediately be evaluated.

(i) p'_μ parallel p_π. This occurs when p_μ assumes its maximum possible value, corresponding to $\theta' = 0$. In this case the muon momentum in the rest frame of the pion, p'_μ is parallel to the muon momentum p_μ in the laboratory frame. Hence the muon spin polarisation along p_μ is the same as in the rest frame, i.e. $\mp 100\%$ for μ^+ or μ^-, respectively.

(ii) p'_μ antiparallel p_π. This situation results when $\theta' = 180°$ and p_μ assumes its minimum possible value in the laboratory frame. We now find that p_μ and p'_μ are antiparallel, though p_μ is still parallel to p_π. Hence the muon spin polarisation in the laboratory frame along p_π is reversed with respect to the first situation, that is we now have $P_\mu = \pm 100\%$ for μ^+ and μ^-, respectively.

For intermediate muon momenta in the laboratory frame one has to resort to the general expression

$$P_{\mu\pm} = \pm p'_\mu \cdot p_\mu/|p'_\mu| \cdot |p_\mu| = \pm \cos \eta \qquad (2.15)$$

(where η is the so called Wigner angle) which shows that $|P_\mu|$ becomes zero for p'_μ perpendicular to p_μ.

By selecting the proper muon momentum (with the second bending magnet in figure 2.4) it is possible to change the spin polarisation of the muon beam at will, and even to reverse its sign. Of course, one would like to have a beam with the largest spin polarisation possible, which implies that one either makes use of the so-called forward emitted muons ($p_\mu \simeq p_{\mu,\max}$) or of the backward emitted ones ($p_\mu \simeq p_{\mu,\min}$). Since the beam line optics only allow for a limited momentum resolution, muons of a certain spread in momentum below $p_{\mu,\max}$ and above $p_{\mu,\min}$ will be transported by the beam line, and this will result in a reduction of the attainable polarisation. There is also a further reduction stemming from the concomitant spread in angles, θ, around zero degrees, introducing a projection factor of $\cos \theta$. The overall kinematic depolarisation may reduce the actual spin polarisation of a decay muon beam to $P_\mu \simeq 60\text{--}80\%$.

2.1.3 Range and stopping distribution in matter

In a μSR experiment, muons have to be brought to rest in the target or sample being investigated. This implies that one has to pay attention to the range in matter of muons with a particular momentum and to the width of their stopping distribution.

Range tables can be found in the annual review of particle properties by the particle data group (e.g. in 1984 *Rev. Mod. Phys.* **56**, No 2, part II s 54) or in the Lawrence Berkeley Laboratory Report UCRL-2426 (Trower 1966). As an example, figure 2.6 displays the measured range of muons in CH_2. If the range R_μ exceeds the target thickness it becomes necessary to degrade the muon's energy first by introducing additional material into the beam in front of the

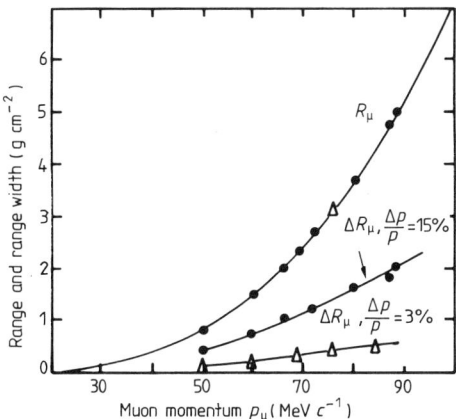

Figure 2.6 Measured range, R_μ, and range width, ΔR_μ, of muons in CH_2 as a function of muon momentum p_μ. The range width has been obtained for two different momentum widths $\Delta p/p$ of the beam. (From the SIN Users handbook 1981.)

target. This ensures that the muons stop in the target rather than just passing through. The amount of degrader thickness needed is usually determined experimentally by measuring the number of stopped muons in the target as a function of degrader thickness.

It is important to consider next the stopping distribution. The width ΔR_μ of the stopping distribution determines the minimum thickness of the target to be investigated so that—after proper adjustment of the degrader—all muons stop in the target. The width of the stopping distribution or range is always finite and is caused, first by range straggling which mostly cannot be avoided, and secondly by the finite momentum resolution $\Delta p_\mu / p_\mu$ of the muon beam. The effect of $\Delta p_\mu / p_\mu$ on the width of the stopping range can be easily seen from figure 2.6, which displays measured data in CH_2 for $\Delta p_\mu / p_\mu = 3\%$ and $\Delta p_\mu / p_\mu = 15\%$. The range straggling, as a rule of thumb, is rather independent of the initial energy and the target material and amounts to 8–9% of the nominal range corresponding to the initial energy, i.e. $\Delta R_\mu / R_\mu \simeq 8\text{–}9\%$. Consulting figure 2.6 it is seen that the attainable range width for $\Delta p_\mu / p_\mu = 3\%$ has already reached the principal lower limit set by range straggling.

If very thin targets have to be investigated, one has to go to lower beam momenta. Using a surface μ^+ beam, targets with a mass per square centimetre of as low as $10\ \mathrm{mg\,cm^{-2}}$ may be studied.

2.1.4 μSR signal

Suppose now that an ensemble of N_0 spin polarised μ^+ or μ^- have been implanted at time $t=0$ in a target placed at the end of a μ beam line. We are now interested in the number of positrons or electrons, dN, emitted at time t

within angle θ (with respect to the initial spin polarisation vector $\boldsymbol{P}_\mu(0) = \langle \boldsymbol{\sigma}_\mu \rangle_{t=0}$ of the incoming muons) and with energy ε in the time bin dt, energy bin dε and solid angle d$\sigma = \sin\theta$ dφ. According to equation (2.5) dN is given by

$$dN = W(\theta, \varepsilon) N(t) \, dt \, d\varepsilon \, d\sigma. \tag{2.16}$$

$N(t)$ is the number of muons still present at time t. This number decreases as (radioactivity decay law)

$$N(t) = N_0 \exp(-t/\tau_\mu). \tag{2.17}$$

Making use of equations (2.5), (2.6) and (2.17) equation (2.16) will become

$$dN^\pm = (N_0/\tau_\mu) \exp(-t/\tau_\mu)[(3-2\varepsilon) \pm P_\mu(0)(2\varepsilon-1)\cos\theta]$$
$$\times \varepsilon^2 \, d\varepsilon \, d\cos\theta \, d\varphi \, dt. \tag{2.18}$$

Note that the actual beam polarisation is taken care of by multiplying the $\cos\theta$ dependent term with $P_\mu(0)$. The number of emitted particles in direction θ is therefore proportional to the projection of $\boldsymbol{P}_\mu(0)$ onto this direction. Integrating equation (2.18) over energy, we arrive at the number of all positrons or electrons emitted at time t in the interval dt and with angle θ into the solid angle dσ

$$dN^\pm(\theta, t) = (N_0/\tau_\mu) \exp(-t/\tau_\mu)\tfrac{1}{2}(1 \pm \tfrac{1}{3}P_\mu(0)\cos\theta) \, dt \, d\cos\theta \, d\varphi. \tag{2.19}$$

To observe this distribution it would be necessary to detect positrons or electrons with an efficiency independent of their energy. In practice, however, the detection efficiency is energy dependent (see also §2.2.1). If the detection efficiency as a function of energy, is given by $D(\varepsilon)$ we would instead have to calculate

$$\frac{dN^\pm(\theta, t)}{N(t) \, dt \, d\sigma} = \int_0^1 W(\theta, \varepsilon) D(\varepsilon) \, d\varepsilon$$
$$= \tau_\mu^{-1} d(1 \pm P_\mu(0) A \cos\theta) \tag{2.20}$$

where

$$d = \int_0^1 (3 - 2\varepsilon)\varepsilon^2 D(\varepsilon) \, d\varepsilon \tag{2.21}$$

$$A = \frac{1}{d}\int (2\varepsilon - 1)\varepsilon^2 D(\varepsilon) \, d\varepsilon. \tag{2.22}$$

A is an average asymmetry and d is a reduction factor less than one. Since a separate determination of $P_\mu(0)$ and A is rather difficult and not necessary in most cases, we will in the following, with a few exceptions, always lump $P_\mu(0)$ and A together by defining

$$\boldsymbol{P} = \boldsymbol{P}_\mu(0) \times A. \tag{2.23}$$

Likewise the factor d will be absorbed into a new normalisation constant

$$N = d \times N_0. \tag{2.24}$$

Hence, generally we have

$$dN^{\pm} = (N/\tau_\mu) \exp(-t/\tau_\mu)(1 \pm |\boldsymbol{P}|\cos\theta)\, dt\, d\sigma. \tag{2.25}$$

Equations (2.16)–(2.25) were derived under the assumption that all muons were stopped at the same time in the target. This situation is actually nearly given at some of the accelerator laboratories where the primary proton beam is pulsed and the secondary beams also still reflect this pulsed structure. The pulse structure must be such that the width of the muon beam bursts are short compared with the muon lifetime and the pulse repetition period long compared with the muon lifetime (see also §2.2.1). If, however, a DC beam is used, muons will enter the target one after the other at more or less random time intervals. If a new time zero is defined with the arrival of each muon, equations (2.15)–(2.25) still hold.

Equation (2.25) is the basic equation on which the μSR technique rests. This allows the polarisation vector $\boldsymbol{P}$ to be determined by monitoring the angular distribution of positrons or electrons. If this distribution is also monitored in time for direction $\boldsymbol{r}_0(|\boldsymbol{r}_0| = 1)$, any time dependence of $\boldsymbol{P}$ inside the target will likewise become observable

$$dN^{\pm}(\boldsymbol{r}_0, t) = (N/\tau_\mu) \exp(-t/\tau_\mu)(1 \pm \boldsymbol{P}(t) \cdot \boldsymbol{r}_0)\, dt\, d\sigma. \tag{2.26}$$

From now on we restrict our discussion to positive muons. It is trivial to adapt the following formulae to negative muons.

A time dependence of $\boldsymbol{P}(t)$ will result if the implanted muons are subjected to magnetic fields either of external (applied) or internal (target) origin. Since the muon possesses a magnetic moment, a torque will result causing the spinning muon to perform Larmor precession. The time evolution of $\boldsymbol{P}(t)$ under the influence of a uniform static magnetic field $\boldsymbol{H}$ can be derived as follows. First we introduce the Hamiltonian operator $\mathcal{H}_Z$, describing the magnetic or Zeeman interaction of the μ^+

$$\mathcal{H}_Z = -\boldsymbol{\mu}_\mu \cdot \boldsymbol{H} = -\gamma_\mu \boldsymbol{I} \cdot \boldsymbol{H}. \tag{2.27}$$

Secondly we need to know the time-dependent spin density operator $\rho(t)$ of the muon which is known to obey the equation

$$-d\rho/dt = (i/\hbar)[\mathcal{H}_Z, \rho]. \tag{2.28}$$

In terms of $\boldsymbol{P}(t)$ ρ can be expressed as

$$\rho(t) = (1 + \boldsymbol{P}(t) \cdot \boldsymbol{\sigma}_\mu) \tag{2.29}$$

where $\boldsymbol{\sigma}_\mu = 2\boldsymbol{I} = (\sigma_x, \sigma_y, \sigma_z)$ and $\sigma_{x,y,z}$ are the Pauli spin matrices. Defining the direction of $\boldsymbol{H}$ as the z axis and inserting equations (2.27) and (2.29) into equation (2.28) we arrive at the following system of differential equations by equating terms of the same component σ_i on the left- and right-hand side of equation (2.28)

$$\dot{P}_x(t) = +\gamma_\mu H_z P_y$$
$$\dot{P}_y(t) = -\gamma_\mu H_z P_x \qquad (2.30)$$
$$\dot{P}_z(t) = 0.$$

This set of equations describes the classical phenomenon of the Larmor precession. Note that the Planck constant $\hbar$ has dropped out of these equations. The solution of equation (2.30) is

$$\boldsymbol{P}(t) = [P_x(0) \cos(\omega t + \varphi), P_y(0) \sin(\omega t + \varphi), P_z(0)]. \qquad (2.31)$$

Here $\omega = \gamma_\mu H$ is the Larmor frequency and $P_x(0) = P(0) \cos \varphi$, $P_y(0) = P(0) \sin \varphi$ and $P_z(0)$ represent the initial μ^+ polarisation at the instant of entering the field H. Equation (2.31) describes the motion of the polarisation vector on the surface of a cone around the field H with an aperture 2ϕ determined by

$$\tan \phi = (P_x^2(0) + P_y^2(0))^{1/2} / P_z(0). \qquad (2.32)$$

In practice two experimental arrangements are usually considered.

(i) The initial polarisation is along the applied field (referred to as the longitudinal field configuration), i.e. $P_x(0) = P_y(0) = 0$, $P_z(0) = |\boldsymbol{P}(0)|$. Inserting equation (2.31) into (2.26) we obtain

$$dN_\parallel(t) = \frac{N}{\tau_\mu} \exp(-t/\tau_\mu)(1 + P_z(0) \cos \varphi) \, dt \, d\sigma \qquad (2.33)$$

where φ is the angle between $\boldsymbol{P}(0)$ and $\boldsymbol{r}_0$.

(ii) The initial polarisation is perpendicular to the applied field (referred to as the transverse field configuration), for example in the x direction: $P_x(0) = |\boldsymbol{P}(0)|$, $P_y(0) = P_z(0) = 0$. The polarisation vector $\boldsymbol{P}(t)$ precesses now in the xy plane. If the direction of the observation, $\boldsymbol{r}_0$, is also confined to the xy plane, as in the conventional transverse field set up, equation (2.26) becomes

$$dN_\perp(t) = \frac{N}{\tau_\mu} \exp(-t/\tau_\mu)(1 + P_x(0) \cos(\omega t + \varphi)) \, dt \, d\sigma \qquad (2.34)$$

where φ is again the angle between $\boldsymbol{P}(0)$ and $\boldsymbol{r}_0$. The Larmor precession will thus lead to an oscillatory behaviour of the positron or electron rate when monitored in a certain direction within the solid angle $d\sigma$.

Both $P_z(0)$ in equation (2.33) and $P_x(0)$ in equation (2.34) may assume a time dependence if depolarisation processes take place inside the target (see §2.3). In that case equations (2.33) and (2.34) should be more properly and completely written as

$$dN_\parallel(t) = \frac{N}{\tau_0} \exp(-t/\tau_\mu)(1 + P_z(0) G_z(t) \cos \varphi) \, dt \, d\sigma \qquad (2.35)$$

$$dN_\perp(t) = \frac{N}{\tau_0} \exp(-t/\tau_\mu)(1 + P_x(0)G_x(t)\cos(\omega t + \varphi)) \, dt \, d\sigma \quad (2.36)$$

where $G_z(t) = G_\parallel(t)$ and $G_x(t) = G_\perp(t)$ are the proper relaxation functions in a longitudinal and transverse field geometry ($G_z(0) = 1$, $G_x(0) = 1$).

For completeness we also list the time dependence of $\boldsymbol{P}(t)$ for arbitrary angles of the uniform field $\boldsymbol{H}$ with respect to $\boldsymbol{P}(0)$ (Schenck 1976):

$$P_x(t) = |\boldsymbol{P}(0)|[\tfrac{1}{2}\sin 2v \cos\phi(G_\parallel(t) - G_\perp(t)\cos\omega t)$$
$$+ G_\perp(t)\sin v \sin\phi \sin\omega t] \quad (2.37)$$

$$P_y(t) = |\boldsymbol{P}(0)|[\tfrac{1}{2}\sin 2v \sin\phi(G_\parallel(t) - G_\perp(t)\cos\omega t)$$
$$- G_\perp(t)\sin v \cos\phi \sin\omega t] \quad (2.38)$$

$$P_z(t) = |\boldsymbol{P}(0)|[G_\parallel(t)\cos^2 v + G_\perp(t)\sin^2 v \cos\omega t]. \quad (2.39)$$

The angles v and ϕ define the direction of $\boldsymbol{H}$ in a coordinate system, the z direction of which is assumed to coincide with $\boldsymbol{P}(0)$ (see figure 2.7).

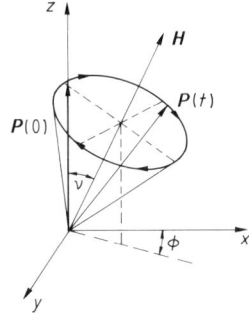

Figure 2.7 Precession of the μ^+ polarisation vector $\boldsymbol{P}(t)$ on the surface of a cone for an arbitrary orentation of the applied field $\boldsymbol{H}$ with respect to $\boldsymbol{P}(0)$. v is the polar angle and ϕ the azimuthal angle of $\boldsymbol{H}$ in the coordinate system x, y, z, the z direction of which is given by $\boldsymbol{P}(0)$.

So far we have only considered the case of one precession frequency ω. However, in general, μ^+ may precess with different frequencies when placed in different environments in the same sample each perhaps associated with a particular relaxation behaviour. These formulae can then be easily extended, depending on the circumstances. Another possibility is the formation of muonium or other paramagnetic compounds containing μ^+ which will lead to a multitude of precession frequencies. Here the evolution of the muon's polarisation has to be considered under the combined action of applied and hyperfine fields (see §7.1.2).

Finally it is instructive to point out that the signals, expressed by equations

(2.35) and (2.36) are also found in other techniques in which the precessional motion of nuclear or electronic spins is observed. For instance equation (2.36) describes the free induction decay signal in nuclear magnetic resonance (NMR) which is observed after application of a so-called 90° radiofrequency (RF) pulse (e.g. Abragam 1975). Equation (2.35) can also be observed in NMR using the spin echo technique, in which it describes the decay of the spin echo amplitude as a function of the delay time between a first 180° RF pulse and a second analysing 90° RF pulse. Analogous expressions are likewise observed in perturbed angular correlation and orientation techniques using nuclear radiation, e.g. $\gamma\gamma$PAC (perturbed angular correlations) (Matthias 1967).

2.2 Experimental Techniques

2.2.1 Time differential spectroscopy in zero, longitudinal and transverse fields

In the preceding section the general expressions for the μSR signal for various field configurations have been derived. Essentially these expressions describe the time dependence of the probability of detecting a positron or electron in a certain direction after a spin polarised $\mu^{\pm}$ has been implanted in a target. To make this time dependence visible, the electron/positron rate in a given direction has to be monitored as a function of elapsed muon lifetime for a large number of stopped muons. But how can this be done?

Figure 2.8 shows a schematic layout of a μSR apparatus. The spin polarised μ beam with $P(0)$ parallel or antiparallel to the beam axis is directed on to a target in which the μ's are brought to a stop. Eventually the implanted μ's will decay and a positron or electron emerges from the target. Various detectors around the target track the incoming μ's and the outgoing positrons or electrons. These detectors are usually plastic scintillators a few mm's thick that can be shaped to any desired form. The light pulse created by a passing charged particle is transmitted via a light pipe (plexiglass) to an electron photomultiplier. The photomultiplier output signal is processed further by electronics, as discussed later. An incoming μ reveals itself by an (almost) simultaneous response of all detectors it is traversing (I, B_2, B_1, M). If the muon is stopped in the target the detectors behind the target (F_1, F_2) will show no response. The logic signature of a stopped muon (μstop) is therefore defined by a coincidence of signals from I, B_2, B_1, M and none from F_1, F_2, in short: $I \cdot B_2 \cdot B_1 \cdot M \cdot \overline{(F_1 + F_2)}$. Likewise an emitted positron or electron in the forward direction will produce coincident signals from F_1 and F_2, and thus is distinguished from a passing through μ^+ by the absence of coincident signals from the dectectors in front of the target. The logic signature could be: $F_1 \cdot F_2 \overline{(I + M)}$. A backward emitted electron or positron could be identified by the logic signature $B_2 \cdot B_1 \overline{(I + F_1 + F_2)}$. Note that the anticoincidence

Experimental techniques   21

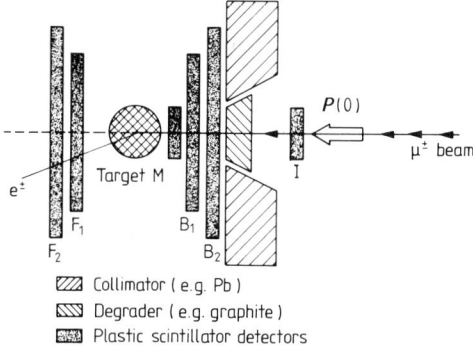

Figure 2.8 Schematic lay out of a μSR apparatus. The incoming $\mu^{\pm}$ are collimated and degraded by a moderator and pass through the plastic scintillator detectors I, B_2, B_1, M, and stop in the target. The outgoing decay $e^{\pm}$ are observed with the scintillator detectors F_1 and F_2, or B_1 and B_2.

or veto signals are combined by logic 'ors' rather than 'ands'. This is to make sure that no veto signal is lost by the inefficiency or different geometries of the corresponding detectors.

To identify stopped muons and emerging positrons or electrons the signals from the various photomultipliers have to be combined by coincidence and anticoincidence circuitry. The actual electronic detection and identification scheme depends now very much on whether a 'DC' or a pulsed muon beam is used. In the following we describe first the principles of the electronics for a DC beam, which show best the essential features of the time differential detection technique. It is also the most commonly used mode of operation (SIN, TRIUMF, CERN, DUBNA). A DC beam here means that muons will arrive at more or less random intervals, the average interval being determined by the total beam rate.

The principal layout of the electronic detection scheme is shown in figure 2.9 and the timing of the various signal pulses in figure 2.10. The signals from the photomultipliers are first transformed to (usually) standard nuclear instrument and measurement (NIM) pulses of a few ns width and then directed on to various coincidence and anticoincidence units. A muon decay event requires that within a certain time interval after a μstop, a positron or electron is detected. This time interval is called the data gate and extends usually over several muon lifetimes (e.g. 10 μs). The data gate is produced by a gate generator upon being triggered by a μstop signal. A good decay positron or electron is then identified by a coincidence of the electron/positron signal with the data gate. Next we want to know the number of such events as a function of the time the muons have spent in the target until their decay. This is facilitated by starting a clock (e.g. a digital clock, or a time-to-amplitude-converter (TAC))

22   The muon spin rotation technique

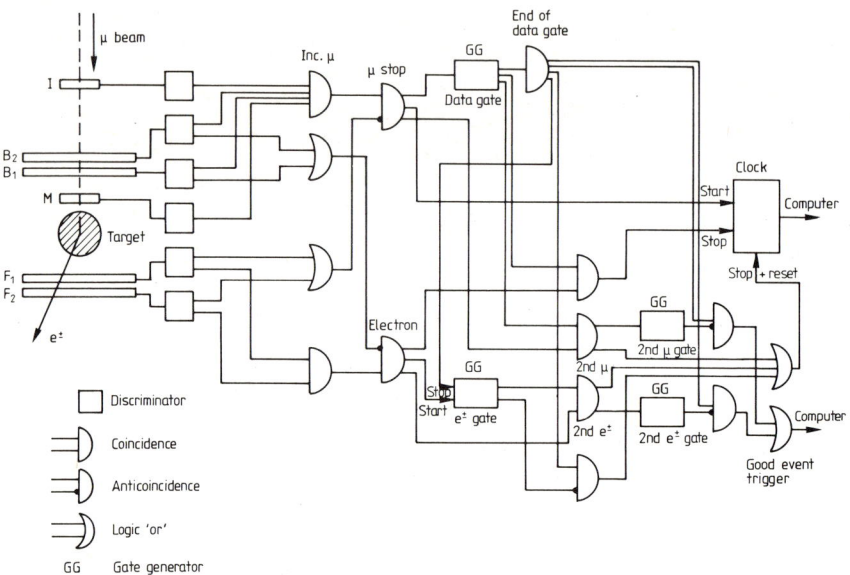

Figure 2.9 Principal components of the electronic detection scheme. The photomultiplier pulses from the detectors are transformed by discriminator units into standard NIM pulses, which are processed further by fast coincidence and anticoincidence units to identify good $\mu^\pm$ stops, good $e^\pm$ decay and to exclude ambiguous events, which involve two stopped $\mu^\pm$ or two decay $e^\pm$ within a certain time interval. For a good event (only one $\mu^\pm$, only one $e^\pm$) the time between the arrival of the $\mu^\pm$ and the appearance of the $e^\pm$ is measured by a clock and subsequently read out to a computer in which the rate-versus-time histogram is formed.

by a μstop signal and stopping the clock by the good decay event signal. The measured time then forms the address of a histogramming memory (e.g. a multichannel analyser or a computer), and the content of the corresponding channel is incremented by one. In this way a rate-versus-time histogram is formed.

In practice the electronics has to check for a number of further details. The most important is the rejection of so-called second muon events (pile up rejection). A second muon event represents a stopped muon which follows a previously stopped first muon within the time interval defined by the data gate (see figure 2.10). There are now possibly two muons in the target and a subsequently emerging positron or electron cannot be related unambiguously to the right parent muon. If such ill-defined events are not excluded from being added to the histogram a reduction in the true μSR signal will result and the uncorrelated (accidental) start–stop background rate is increased. It is important now that not only events of the sequence μstop$_1$–μstop$_2$–positron (electron) are rejected but also those with the sequence μstop$_1$–positron (electron)–μstop$_2$ (all within the data gate set up by the μstop$_1$). Otherwise

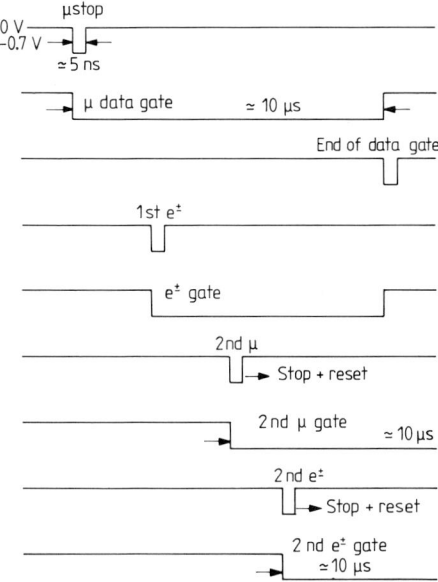

Figure 2.10 Schematic representation of the time sequence of possible signal pulses after a first $\mu^\pm$ stop signal.

there will be another spectrum distortion because the observation of long lived muons will be increasingly suppressed by second muons, while short lived muons are decreasingly less affected by second muons. By rejecting all second muon events regardless of the relative timing of the second muon with respect to the positron or electron, this kind of bias is removed.

Another ill-defined event is one in which a μstop is followed by two positrons or electrons during the same data gate. Again there would be an ambiguity which, if not rejected, would increase the uncorrelated background rate and decrease the true μSR signal. For an in depth discussion of these, and more subtle effects see Garner (1979). If no second muon and second electron signals are detected, a good event signal can be formed which tells the data acquisition system to accept the measured time interval, and update the rate-versus-time histogram.

As an example of such a histogram, figure 2.11(a) shows the μSR spectrum obtained from a Cu sample at low temperatures in a transverse field of 100 G. We notice the exponential decay due to the μ^+ lifetime and superimposed on it the oscillatory pattern due to the μ^+ precession in the 100 G field. Figure 2.11(b) shows the same data with the exponential decay and the non-oscillatory part removed. Only $P_x(t)$ is now displayed, which clearly shows a strong damping.

The rate against time histogram will contain not only the μSR signal (see equation (2.26), but also some uncorrelated background due to accidental

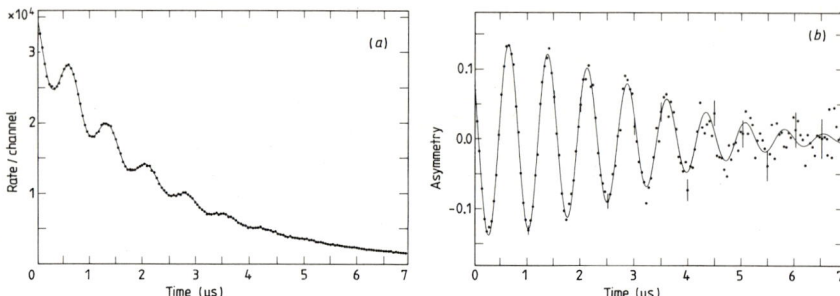

Figure 2.11 (*a*) Example of a transverse field μ^+SR histogram obtained in Cu in a field of 100 G; (*b*) the same histogram but with the exponential decay removed. The signal essentially reflects the polarisation $P_x(t)$.

start–stop events. Such events, as was just mentioned, could consist of uncorrelated stopped muons and decay positrons/electrons, originating from other places than the target as well. In addition, background radiation could simulate μstop and positron/electron signals, totally uncorrelated to each other and true μstop or true positron/electron signals. This will produce a flat background spectrum that is, in principle, time independent.

What are the principal limitations in measuring frequencies and relaxation rates? In order to be able to determine a frequency with some precision, at least one complete oscillation should fall within the data gate interval. Assuming a typical interval of 10 μs the lowest frequency is of the order of 100 KHz, or about 7 G. The highest frequency to be resolved correlates with the intrinsic time resolution of the detection electronics, as well as the time resolution of the clock. As it turns out the former is always the bottleneck. The intrinsic time resolution is determined mainly by the spread in passing times of the light from the scintillators through the light pipes to the photo-tubes, and the transit times inside the photo-tubes. Just a few cm difference in the light path results in a time jitter of the order of 0.1 ns. The intrinsic time resolution can be determined by measuring the time distribution of so-called prompt events, which consist of muons passing through the muon detectors as well as the positron ones. The obtained time distribution usually has a width of the order of 1 ns, thus limiting the highest resolvable frequencies to several hundred MHz. The time resolution can be considerably improved by attaching the scintillators directly to the photo-tube, using specially selected photo-tubes and fast scintillator material (NE 111). In this way Holzschuh *et al* (1981) succeeded in resolving the muonium hyperfine splitting frequency of 4.6 GHz in quartz in zero applied field (see also §7.3.2). The scintillator and photo-tube arrangement is shown in figure 2.12.

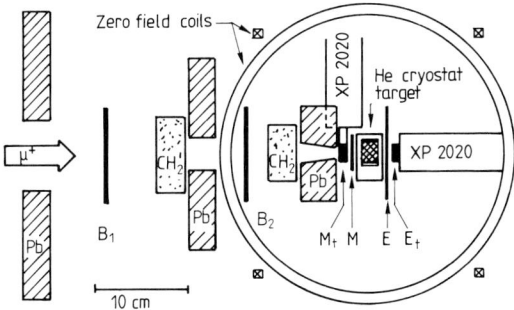

Figure 2.12 Schematic view of the detector (scintillator and photomultiplier) arrangement in the high resolution set up of Holzschuh et al (1981).

The fastest relaxation rates or shortest damping times of the μSR signal that can be observed are limited by the dead time of the electronics following a μstop signal. The dead time is introduced by the veto signal on the positron coincidence derived from the incoming μ, which usually has a width of 10–20 ns. Muons relaxing within times shorter than the dead time will appear totally depolarised. In practice relaxation times should not be much less than 50 ns in order to be reliably tracked down. The smallest relaxation rates accessible by μSR, on the other hand, are limited principally by the $\mu^{\pm}$ lifetime, and practically by the length of data gate and the time independent background due to accidental start–stop events. As a rule of thumb, relaxation times exceeding 50 μs can no longer be reliably determined. Compared with nuclear magnetic resonance (NMR) we find that μSR allows us to extend the range of relaxation rates by 1–3 orders of magnitude. However, for small relaxation rates μSR is no match for NMR.

The necessity to reject second muon events etc has important consequences for the total acceptable rate of stopped muons. If the μstop rate is denoted by I_s and the data gate interval by ΔT, the average rate of second μ events is given by

$$I_{2\mathrm{nd}\mu} = \Delta T I_s^2 \quad \text{or} \quad I_{2\mathrm{nd}\mu}/I_s = \Delta T/I_s. \tag{2.40}$$

If, for example, we use a data gate interval of $T = 10$ μs, and require that $I_{2\mathrm{nd}\mu}/I_s \leq 10\%$, we get, as an upper limit for the total acceptable stop rate, $I_s \leq 10^{-1}/10^{-5} = 10^4$ s^{-1}. In view of the much larger stop rates available at the modern meson factories, this limitation is quite a nuisance. Ways out of this dilemma are the integral methods of observation that will be dealt with in the next section.

The μ stopping rate is of no concern in principal, if the muon beam has a pulsed structure, such that beam bursts are short in comparison to the μ lifetime, and the separation between two bursts is long in comparison with the μ lifetime. Such a condition is given at the μSR facility BOOM at KEK with a burst width of 50 ns and a pulse separation of 50 ms. Here the time measurement is started by the arrival of the muon burst (consisting of up to 10^4 μ^+ at BOOM) and detected by a single plastic scintillator, for example, in front of the target. Positrons are detected individually as before in a telescope. The instantaneous positron rate might be considerable, depending on the solid angle covered by the e^+ telescope. It is therefore advantageous to split the positron telescope into many small ones, each covering a small solid angle, so as to keep each telescope's rate down to a manageable level. To measure the time interval between the arrival of the muon burst and the individual positrons, a clock with a multistop option has to be used (see Nagamine and Yamazaki 1981, 1982). The number of events that can be handled in this fashion is limited by the technical capacity of the clock. In any case the instantaneous rate that can be handled is much larger than in the DC beam technique.

To take unlimited advantage of the high instantaneous μ^+ stop rate, the BOOM μSR group has also developed a so-called macroscopic detection technique, by which the individual positrons are no longer recorded, but the decay positrons (up to ten per burst) are monitored by a large Čerenkov counter. The light output from this counter is proportional to the rate of impinging positrons, and is therefore also represented as a function of time by equation (2.36). Zero time is again defined by the arrival of the muon burst. The Čerenkov light is detected by a photomultiplier, the output of which is further processed by a transient digitiser (Nakayama *et al* 1981). A disadvantage of this method is the inherently poor ability to distinguish between positrons originating from the sample under investigation and other sources in the target's vicinity (e.g. target holders, cryostate walls etc). Such 'analogue' detection techniques may become very important in the future in view of the prospect of even higher beam intensities.

Although pulsed muon beams appear advantageous with respect to the ability to handle high stopping rates, there is an important limitation on the time resolution to be achieved. This limitation is related to the width, Δ, of the individual muon bursts. This introduces an uncertainty of the same magnitude in the time measurement as this measurement is started each time by the first muons in a burst. This has the effect of averaging out any time dependence of the μSR signal which proceeds on a scale comparable or faster than Δ. This leads, in particular to the amplitude of the transverse field μSR signal decreasing (equation (2.36)) with increasing precession frequency. If the distribution of muons in a burst is assumed to be Gaussian with total width T_W, i.e. given by $\exp(-2t^2/T_W)$, then an effective signal amplitude of

$$A_{\text{eff}} = P_x(0) \exp(-T_W\omega)^2/8 \qquad (2.41)$$

is obtained by folding equation (2.36) with the Gaussian distribution (Nagamine and Yamazaki 1982). Figure 2.13 shows the reduction of the

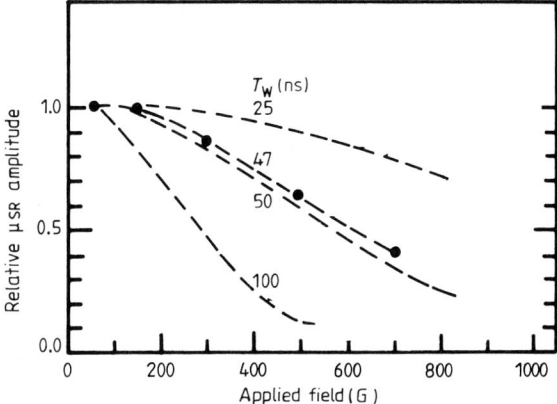

Figure 2.13 Reduction of the effective precession amplitude as a function of applied field for various burst widths T_w of a pulsed muon beam (Nagamine and Yamazaki 1982).

amplitude with rising field (or Larmor frequency) for various values of T_W. It has been realised that muon precession in fields much greater than 800 G cannot be observed for burst widths of 50 ns.

With respect to measuring relaxation rates and, perhaps more importantly, the shape of the relaxation functions, the pulsed beam time differential μSR allows the range of accessible relaxation rates to be extended to much smaller values. This is because of the virtual absence of any background during the off beam period. Relaxation rates below 0.01 μs^{-1} seem to appear observable (Nagamine and Yamazaki 1982).

As already indicated, the time differential signals obtained by these techniques are described by equations (2.35) and (2.36) or the modified, more complicated versions, which now include a time independent background term. The analysis of experimentally obtained histograms will consist of fitting the appropriate equations to the data, and thereby determining the parameters of interest, such as relaxation times, amplitudes, precession frequencies and phases. In the presence of several frequencies a Fourier analysis of the histograms will be very useful. The Fourier technique may also be used to determine relaxation rates from the width of the Fourier peaks. The application of Fourier analysis to transverse field μSR spectra has been discussed in detail by Brewer et al (1982) and Burkhard et al (1984).

2.2.2 Integral techniques: stroboscopic and magnetic resonance spectroscopy

The two methods discussed here do not involve the measurement of time intervals and are therefore free of the various inherent limitations of time differential techniques. These methods inevitably have their own limitations and these restrict their use to selected areas of applications. Common to both methods is that their measurements are made in frequency space rather than time space. Consequently they yield the Fourier transform of the time differential μSR signal as defined in equation (2.36). Both methods require the presence of sufficiently strong magnetic fields in the muon's environment (either of external or internal origin). In the stroboscopic method the field is orientated perpendicular to the initial muon polarisation. Hence what is observed is the muons undergoing Larmor precession, as in the case of the transverse field time differential detection scheme. The magnetic resonance technique, on the other hand, involves a longitudinal magnetic field and is just an extension of the NMR technique to muons. Here the resonance is detected via a change in the angular distribution of the electrons or positrons (NMR is detected by nuclear radiation).

(1) Stroboscopic μSR spectroscopy

A prerequisite for this method is a fast (MHz) periodic intensity modulation of the muon beam. Indeed, at many accelerators the primary proton beam possesses a microscopic time structure that derives from cyclic acceleration by radiofrequency (RF) fields. The beam consists of small bunches, with repetition cycles given by the angular frequency Ω of the accelerator RF. This time structure is repeated in the secondary pion beams, and muon beams formed from pions decaying in flight can still reflect this time structure, although with an increased bunch width. This will occur if the difference between the pion and muon flight time, over the decay section of the beam, is kept well below the bunch repetition period $2\pi/\Omega$. A different situation arises for surface muon beams. Here the pion lifetime τ_π usually determines the shape and width of the muon bunches and, if the repetition cycle Ω^{-1} is not long compared with τ_π, it may completely wash out all time structure.

The principle consists of letting the muons precess with a frequency close to an integral multiple of the frequency Ω with which the intensity of the muon beam is being modulated.

Thus, muons entering the target have almost the same spin phase as those that have already spent some time inside the target. In effect, the muons, independent of their arrival time, will precess more or less coherently. The coherence is strongest when

$$\omega = n\Omega \qquad (n = 1, 2, \ldots)$$

and vanishes for

$$|\omega - n\Omega| > \tau_\mu^{-1}.$$

The degree of coherence is detected by simply measuring the positron rate in a time window (gate) of suitable length and phase, fixed with respect to the beam intensity modulation. This will be called the stroboscopic signal.

Such a method was first used by Christiansen et al (1970) in a γγPAC experiment, and first applied to muons by Camani et al (1978) (see also Schenck 1976). We will now derive the shape of the stroboscopic signal as a function of either the muon precession frequency, ω, or the applied field. A schematic representation of the stroboscopic measurement is given in figure 2.14 (Klempt et al 1982).

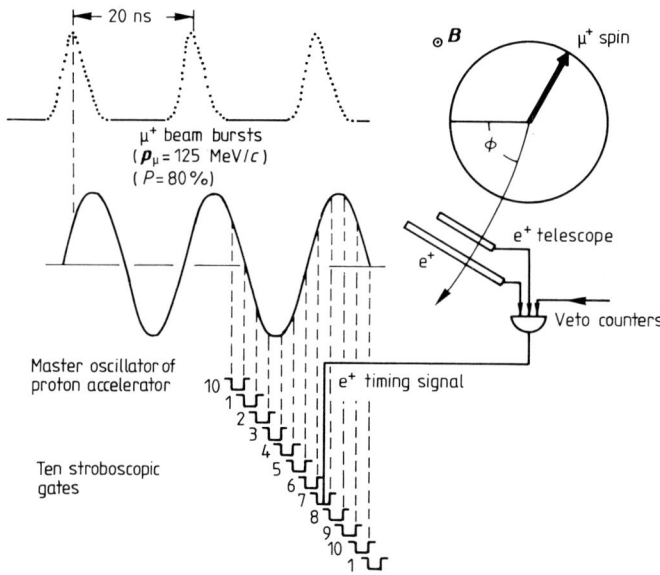

Figure 2.14 Schematic illustration of the stroboscopic technique. See text for details (Klempt et al 1982).

Consider a beam of polarised muons which has been stopped in a magnetic field. The polarisation of the muons is transverse to the magnetic field, and denoted $P_x(0)$. The incoming muon beam exhibits a periodic time structure of frequency $\Omega = 2\pi/T$. Therefore, the muon stopping rate, $B(t)$, is modulated as well and satisfies $B(t) = B(t + nT)$. Positrons emitted in the decay of muons are detected in the positron telescope, centred at the angle ϕ in the plane of precession and covering a finite solid angle. The telescope is gated by the electronically defined gate function $G(t)$, which is periodic in time and phase locked to the master oscillator of frequency Ω. We now need to know the total number of positron counts in the gate $G(t)$ during the time interval (t_i, t_f) as a function of either the magnetic field or the muon spin precession frequency ω. The decay probability of a muon stopped at time t_s in the target, is given by $\tau_\mu^{-1} \exp(-(t-t_s)/\tau_\mu)$. If the decay of polarised muons into direction ϕ' is

30 The muon spin rotation technique

considered, following equation (2.36) the decay probability for $t \geq t_s$ has to be multiplied by

$$1 + P_x(0)G_x(t) \cos[\omega(t-t_s) + \phi']. \qquad (2.42)$$

In order to calculate the number of counts N in the stroboscopic gate $G(t)$, the decay probability has to be integrated over the finite solid angle, σ, covered by the positron telescope, over all beam particles which have entered the target and over the time when the stroboscopic gate $G(t)$ is active during the data taking time interval (t_i, t_f). Therefore

$$N = \int_{t_i}^{t_f} dt\, G(t) \int_{-\infty}^{t} dt_s B(t_s) \int_\sigma D'(\phi', \varphi')\, d\sigma$$
$$\times \tau_\mu^{-1} \exp{-(t-t_s)\tau_\mu}\{1 + P_x(0)G_x(t) \cos[\omega(t-t_s) + \phi']\} \qquad (2.43)$$

$D'(\phi', \varphi')$ is the geometrical detection efficiency in the direction ϕ' (polar angle) and φ' (azimuthal angle). The integration over σ leads to an average geometrical detection efficiency D

$$D = \int_\sigma D'(\phi', \varphi')\, d\sigma. \qquad (2.44)$$

The average decay asymmetry A, as observed by the positron telescope, is defined by

$$AD \cos(\alpha + \phi) = \int_\sigma d\sigma P_x(0) D'(\phi', \varphi') \cos(\alpha + \phi'). \qquad (2.45)$$

Changing the variable t_s to $t' = t - t_s$ leads to

$$N = D \int_0^\infty \tau_\mu^{-1}\, dt'\{\exp(-t'/\tau_\mu)[1 + AG_x(t') \cos(\omega t' + \phi)] \int_{t_i}^{t_f} dt\, B(t-t') G(t)\}. \qquad (2.46)$$

We define

$$S(t') = (t_f - t_i)^{-1} \int_{t_i}^{t_f} B(t-t') G(t)\, dt \qquad (2.47)$$

which we call the stroboscopic structure function. It contains all the details of a certain measurement, like the structural features of a particular beam burst shape and the electronic gate, as well as transient features due to singular distortions. In the limit of large data taking intervals, $t_f - t_i$ transient features are averaged out and only the periodic features of $S(t')$ remain.

The muon beam can be described as a superposition of individual bunches of muons, the so-called muon bursts

$$B(t) = B \sum_{n=-\infty}^{+\infty} b(t - nT - t_B) \qquad (2.48)$$

where $b(t)$ describes the time structure of an individual burst, and t_B defines the time offset of the bursts relative to an arbitrary time synchronisation signal, the latter being periodic with the accelerator frequency $\Omega/2\pi$. If $b(t)$ is normalised by $\int_{-\infty}^{+\infty} b(t)\, dt = 1$, then B is the average number of stopped muons per burst. $N_t = D \times B \times (t_f - t_i)/T$ is the total number of positrons in the time period $(t_f - t_i)$ traversing the detector telescope.

Similarly one can define

$$G(t) = \sum_{-\infty}^{+\infty} g(t - mT - t_G). \qquad (2.49)$$

The gate function $g(t)$ is normalised by $\int_{-\infty}^{+\infty} g(t)\, dt = \lambda T$, where λ is the fractional width of the gate, G.

Clearly $B(t)$ and $G(t)$ are periodic functions of time, and we can express them by Fourier sums

$$B(t) = (B/T) \sum_{m=-\infty}^{\infty} \hat{b}(m\Omega) \exp im\Omega(t - t_B) \qquad (2.50)$$

with

$$\hat{b}(\omega) = \int_{-\infty}^{+\infty} b(t) \exp i\omega t\, dt \qquad (2.51)$$

and

$$G(t) = T^{-1} \sum_{n=-\infty}^{\infty} \hat{g}(n\Omega) \exp in\Omega(t - t_G) \qquad (2.52)$$

with

$$\hat{g}(\omega) = \int_{-\infty}^{+\infty} g(t) \exp i\omega t\, dt. \qquad (2.53)$$

Inserting equations (2.50) and (2.52) into equation (2.47), we find that in the limit of long data taking intervals, $t_f - t_i$, the integration will yield non-vanishing contributions only for those products with $m = -n$, so that

$$S(t') \rightarrow \frac{B}{T} \sum_{n=-\infty}^{\infty} T^{-1} \hat{b}^*(n\Omega) \hat{g}(n\Omega) \exp in(t' + t_B - t_G). \qquad (2.54)$$

We define

$$\hat{s}(\omega) = \hat{b}(\omega)\hat{g}(\omega) \qquad (2.55)$$

which is the Fourier transform of

$$S(t') = \int_{-\infty}^{+\infty} dt\, b(t - t')g(t). \qquad (2.56)$$

Note that $s(t)$ is a real function, so that $\hat{s}(-\omega) = \hat{s}^*(\omega)$.

The stroboscopic signal now takes the form

$$N = N_t \sum_{m=-\infty}^{\infty} T^{-1} \hat{s}(m\Omega) \exp im\Omega \, \Delta t \int dt' \tau_\mu^{-1} \exp -t'/\tau_\mu$$
$$\times [1 + A G_x(t) \cos(\omega t' + \phi)] \exp im\Omega \quad (2.57)$$

with $\Delta t = t_B - t_G$.

For further evaluation, we need to know the functional form of $G_x(t)$, the transverse field relaxation function. Assuming $G_x(t)$ to be given by a simple exponential decay, $\exp(-t/T_2)$ with relaxation time T_2, equation (2.57) can be straightforwardly evaluated as follows

$$N = \frac{N_t}{T} \sum_{m=-\infty}^{\infty} \left[\frac{\text{Re}[\hat{s}(m\Omega) \exp im\Omega \, \Delta t] + m\Omega\tau_\mu \, \text{Im}[\hat{s}(m\Omega) \exp im\Omega \, \Delta t]}{1 + m^2 \Omega^2 \tau_\mu^2} \right.$$

$$\times A \frac{\tau}{\tau_\mu} \left(\frac{\text{Re}[\hat{s}(m\Omega) \exp i(m\Omega \, \Delta t + \phi)]}{1 + (\omega - m\Omega)^2 \tau^2} \right.$$

$$\left. \left. + \frac{(\omega - m\Omega)\tau \, \text{Im}[s(m\Omega) \exp i(m\Omega \, \Delta t + \phi)]}{1 + (\omega - m\Omega)^2 \tau^2} \right) \right] \quad (2.58)$$

$$\simeq \frac{N_t}{T} \left(\text{Re } \hat{s}(0) + \sum_{m=-\infty}^{\infty} A \frac{\tau}{\tau_\mu} |\hat{s}(m\Omega)| \frac{\cos \psi + (\omega - m\Omega)\tau \sin \psi}{1 + (\omega - m\Omega)^2 \tau^2} \right) \quad (2.59)$$

with

$$\tau^{-1} = \tau_\mu^{-1} + T_2^{-1} \quad (2.60)$$

$$\psi = m\Omega \, \Delta t + \phi + \arg \hat{s}(m\Omega). \quad (2.61)$$

We see that the stroboscopic signal consists of three parts. The first one is just a constant, corresponding to the isotropic part in the positron decay distribution. It is dominated by the $m=0$ term as long as $\Omega\tau_\mu \gg 1$, and this was assumed by going from equation (2.58) to (2.59). The two other parts show resonances at $\omega = m\Omega$, $m = 1, 2, \ldots$. The first of the two exhibits a Lorentzian shape, whilst the second a dispersion-like shape. All terms with m negative are negligibly small provided that $(\omega - m\Omega) \gg 1$ which is usually the case. It is interesting to note that these terms lead to a small shift in the true resonance frequency analogous to the Bloch–Siegert shift in RF spectroscopy when using linearly polarised radiation. The presence of these terms in equation (2.59) follows from the fact that the stroboscopic detection arrangement can be viewed as a left and a right turning stroboscope. Another interesting term is the one when $m=0$. Here the resonance condition is fulfilled for $\omega=0$. The corresponding effect in level crossing spectroscopy is known as the Hanle effect. If ω is close to a particular resonance, i.e. $\omega \simeq m'\Omega$, all other terms with $m \neq m'$ can be neglected. We also find that the width of the resonance signal is given solely by the μ lifetime and relaxation time T_2 combined. All the details concerning the beam burst shape, the data gate width and relative position, and the positron telescope direction are absorbed into the phase ψ. This determines the relative weight of the Lorentzian and the dispersion signals for a given m.

The amplitude of the resonance at a particular m is primarily given by $|\hat{s}(m\Omega)|$ or

$$|\hat{s}(m\Omega)| = |\hat{b}^*(m\Omega)\hat{g}(m\Omega)| \qquad (2.62)$$

For rectangular gates $g(t) = \theta(t - \frac{1}{2}\lambda T)\theta(\frac{1}{2}\lambda T - t)$, where θ is the Heaviside function, the Fourier transform of $g(t)$ is given by

$$\hat{g}(\omega) = \lambda T \sin(\lambda \omega T/2)/(\lambda \omega T/2). \qquad (2.63)$$

The muon beam burst can be approximately described by a gaussian function

$$b(t) = (\kappa T \pi^{1/2})^{-1} \exp\left[-(t/\kappa T)^2\right] \qquad (2.64)$$

where κ is a measure of the width of the beam bursts. The Fourier transform of $b(t)$ is then

$$\hat{b}(\omega) = \exp -(\kappa \omega T/2)^2. \qquad (2.65)$$

Therefore

$$|\hat{s}(m\Omega)| = \exp\left[-(\kappa m \pi)^2 T\right] \frac{\sin \lambda m \pi}{m \pi}. \qquad (2.66)$$

This formula shows that the resonance amplitude quickly decreases with m and the fractional burst width, κ. The influence of the fractional gate width λ is better seen by rewriting equation (2.59) in the form

$$N = N_t \lambda \left(1 + \sum_{m=-\infty}^{\infty} A_{\text{eff}}(m\Omega) \frac{\cos\psi + (\omega - m\Omega)\sin\psi}{1 + (\omega - m\Omega)^2 \tau^2}\right) \qquad (2.67)$$

with the effective signal strength given by

$$A_{\text{eff}}(m\Omega) = A \cdot \frac{\tau}{\tau_\mu} \frac{\sin \lambda m \pi}{\lambda m \pi} \exp -(\kappa m \pi)^2. \qquad (2.68)$$

We find that the signal strength is largest when λ is vanishingly small, and decreases roughly with $1/\lambda$. This immediately shows us that the largest observable precession frequency, $\omega = n\Omega$, is limited by the fractional width of the beam burst and the gate. The smallest frequency, on the other hand, is given by $\omega = \Omega$. Since the beam burst frequency Ω is usually fixed, the stroboscopic method only works for a limited number of well defined ω (including $\omega = 0$, if the Hanle effect is studied).

This restriction in frequency (or field) range implies that the stroboscopic method is only applicable in selected study fields. If applicable, however, this method proves very powerful in measuring ω as well as τ or T_2 with extreme precision. This is not only a result of the number of events that can be accumulated in a short time because the data taking rate is unrestricted, but also because, contrary to the time differential technique, an essentially infinite time range can be covered (no data gate!). The precision in determining ω is also related to the fact that it results from a direct comparison with the beam burst repetition rate Ω. This usually derives from a master quartz oscillator and is typically known with an accuracy of a few parts in 10^8.

So far we have only discussed the stroboscopic signal in the presence of an exponential relaxation function $G_x(t)$. For other functional forms of $G_x(t)$ the situation is more complex and no analytical expressions for N can be derived (e.g. for $G_x(t) = \exp(-\sigma^2 t^2)$). In principle one has to fold the stroboscopic signal (where $\tau = \tau_\mu$) with the frequency distribution $F(\omega)$ which is the Fourier transformation of $G_x(t)$

$$F(\tilde{\omega}) = \int_0^\infty G_x(t) \exp(i\tilde{\omega} t) \, \mathrm{d}t. \tag{2.69}$$

The most general form of the stroboscopic signal is then given by

$$N = N_t \lambda \left(1 + \sum_{m=-\infty}^{+\infty} A_{\mathrm{eff}}(m\Omega) \int_{-\infty}^{+\infty} \mathrm{d}\tilde{\omega} F(\tilde{\omega}) \frac{\cos\psi + (\omega + \tilde{\omega} - m\Omega)\tau_\mu \sin\psi}{1 + (\omega + \tilde{\omega} - m\Omega)^2 \tau_\mu^2} \right). \tag{2.70}$$

Figure 2.15 shows a schematic diagram of the particle detection electronics of the stroboscopic μSR set-up at SIN (Klempt et al 1982). Positrons from the target are detected in the telescopes P_1, P_2. All other counters around the target serve as veto counters for the rejection of muons, and positrons from sources other than the target passing through P_1, P_2. Nevertheless there are a lot of 'good' positron events in P_1, P_2 that do not originate from the target. To keep this rate as low as possible one has to optimise the solid angle covered by P_1, P_2 with respect to positrons from the target, and minimise the solid angle with respect to other positron sources. This primarily means the positron telescope is mounted as close as possible to the target. It should be noted that this kind of background problem is much reduced in the time differential detection scheme, as there one can make sure that only muons actually entering the target lead to accepted events.

An example of the stroboscope electronic circuitry itself is shown in figure 2.16. It is built in such a way as to cover the burst period $T = 2\pi/\Omega$ by ten contiguous non-overlapping time windows (gates), in each of which the stroboscopic signal is observed.

In principle, as emphasised before, the stroboscopic method is not limited by rate. In practice, however, saturation problems arise at very high positron rates because of the concomitant high veto rates. 'Good' positron rates of, for example $10^5 \, \mathrm{s}^{-1}$ can be achieved in the SIN apparatus using one telescope and ten time windows. This number is almost two orders of magnitude higher than

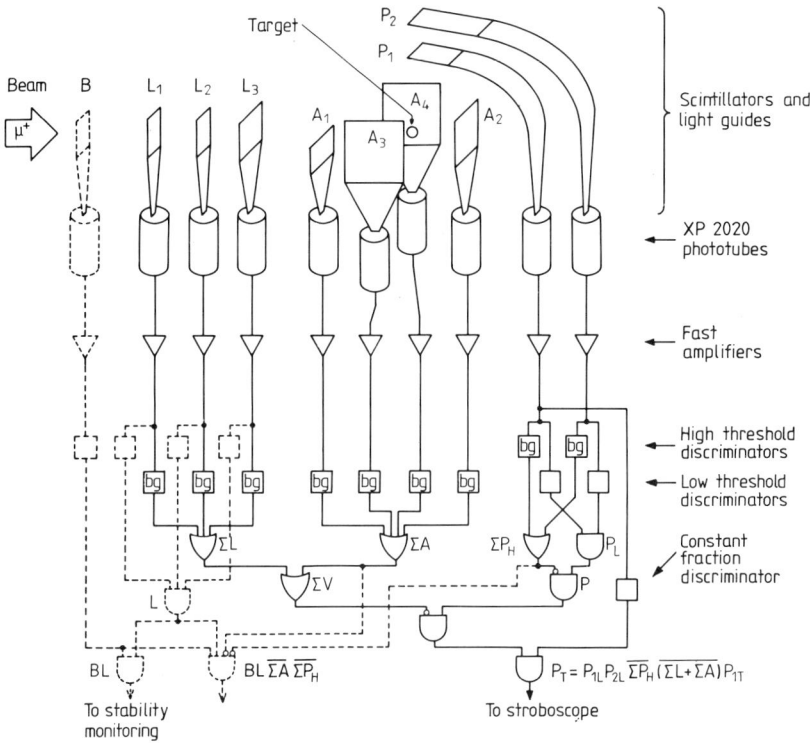

Figure 2.15 Schematic arrangement of detectors and electronics in the SIN stroboscopic μ^+ SR spectrometer. The essential part (solid lines) selects and prepares the accepted positron signals for the stroboscope. Additional components (dashed lines) are used for stability checks (Klempt et al 1982).

that obtained with the time differential technique in a similar geometrical set-up.

Figure 2.17 shows, as an example, five stroboscopic signals from the first five gates, out of a total of ten gates (Camani et al 1978).

(2) Magnetic resonance spectroscopy

The next method to be discussed is the application of nuclear magnetic resonance to spin polarised muons. The resonance is detected via its effect on the angular distribution of $e^{\pm}$. The technique of detecting the resonance by nuclear radiation is well known in conjunction with $\gamma\gamma$PAC, polarised neutron capture etc applications. Compared with classical NMR, muon magnetic resonance, (MMR) is an extremely sensitive technique, which can be performed with only a few million decay events.

Usually the resonance signal is derived by considering the change in occupation numbers of the Zeeman levels which has been induced by the

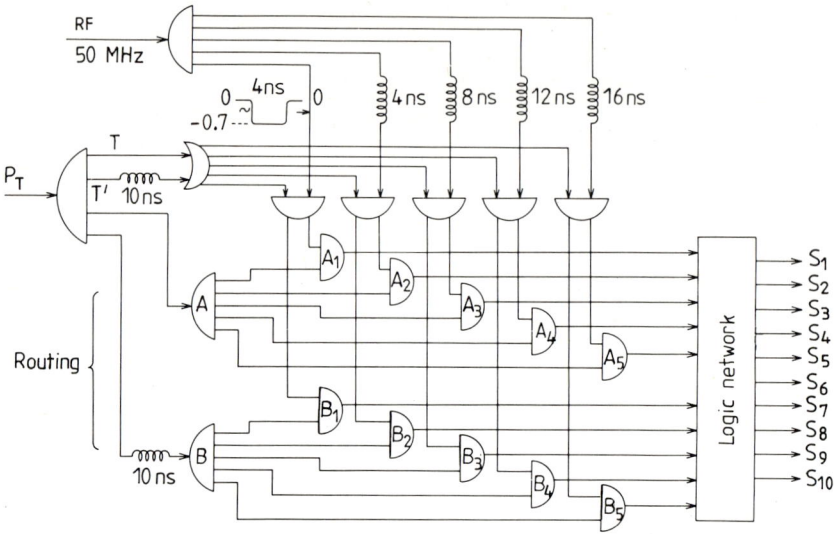

Figure 2.16 Schematic circuitry of the stroboscope adapted to conditions at SIN ($\Omega/2\pi = 50.6$ MHz, $T \simeq 20$ ns). Two sets of five gates delayed by 10 ns relative to each other, each set covering roughly the 20 ns period between two beam bursts, are derived from the 50 MHz master oscillator. The accepted positron signals, P_T, are brought into ten coincidences with these gates. The outputs are combined in such a way by a logic network to produce ten positron signal channels, corresponding to ten nonoverlapping and gap-free time windows which exactly cover the beam burst repetition period.

magnetic radio frequency (RF) radiation. Here we will follow a different procedure by directly treating the effect of an RF field on the time evolution of the muons polarisation.

For simplicity let us assume that a strong static magnetic field (H_z) is applied along the z-direction and that a weak oscillatory magnetic field, circularly polarised in the xy plane, is superimposed. Other cases can always be reduced to such an arrangement. The total field acting on the muon is therefore given by

$$H(t) = H_z \hat{z}_0 + H_1(\cos{(\omega t + \varphi)}\hat{x}_0 + \sin{(\omega t + \varphi)}\hat{y}_0). \quad (2.71)$$

Note that the phase φ of the RF field is not specified. The time evolution of the muon spin operator I is then given by

$$\frac{\hbar}{i} \frac{dI}{dt} = [\mathcal{H}, I] \quad (2.72)$$

with the Hamiltonian $\mathcal{H} = \gamma_\mu H(t) \cdot I$. Equation (2.72) can be reduced to the classical equation describing the Larmor precession

$$dI/dt = \gamma_\mu [I \times H(t)]. \quad (2.73)$$

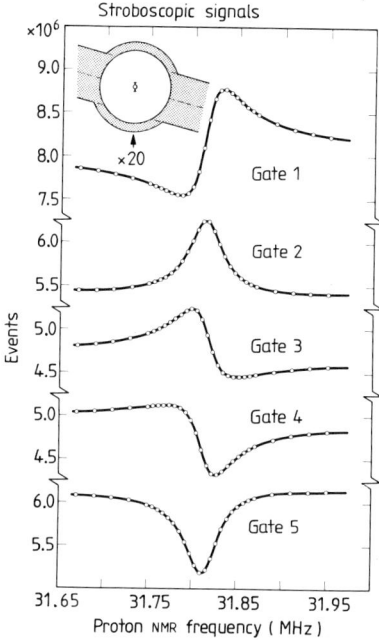

Figure 2.17 Example of stroboscopic signals from five sequentially ordered time gates (Camani et al (1978)).

Since the polarisation $P(t)$ is proportional to $I(t)$, the evolution of $P(t)$ is given by the same equation. In the static case, the solution of equation (2.73) for arbitrarily oriented H is given by the set of equations (2.37)–(2.39). To reduce the problem of solving equation (2.73) to the static case, we apply the well known trick of considering the time evolution of $P(t)$ in the so-called rotating coordinate system $(\hat{x}', \hat{y}', \hat{z}')$ which is related to the laboratory frame $(\hat{x}, \hat{y}, \hat{z})$ by

$$\hat{x}' = \hat{x} \cos \omega t + \hat{y} \sin \omega t$$
$$\hat{y}' = -\hat{x} \sin \omega t + \hat{y} \cos \omega t \qquad (2.74)$$
$$\hat{z}' = \hat{z}$$

that is, the new frame rotates around the z axis with angular velocity ω, chosen to be identical with the RF field frequency. Hence $H_1(t)$ appears static in the rotating frame ($H_1' = H_1(\hat{x}_0' \cos \varphi + \hat{y}_0' \sin \varphi)$ and $\hat{x}_0, \hat{y}_0, \hat{z}_0, \hat{x}_0', \hat{y}_0', \hat{z}_0'$ are unit vectors). The time derivative of $P(t)$ with respect to the rotating frame is then given by the well known expression (e.g. Slichter 1978)

$$\partial P(t)/\partial t = \gamma_\mu P(t) \times [(H_z + \omega/\gamma_\mu)\hat{z}_0' + H_1 \hat{x}_0' \cos \varphi + H_1 \hat{y}_0' \sin \varphi]. \qquad (2.75)$$

This equation is in the same form as equation (2.30), with the desired time independence of the effective field acting on the muon in the rotating frame

$$\boldsymbol{H}_{\text{eff}} = (H_z + \omega/\gamma_\mu)\hat{z}_0' + H_1\hat{x}_0'\cos\varphi + H_1\hat{y}_0'\sin\varphi$$

or

$$|\boldsymbol{H}_{\text{eff}}| = [(H_z + \omega/\gamma_\mu)^2 + H_1^2]^{1/2}. \tag{2.76}$$

For simplicity let us assume that the initial muon polarisation $P(0)$ is along the z axis. The angle between $\hat{z}_0$ and $\boldsymbol{H}_{\text{eff}}$ is denoted by v, with

$$\sin v = H_1/H_{\text{eff}}. \tag{2.77}$$

The azimuthal angle ϕ of $\boldsymbol{H}_{\text{eff}}$ is simply given by the RF field phase φ. The solution of equation (2.75), neglecting any relaxation in the rotating frame, is then given by equations (2.37)–(2.39)

$$P_x'(t) = P(0)[\tfrac{1}{2}\sin 2v \cos\varphi(1 - \cos\omega_{\text{eff}}t) + \sin v \sin\varphi \sin\omega_{\text{eff}}t] \tag{2.78}$$

$$P_y'(t) = P(0)[\tfrac{1}{2}\sin 2v \sin\varphi(1 - \cos\omega_{\text{eff}}t) - \sin v \cos\varphi \sin\omega_{\text{eff}}t] \tag{2.79}$$

$$P_z'(t) = P(0)[\cos^2 v + \sin^2 v \cos\omega_{\text{eff}}t] \tag{2.80}$$

with $\omega_{\text{eff}} = \gamma_\mu H_{\text{eff}}$.

Finally it remains to transform $P(t)$ back into the laboratory frame, with the results:

$$\begin{aligned}P_x(t) = P(0)\{&[\tfrac{1}{2}\sin 2v \cos\varphi(1-\cos\omega_{\text{eff}}t) + \sin v \sin\varphi \sin\omega_{\text{eff}}t]\\&\times \cos\omega t - [\tfrac{1}{2}\sin 2v \sin\varphi(1-\cos\omega_{\text{eff}}t)\\&- \sin v \cos\varphi \sin\omega_{\text{eff}}t]\sin\omega t\}\end{aligned} \tag{2.81}$$

$$\begin{aligned}P_y(t) = P(0)\{&[\tfrac{1}{2}\sin 2v \cos\varphi(1-\cos\omega_{\text{eff}}t) + \sin v \sin\varphi \sin\omega_{\text{eff}}t]\\&\times \sin\omega t + [\tfrac{1}{2}\sin 2v \sin\varphi(1-\cos\omega_{\text{eff}}t)\\&- \sin v \cos\varphi \sin\omega_{\text{eff}}t]\cos\omega t\}\end{aligned} \tag{2.82}$$

$$P_z(t) = P(0)\{\cos^2 v + \sin^2 v \cos\omega_{\text{eff}}t\}. \tag{2.83}$$

These equations assume a particularly simple form for the resonance condition $\omega_0 = -\gamma_\mu H_z$. Let us also specify the RF phase as $\varphi = 0$. Then

$$\begin{aligned}P_x(t) &= P(0)\sin\omega_1 y \sin\omega_0 t\\ P_y(t) &= P(0)\sin\omega_1 t \cos\omega_0 t\\ P_z(t) &= P(0)\cos\omega_1 t\end{aligned} \tag{2.84}$$

with $\omega_1 = \gamma_\mu H_1$.

$P_z(t)$ oscillates with the frequency ω_1 solely determined by the RF field strength H_1. $P_x(t)$ and $P_y(t)$ oscillate at the frequency ω_0 determined by the applied field and are modulated with the frequency ω_1. Observation of the time dependence of $P_x(t)$ and $P_y(t)$ requires a fixed phase between the initial polarisation $P(0)$ and the RF field. If the phase is not fixed, averaging leads to $\langle P_x(t) \rangle = 0, \langle P_y(t) \rangle = 0$. Note that $P_z(t)$ does not involve the RF phase at all. Similar equations can be derived for other relative directions of initial muon polarisation $P(0)$, static field H_0 and RF field $H_1(0)$.

In summary we find that the evolution of the muon's polarisation in the presence of an RF field assumes a complex and interesting behaviour. This can be monitored by the time differential technique as discussed in §2.2.1. To explore all the details in this way, however, as previously pointed out, a fixed phase is required between $P(0)$ and $H_1(0)$. In addition H_1 has to be strong enough to yield a $\omega_1/2\pi$ comparable or larger than the μ decay rate $(1/\tau_\mu)$. This requires RF field amplitudes above 10 G or about 100 kW RF peak power. These requirements can more easily be met when working with a pulsed muon beam, since the RF generator can then be triggered on by the arrival of the muon beam burst and need not be running in a CW mode. As an example figure 2.18 shows data obtained by the BOOM group using grafoil as a target. Here initial muon polarisation and the applied static field were in the same direction, (z), a situation described by equations (2.81)–(2.83). The polarisation was monitored in the z direction and showed clearly an oscillation with $\omega_1 = \gamma_\mu H_1$ at the resonance, in accordance with equation (2.84).

A time differential observation is, however, not necessary to observe the resonance. Rather it is sufficient to just monitor the $e^\pm$ rates in a telescope in the x, y or z direction as a function of RF frequency ω. The rates are obtained

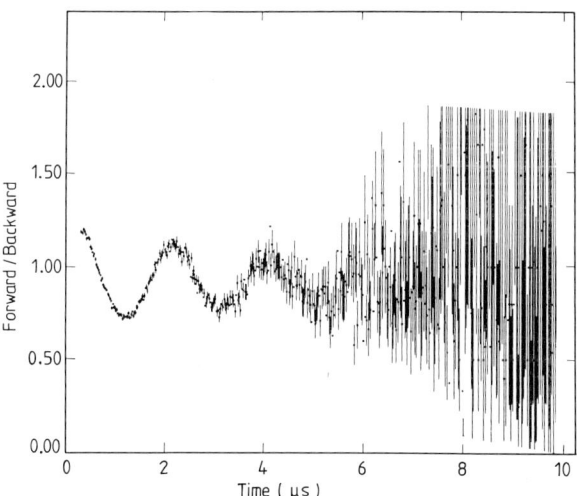

Figure 2.18 Precession of the μ^+ in the RF field H_1 at the resonance condition $\omega_{RF} = \omega_0 = 2\pi\gamma_\mu H_0$, where H_0 is the static field, applied along $P(0)$ (from Nagamine and Yamazaki 1981).

from the expressions for the time-dependent polarisation and equation (2.26) by integrating over time:

$$\bar{N}_i = \frac{N}{\tau_\mu} \int_0^\infty (1+P_i(t)) \exp{-(t/\tau_\mu)} \, dt \qquad i=x,y,z. \qquad (2.85)$$

Inserting the expressions equations (2.81)–(2.83) for $P_i(t)$, we arrive at the following formulae for the integrated $e^\pm$ rates:

$$\bar{N}_x = N \left\{ 1 + P(0) \left[\frac{\omega_1(\omega_0-\omega)}{(\omega_0-\omega)^2+\omega_1^2} \left(\cos\varphi \frac{1}{1+\omega^2\tau_\mu^2} - \sin\varphi \frac{\tau_\mu\omega}{1+\omega^2\tau_\mu^2} \right) \right. \right.$$

$$+ \sin\varphi \frac{\Delta\omega\tau_\mu}{2(1+\Delta\omega^2\tau_\mu^2)} \frac{\omega_1(\omega_0-\omega)+\omega_1\omega_{\text{eff}}}{(\omega_0-\omega)^2+\omega_1^2}$$

$$\left. \left. - \cos\varphi \frac{1}{2(1+\Delta\omega^2\tau_\mu^2)} \frac{\omega_1(\omega_0-\omega)-\omega_1\omega_{\text{eff}}}{(\omega_0-\omega)^2+\omega_1^2} \right] \right\} \qquad (2.86)$$

$$\bar{N}_y = N \left\{ 1 + \dot{P}(0) \left[\frac{\omega_1(\omega_0-\omega)}{(\omega_0-\omega)^2+\omega_1^2} \left(\cos\varphi \frac{\tau_\mu\omega}{1+\omega^2\tau_\mu^2} + \sin\varphi \frac{1}{1+\omega^2\tau_\mu^2} \right) \right. \right.$$

$$- \sin\varphi \frac{1}{2(1+\Delta\omega^2\tau_\mu^2)} \frac{\omega_1(\omega_0-\omega)-\omega_1\omega_{\text{eff}}}{(\omega_0-\omega)^2+\omega_1^2}$$

$$\left. \left. - \cos\varphi \frac{\Delta\omega\tau_\mu}{2(1+\Delta\omega^2\tau_\mu^2)} \frac{\omega_1(\omega_0-\omega)+\omega_1\omega_{\text{eff}}}{(\omega_0-\omega)^2+\omega_1^2} \right] \right\} \qquad (2.87)$$

$$\bar{N}_z = N \left\{ 1 + P(0) \left[1 - \frac{\omega_1^2\tau_\mu^2}{1+\tau_\mu^2[(\omega_0-\omega)^2+\omega_1^2]} \right] \right\} \qquad (2.88)$$

where

$$\omega_{\text{eff}} = [(\omega_0-\omega)^2+\omega_1^2]^{1/2} \qquad \Delta\omega = \omega_{\text{eff}} - \omega. \qquad (2.89)$$

The rates in the x and y directions show a very weak dependence on ω, as is to be expected, since $P_x(t)$ and $P_y(t)$ oscillate around zero. The signal will only show up significantly if ω_0 and ω_1 are of comparable magnitude. The signals consist, as in the stroboscopic case, of a Hanle term and higher field resonance terms.

The rate in the z direction, on the other hand, shows a very pronounced resonance behaviour at $\omega = \omega_0$, the strength of which is directly proportional to ω_1^2, i.e. to H_1^2. On closer inspection one realises that the analytical form of equation (2.88) corresponds to the absorptive part of the CW NMR signal as one might have expected from the beginning.

It is also possible to observe a dispersive signal by starting with the initial condition

$$\boldsymbol{P}(0) = P(0)\hat{\boldsymbol{x}}_0 \parallel \boldsymbol{H}_1(0) = H_1(0)\hat{\boldsymbol{x}}_0 \perp \boldsymbol{H}_0 = H_0\hat{\boldsymbol{z}}_0.$$

The counting rate in the z direction is then computed to be

$$\bar{N}_z = N\left(1 + P(0)\frac{(\omega_0-\omega)\tau_\mu^2\omega_1}{1+\tau_\mu^2[(\omega_0-\omega)^2+\omega_1^2]}\right). \qquad (2.90)$$

So far we have neglected the possibility of relaxation. Its influence may be accounted for by introducing damping terms into equation (2.75), which then looks identical to the phenomenological Bloch equations in the rotating frame

$$\frac{\partial \boldsymbol{P}(t)}{\partial t} = \gamma_\mu \boldsymbol{P}(t)[(H_z + \omega/\gamma_\mu)\hat{z}'_0 + H_1(\cos\varphi\hat{x}'_0 + \sin\varphi\hat{y}'_0)]$$

$$-\frac{P_x(t)\hat{x}_0}{T_2} - \frac{P_y(t)\hat{y}_0}{T_2} - \frac{P_z(t)\hat{z}_0}{T_1}. \qquad (2.91)$$

It is easy to see that this corresponds to an exponential relaxation, both for the longitudinal component P_z, and for the transverse components P_x and P_y. As mentioned before, T_1 is usually very long compared with the muon lifetime, and the corresponding term in equation (2.91) may be neglected. The effect of the remaining damping terms for the transverse components is mainly to broaden the integral signals given by equations (2.85)–(2.88). That is, the resonance term in equation (2.88) is replaced by (Slichter 1978)

$$\omega_1^2\tau^2\{1+\tau^2[(\omega_0-\omega)^2+\omega_1^2]\}^{-1} \qquad (2.92)$$

with

$$1/\tau = 1/\tau_\mu + 1/T_2. \qquad (2.93)$$

From a different point of view, application of a magnetic radiofrequency field induces transitions between the two Zeeman states of a muon, whilst a static magnetic field is being applied along the initial muon polarisation. The Zeeman splitting is given by

$$\Delta E = 2\langle \boldsymbol{\mu}_\mu \cdot \boldsymbol{H}_0\rangle = \hbar\gamma_\mu H_z = \omega_0\hbar. \qquad (2.94)$$

The resonance condition implies that $\omega = \omega_0$ as before.

The first application of magnetic resonance to the measurement of the Zeeman splitting of μ^+ was performed by Coffin et al (1958), with the aim to determine the magnetic moment of the muon. The applied static field amounted to 2.8 kG. More recently the μSR group at KEK (BOOM) has adapted the magnetic resonance technique as a routine method in μSR spectroscopy. The schematic layout of their experimental arrangement is shown in figure 2.19. Figure 2.20 shows the magnetic resonance signal obtained from μ^+ stopped in water, and equation (2.88) describes this well.

Relaxation of the transverse polarisation components may also be induced by a spread of the static field, $\boldsymbol{H}_0 = H_z\hat{z}_0$, over the various muon sites. If the spread is described by a distribution function $f(\omega_0)$ with $\int_{-\infty}^{+\infty} f d\omega_0 = 1$,

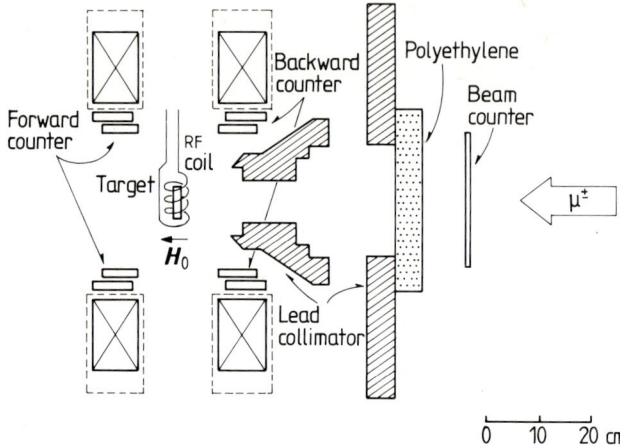

Figure 2.19 Schematic view of the experimental arrangement for a muon spin resonance experiment (Kitaoka *et al* 1982).

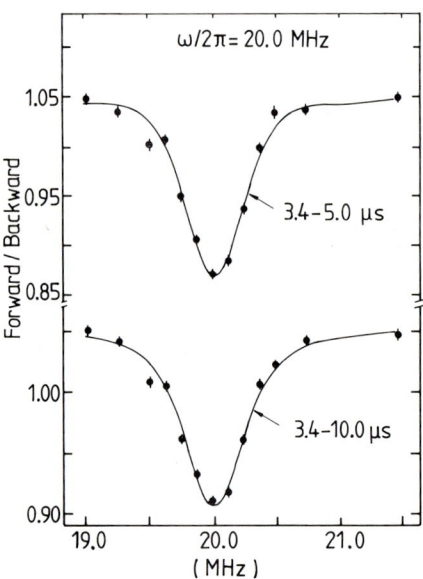

Figure 2.20 Time integrated μ^+ spin resonance curve obtained from the diamagnetic fraction in water (Kitaoka *et al* 1982). The points represent the experimental data, and the curve the best fit to these data.

equation (2.88) may be rewritten as follows

$$\bar{N}_z = N_0 \left[1 + P(0) \left(1 + \omega_1^2 \tau_\mu^2 \int_{-\infty}^{+\infty} \frac{f(\omega_0) \, d\omega_0}{1 + \tau_\mu^2[(\omega_0 - \omega)^2 + \omega_1^2]} \right) \right]. \quad (2.95)$$

Very often (see §2.3) $f(\omega_0)$ can be expressed by a gaussian function, centred at a frequency ω_0^0, with RMS width 2σ:

$$f(\omega_0) = [2/(\pi\sigma^2)^{1/2}] \exp[-(\omega_0 - \omega_0^0)^2/\sigma^2]. \quad (2.96)$$

In the limit $\sigma \gg 1/\tau_\mu$, the resonance term of equation (2.95) will itself assume a gaussian shape given by equation (2.96). This corresponds to the usual situation in NMR when one uses stable nuclei. We see that the situation is more difficult in μSR, where generally $\sigma < 1/\tau_\mu$ because of the relatively short lifetime of the $\mu^\pm$. Finally, it is instructive to point out the equivalence between the stroboscopic signal and the integrated magnetic resonance signal. Both methods essentially perform a Fourier analysis of the time dependent polarisation.

Magnetic resonance has also been applied to the study of muonium states (Brown *et al* 1979), by a special technique called Double Electron Muon Resonance (DEMUR) analogous to the electron–nuclear double resonance (ENDOR) technique. This method will be described in §7.3.4.

2.3 Dipolar Relaxation

The interaction of the muon's magnetic dipole moment (or spin) with other magnetic moments (or spins) of nuclear or electronic origin in its environment, is one of the prime sources of important information on properties of either the host material or μ^+ specific phenomena (Schenck 1970, Schenck *et al* 1972). In this respect the dipolar interaction of the μ^+ with its environment constitutes an important element in the basic principles of μSR. In this section we will be concerned only with the effect of many static, randomly oriented spins on the evolution of the muon's ensemble polarisation. The net effect will lead to a depolarisation by dephasing which will become visible in transverse and zero field experiments. The effect of ordered electronic moments will be deferred to §4.6.

2.3.1 Transverse fields

The interaction Hamiltonian between two magnetic dipole moments, say of a μ^+ and a nucleus with magnetic moments μ_μ, μ_N and spins S, I ($\mu_\mu = \hbar\gamma_\mu S$, $\mu_N = \hbar\gamma_N I$) is given by

$$\mathcal{H}_{dip} = \frac{\hbar^2 \gamma_\mu \gamma_N}{r^3} \left(\mathbf{S} \cdot \mathbf{I} - \frac{3(\mathbf{S} \cdot \mathbf{r})(\mathbf{I} \cdot \mathbf{r})}{r^2} \right) \quad (2.97)$$

where r is the vector connecting the two dipoles. By introducing polar coordinates $\mathcal{H}_{dip}$ can be rewritten

$$\mathcal{H}_{dip} = (\hbar^2 \gamma_\mu \gamma_N / r^3)(A + B + C + D + E + F) \quad (2.98)$$

with

$$A = S_z I_z (1 - \cos^2 \theta)$$
$$B = -\tfrac{1}{4}(S_+ I_- + S_- I_+)(1 - 3\cos^2 \theta)$$
$$C = -\tfrac{3}{2}(S_+ I_z + S_z I_+) \sin \theta \cos \theta \, e^{-i\phi} = D^+$$
$$E = -\tfrac{3}{4}(S_+ I_+) \sin \theta \, e^{-2i\phi} = F^+. \quad (2.99)$$

θ is the angle between the radius vector r and the z axis, which is chosen as the axis of quantisation (Abragam 1970).

In addition, we now assume that the two dipoles are interacting with a common external static field $\boldsymbol{B}$, which is directed along the z axis: $|\boldsymbol{B}| = B_z$. Hence the total Hamiltonian reads

$$\mathcal{H} = \mathcal{H}_{dip} - \hbar \gamma_\mu \boldsymbol{S} \cdot \boldsymbol{B} - \hbar \gamma_N \boldsymbol{I} \cdot \boldsymbol{B}. \quad (2.100)$$

For the case of two unlike spins, such as a muon and a nuclear spin, the B, C, D, E, F terms do not contribute because they involve spin-flip processes which are forbidden for reasons of energy conservation. Naturally this is not true for zero external magnetic field. The terms B–F are also very important if time dependent perturbing fields are present, as these may induce transitions between different spin states.

Assuming $\boldsymbol{B}$ to be sufficiently strong, (i.e. $|\boldsymbol{B}| \gg |H_{dip}|$) only term A, the secular term with respect to the Zeeman interactions, of $\mathcal{H}_{dip}$ needs to be further considered. This corresponds to the classical case of two interacting dipoles.

We will now consider the situation of many dipole moments interacting jointly with the μ^+ spin. This is, for example, the case for μ^+ implanted in copper, where all the host nuclei carry spin and a non-zero magnetic moment. Neglecting the nuclear Zeeman energies and the dipole–dipole interactions among the nuclear spins, I, the total Hamiltonian assumes the form

$$\mathcal{H} = \sum_{j=1}^{N} \alpha_j S_z I_{z,j} - \gamma_\mu S_z B_z = \mathcal{H}_{dip} + \mathcal{H}_z \quad (2.101)$$

where

$$\alpha_j = (\hbar^2 / r_j^3) \gamma_N \gamma_\mu (1 - 3\cos^2 \theta_j). \quad (2.102)$$

The sum extends, in principle, over all nuclei in the sample. However it will now be assumed that they are of one type only and the evolution of the polarisation $\boldsymbol{P}(t)$ of a μ^+ ensemble, subject to the Hamiltonian equation (2.101) will be calculated. We assume that the initial μ^+ polarisation $\boldsymbol{P}_\mu(0)$ is along the x axis, i.e. perpendicular to $\boldsymbol{B}$, and that the nuclei are unpolarised $|\boldsymbol{P}_{n,j}(0)| = 0$.

We are interested in the quantity $P_x(t) = 2\langle S_x(t)\rangle = \langle \sigma_x(t)\rangle$. The ensemble average, $\langle \sigma_x(t)\rangle$ can be calculated from the well known expression

$$\langle \sigma_x(t)\rangle = \text{Tr}\,\rho\sigma_x(t) = P_x(t). \tag{2.103}$$

ρ is a spin density matrix, representing the initial spin configuration of the μ^+ and the nuclei

$$\rho = 2(2I+1)^{-N}\left[1 + P_\mu(0)\sigma_\mu + \sum P_{n,j}\sigma_j\right] \tag{2.104}$$

$$= 2(2I+1)^{-N}\left[1 + P_x(0)\sigma_x\right] \tag{2.105}$$

In order to evaluate the trace in equation (2.103), we have to select a convenient complete set of basis states. In view of the assumption $|B| \gg |H_{\text{dip}}|$ the nuclear and muon spin states are naturally quantised with respect to the z axis and hence the Zeeman eigenstates $|n\rangle = |m_\mu\rangle \prod_{j=1}^{N}|m_j\rangle$ constitute the appropriate set of basis states. They span a $2(2I+1)^N$ dimensional space, representing all possible spin configurations. $|m_j\rangle$ describes the spin state of the jth nucleus with $\langle I_{z,j}\rangle = m_j$, and $|m_\mu\rangle$ the spin state of the μ^+ with $\langle S_z\rangle = m_\mu$.

The time evolution of $\sigma_x(t)$ obeys the equation

$$\dot{\sigma}_x(t) = (i/\hbar)[\mathcal{H}, \sigma_x(t)] \tag{2.106}$$

with the formal solution

$$\sigma_x(t) = \exp(i\mathcal{H}t/\hbar)\sigma_x \exp(-i\mathcal{H}t/\hbar). \tag{2.107}$$

Inserting equations (2.105) and (2.107) into equation (2.103) one arrives at

$$P_x(t) = \frac{1}{2(2I+1)^N}\sum_{n,n'}\langle n|1 + P_x(0)\sigma_x|n'\rangle$$

$$\times \langle n'|\exp(i\mathcal{H}t/\hbar)\sigma_x \exp(-i\mathcal{H}t/\hbar)|n\rangle. \tag{2.108}$$

Since $[\mathcal{H}_z, \mathcal{H}_{\text{dip}}] = 0$, we can write

$$\exp(i\mathcal{H}t/\hbar) = \exp(i\mathcal{H}_{\text{dip}}t/\hbar)\exp(i\mathcal{H}_z t/\hbar)$$

$$= \exp(i\mathcal{H}_{\text{dip}}t/\hbar)\exp(-\tfrac{1}{2}i\omega\sigma_z t) \tag{2.109}$$

The operator $\exp(-\tfrac{1}{2}i\omega\sigma_z t)$ represents a rotation of angle ωt about the z axis. We can therefore write

$$\exp(-\tfrac{1}{2}i\omega t\sigma_z)\sigma_x \exp(\tfrac{1}{2}i\omega t\sigma_z) = \sigma_x \cos\omega t + \sigma_y \sin\omega t \tag{2.110}$$

and

$$P_x(t) = Q(t)\cos\omega t + Q'(t)\sin\omega t \tag{2.111}$$

where

$$Q(t) = \frac{1}{2(2I+1)^N}\sum_{n,n'}\langle n|1 + P_x(0)\sigma_x|n'\rangle$$

$$\times \langle n'|\exp\left(\frac{i}{\hbar}\mathcal{H}_{\text{dip}}t\right)\sigma_x \exp\left(\frac{-i}{\hbar}\mathcal{H}_{\text{dip}}t\right)|n\rangle. \tag{2.112}$$

The second term in equation (2.111) vanishes, since $\mathcal{H}_{dip}$ is invariant under time reversal, and consequently $P_x(t)$ also has to be invariant under time reversal.

In order to evaluate $Q(t)$ we invoke the concept of moments. By a Fourier transformation of $Q(t)$, we obtain a frequency distribution $f(\omega)$ in frequency space

$$f(\omega) = \int Q(t) \cos \omega t \, dt \qquad (2.113)$$

or

$$Q(t) = (2/\pi) \int f(\omega) \cos \omega t \, d\omega. \qquad (2.114)$$

The nth moment is now defined as

$$M_n = \int_{-\infty}^{+\infty} f(\omega) \omega^n \, d\omega \qquad \neq 0 \text{ only if } n \text{ is even} \qquad (2.115)$$

from which we derive

$$M_{2n} = (-1)^n [d^{2n} Q(t)/dt^{2n}]_{t=0}. \qquad (2.116)$$

The moment M_2 is the most important (see Abragam 1970) and we will restrict ourselves to its consideration. Inserting $Q(t)$ from equation (2.112) into equation (2.116) one obtains

$$M_2 = -\frac{1}{\hbar^2} \frac{1}{2(2I+1)^N} \sum_{n,n'} \langle n|1 + P_x(0)\sigma_x|n'\rangle$$
$$\times \langle n'|[\mathcal{H}_{dip},[\mathcal{H}_{dip},\sigma_x]]|n\rangle. \qquad (2.117)$$

A straightforward calculation leads to

$$M_2 = \frac{1}{\hbar^2(2I+1)} \sum_{j=1}^{N} \alpha_j^2 \, \text{Tr}\,(I_z^2). \qquad (2.118)$$

By noting that

$$\text{Tr}\,(I_x^2) = \text{Tr}\,(I_y^2) = \text{Tr}\,(I_z^2) = \tfrac{1}{3} \text{Tr}\,(I^2) = \tfrac{1}{3} I(I+1)(2I+1)$$

one finally obtains (Van Vleck 1948)

$$M_2^{VV} = \frac{I(I+1)}{3\hbar^2} \sum_{j=1}^{N} \alpha_j^2 = \gamma_\mu^2 \sum_{j=1}^{N} \overline{(H_{dip,z}^{(j)})^2} \qquad (2.119)$$

where $\overline{(H_{dip,z}^{(j)})^2}$ is the average of the squared dipole field in the z direction at the μ^+ position originating from the jth nucleus. Hence, the second moment is a measure of the spread of the z component of the nuclear dipole fields at the μ^+.

A gaussian distribution $f(H_{dip,z})$ of the dipolar fields at the μ^+ would be completely characterised by the second moment M_2/γ_μ^2, as can be deduced from equation (2.115). The second moment is then identical to the gaussian distribution width (explicitly: $\exp[-(\gamma_\mu H_{dip,z})^2/2M_2]$). In practice therefore, we may approximate the unknown $f(\omega)$ by a gaussian distribution with a width determined by the second moment of the true distribution $f(\omega)$. In this approximation we use for $Q(t)$ the ansatz

$$Q(t) = \exp(-\sigma^2 t^2) \qquad (2.120)$$

and obtain from equation (2.116)

$$M_2 = 2\sigma^2. \qquad (2.121)$$

The precession of muons in the presence of many magnetic dipoles is thus characterised by a gaussian damping of the precession amplitude.

From the second moment the effective RMS field spread ΔB_z felt by the μ^+ can be calculated:

$$\Delta B_z = 2(M_2/\gamma_\mu^2)^{1/2} = 2^{3/2}(\sigma/\gamma_\mu). \qquad (2.122)$$

Note that the ΔB_z would be increased by a factor of $\frac{3}{2}$ for the like spin case. For a polycrystalline sample, integrating over all orientations, θ_j, equation (2.119) becomes:

$$M_{2,\text{polyc}}^{VV} = \frac{4}{5} \frac{I(I+1)}{3} \hbar^2 \gamma_\mu^2 \gamma_N^2 \sum_{j=1}^{N} \frac{1}{r_j^6}. \qquad (2.123)$$

The ansatz equation (2.120) is well justified if a large number of dipole field sources are around. The situation is different in systems with only a dilute and random distribution of magnetic moments as, for example, in substances with paramagnetic impurities, and in spin glasses, (see §6.2). Walstedt and Walker (1974) have shown that in such systems the internal field distribution is essentially Lorentzian, and consequently the relaxation function will turn out to be exponential.

2.3.2 Inclusion of quadrupole effects

So far it has been implied that the axis defined by the applied field provides a natural axis of quantisation for the nuclear spins I. This is, however, no longer valid of the nuclear spin system is subject to an additional interaction which is not negligible in comparison with the Zeeman interaction. The situation arises in the presence of strong electric field gradients and non-vanishing nuclear quadrupole moments, Q $(I > 1/2)$. An electric quadrupole interaction results in competition with the magnetic Zeeman interaction. The resulting effect on the second moment of the dipolar field distribution was first considered by Hartmann (1977). In essence one not only obtains a non-zero expectation value of I in the z direction, $\langle I_z \rangle$, but also a non-zero component in the perpendicular direction chosen to be the x direction $\langle I_x \rangle$.

The second moment then reads:

$$M_2^Q = \hbar^2 \gamma_\mu^2 \gamma_N^2 \sum_{i=1}^{N} \left[\left(\langle I_z \rangle_i \frac{(1 - 3\cos^2\theta_i)}{r_i^3} - \langle I_x \rangle_i \frac{3\sin\theta_i \cos\theta_i}{r_i^3} \right)^2 \right]. \qquad (2.124)$$

The second term under the sum represents the dipole field in the z direction originating from the static spin component in the x direction. The square of the total field component in the z direction has to be averaged over all eigenstates of the ith nuclear spin. $\langle I_z \rangle_i$ and $\langle I_x \rangle_i$ will now depend on the strength of the external magnetic field. For very strong external magnetic fields the quadrupole interaction will become quenched and one obtains

$$\overline{\langle I_z \rangle_i^2} = \tfrac{1}{3} I(I+1) \qquad \overline{\langle I_x \rangle_i^2} = 0. \qquad (2.125)$$

This reduces equation (2.123) to the Van Vleck expression equation (2.119).

For the actual calculation of $\langle I_x \rangle_i$ and $\langle I_z \rangle_i$, one needs the eigenstates of the total nuclear hyperfine Hamiltonian, which reads, for the ith nucleus:

$$\mathcal{H}_N^{(i)} = \mathcal{H}_Q^{(i)} + \mathcal{H}_z^{(i)}$$

$$= \sum_{j,k} \frac{\partial^2 V^{(i)}}{\partial x_j \, \partial x_k} \mathbf{Q}_{jk} + \gamma_N \hbar I_{z,i} B_z. \qquad (2.126)$$

$\partial^2 V^{(i)} / \partial x_j \, \partial x_k = \mathbf{V}_{jk}^{(i)}$ is the electric field gradient (EFG) tensor,

$$\mathbf{Q}_{jk} = \frac{eQ}{6I(2I-1)} [\tfrac{3}{2}(I_j I_k + I_k I_j) - \delta_{jk} I(I+1)] \qquad (2.127)$$

is the quadrupole tensor and Q, the nuclear spectroscopic quadrupole moment. Note that the EFG tensor may be different from site to site.

The origin of the EFG may be the μ^+ itself and its screening charge, in which case the EFG will be radially directed away from the μ^+. We would then expect the interaction strength to rapidly decrease with distance from the μ^+, so that only the quadrupole interaction of the nearest neighbour host nuclei will need to be considered, and the more distant neighbour nuclei can be treated in the Van Vleck limit. Examples for the field dependence of σ for the tetrahedral and octahedral interstital μ^+ positrons in the rigid lattices of Cu, V, Nb and Ta are displayed in figures 2.21(a)–(d).

2.3.3 Zero and longitudinal fields

Even in the absence of an external magnetic field the μ^+ will precess around the internal dipolar fields H_i. The situation is pictured in figure 2.7. If θ is the angle between $P(0) = P_z(0)\hat{z}_0$ and H_i, the time evolution of $P_z(t)$ will be (see also equation (2.39))

$$P_z^\theta(t) = \cos^2 \theta + \sin^2 \theta \cos \gamma_\mu H_i t \qquad (2.128)$$

with

$$\cos^2 \theta = H_{i,z}^2 / H_i^2 \qquad \sin^2 \theta = (H_{i,x}^2 + H_{i,y}^2)/H_i^2. \qquad (2.129)$$

Dipolar relaxation 49

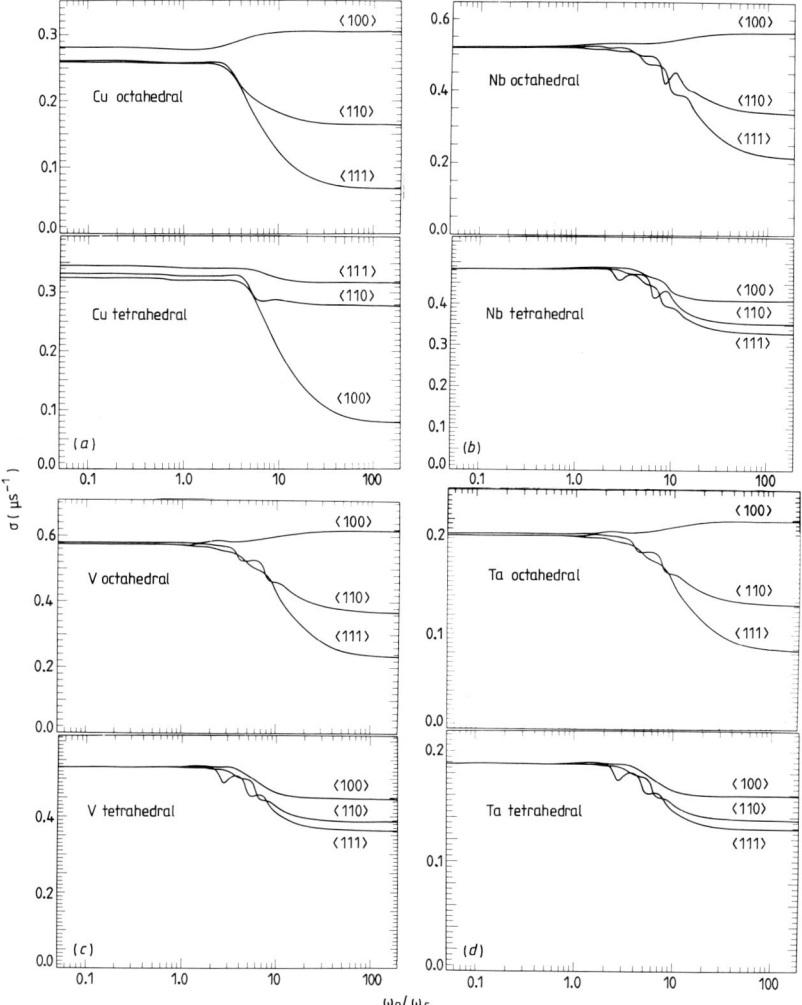

Figure 2.21 Dependence on σ on the relative strength, ω_B/ω_E, of the Zeeman interaction, $\omega_B = \gamma_\mu H_0$, and the quadrupole interaction, $\omega_E = \frac{1}{2}\{eQq/\hbar[2I(I+1)]\}$, where $q \sim 1/r^3$ is the EFG at the nucleus in (a) Cu ($I=3/2$), (b) Nb ($I=9/2$), (c) V ($I=7/2$) and (d) Ta ($I=7/2$). The nuclear moments are taken from Fuller (1976). ω_E is given by the interaction strength at a distance $a/2$ from the μ^+. $\langle 100 \rangle, \langle 110 \rangle$ and $\langle 111 \rangle$ indicates the direction of the external field. Note that for the $\langle 100 \rangle$ and $\langle 110 \rangle$ orientations in the BCC metals two magnetically non-equivalent tetrahedral and octahedral interstitial sites exist, with relative weights of 2:1 (Hartmann 1977).

We will assume that the internal field components are distributed isotropically according to a gaussian function (random local field approximation) with width Δ, i.e.

$$f(H_{i,x}) = f(H_{i,y}) = f(H_{i,z}) = \frac{1}{(2\pi)^{1/2} \Delta} \exp\left(\frac{-H_{i,\rho}^2}{2\Delta^2}\right)$$

$$\rho = x, y, z. \tag{2.130}$$

The ensemble average of the polarisation in the z direction follows from

$$P_z(t) = \int f(H_{i,x}) f(H_{i,y}) f(H_{i,z}) (\cos^2 \theta + \sin^2 \theta \cos \gamma_\mu H_i t) \, d^3 H_i \tag{2.131}$$

where

$$d^3 H_i = H_i^2 \sin \theta \, d\theta \, d\phi \, dH_i = dH_{i,x} \, dH_{i,y} \, dH_{i,z}.$$

Evaluation of equation (2.131) leads to

$$P_z(t) = \tfrac{1}{3} + \tfrac{2}{3}(1 - \gamma_\mu^2 \Delta^2 t^2) \exp(-\tfrac{1}{2}\gamma_\mu^2 \Delta^2 t^2). \tag{2.132}$$

This equation was first derived by Kubo and Toyabe (1967) and has been rediscovered and discussed with respect to μSR by Yamazaki (1979) and Hayano et al (1979). In fact, at the time of its first derivation, it was rather an academic endeavour since its experimental verification by NMR (in zero field!) was out of the question. The predicted time dependence is shown in figure 2.22. The early time behaviour follows a gaussian curve $\exp(-\gamma_\mu^2 \Delta^2 t^2)$, $P_z(t)$ drops to a minimum at $t\Delta^2 = 1 + (1+\Delta^2)^{1/2}$ and later shows a recovery to $\tfrac{1}{3}$. This recovery to $\tfrac{1}{3}$ can be understood intuitively if one considers the random distribution of H_i, which implies that on average $\tfrac{1}{3}$ of the μ^+ will not precess, since for them H_i and $P(0)$ are parallel. This also follows from equation (2.39), by noting that $\langle \cos^2 \theta \rangle = \tfrac{1}{3}$.

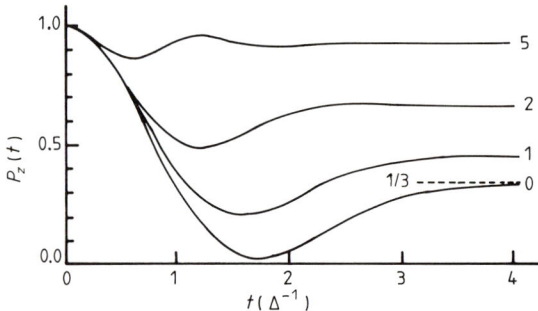

Figure 2.22 Static Kubo–Toyabe function $P_z(t)$, (see equations (2.132) and (2.133)) for different external field strengths, $\gamma_\mu B/\Delta H$, in the longitudinal geometry using a time unit of (Δ^{-1}) (Yamazaki 1979).

If an external field B is applied parallel to $P(0)$, one finds (Hayano et al 1979)

$$P_z(t) = P_\parallel(t) = 1 - (2\gamma_\mu^2 \Delta^2/\omega_0^2)[1 - \exp(-\tfrac{1}{2}\gamma_\mu^2 \Delta^2 t^2) \cos \omega_0 t]$$

$$+ (2\Delta^4/\omega_0^3) \int_0^t \exp(-\tfrac{1}{2}\Delta^2 \tau^2) \sin \omega_0 \tau \, d\tau \qquad (2.133)$$

where $\omega_0 = \gamma_\mu B$. This function is also displayed in figure 2.24 for various external fields ($\gamma_\mu B/\Delta$). We notice that the plateau of recovery increases with increasing field and that the overall time dependence tends to become weaker and weaker. This is a consequence of the increasing decoupling of the μ^+ spin from the internal fields, because of the external field. In other words, this is caused by the cone of precession around the external field becoming more and more narrow.

We have introduced, rather phenomenologically, the width Δ, the second moment of the assumed isotropic distribution of the dipolar fields H_j. We now want to determine Δ for the zero field case from the arrangement of nuclear spins. We first consider the case without any quadrupole interaction. In this instance the μ^+ and nuclear Zeeman levels in zero external field are degenerate, and the full Hamiltonian, equation (2.98), will come into play. Assuming the initial μ^+ polarisation to be along the z axis, as before, the relevant second moment equation (2.117) can be evaluated analytically with the result

$$M_2^{ZF} = -\frac{1}{\hbar^2} \frac{1}{2(2I+1)^N} \sum \langle n|1 + P_z(0)\sigma_z|n' \rangle$$

$$\times \langle n'|[\mathcal{H}_{dip},[\mathcal{H}_{dip},\sigma_z]]|n \rangle$$

$$= \tfrac{1}{3} I(I+1) \sum_{j=1}^{N} \frac{(\hbar\gamma_\mu\gamma_N)^2}{r_j^6} \left(2 + 3\frac{x_j^2 + y_j^2}{r_j^2}\right). \qquad (2.134)$$

Denoting the angle between $P_z(0)\hat{z}_0$ and $\mathbf{r}_j$ by Ω_j we further obtain

$$M_2^{ZF} = \tfrac{1}{3} I(I+1) \sum_{j=1}^{N} \frac{(\hbar\gamma_\mu\gamma_N)^2}{r_j^6} (2 + 3\sin^2 \Omega_j) \qquad (2.135)$$

and the powder average becomes

$$M_{2,\text{polyc}}^{ZF} = \tfrac{4}{3} I(I+1) \sum_{j=1}^{N} \left(\frac{\hbar\gamma_\mu\gamma_N}{r_j^3}\right)^2. \qquad (2.136)$$

Here, we have to take notice that the thus derived M_2^{ZF} is a measure of the field spread perpendicular to $P(0)\hat{z}_0$, that is, it contains contributions from the spread in the x and the y direction. Since in a polycrystalline sample the average internal field is isotropic, we can make the identification

$$\Delta = \tfrac{1}{2} M_{2,\text{polyc}}^{ZF} = \tfrac{2}{3} I(I+1)(\hbar\gamma_\mu\gamma_N)^2 \sum_{j=1}^{N} 1/r_j^6 \qquad (2.137)$$

in accordance with equation (2.130). Comparing this expression with the Van Vleck expression, equation (2.123), we find

$$\Delta = \tfrac{5}{2} M_{2,\text{polyc}}^{VV}. \tag{2.138}$$

Next, we take the quadrupole interaction into account. If the nuclear spin is integer, the degeneracy between the μ^+ and the nuclear Zeeman levels is completely lifted. If the nuclear spin is an odd half-integer there exists a partial degeneracy which originates from the nuclear substates, $m = \pm\tfrac{1}{2}$. A simultaneous flip of μ^+ spin and nuclear spin between the $m = \pm\tfrac{1}{2}$ states without any energy exchange is still possible.

It is possible to derive an expression for the second moment for integer nuclear spins from purely classical arguments if the quadrupole coupling of the nuclear spins is much stronger than either their mutual dipolar coupling, or their dipolar coupling to the μ^+. Assuming, as before, an electric field gradient extending radially from the μ^+, the quadrupole interaction will, in this case, lead to a rapid precession of the jth nuclear spin around the radius vector, r_j (see figure 2.23). The projection, I_{r_j} of I_j, on r_j is time independent with an average value of

$$\overline{\langle I_{r_j} \rangle^2} = \tfrac{1}{3} I(I+1). \tag{2.139}$$

The second moment is then obtained as follows:

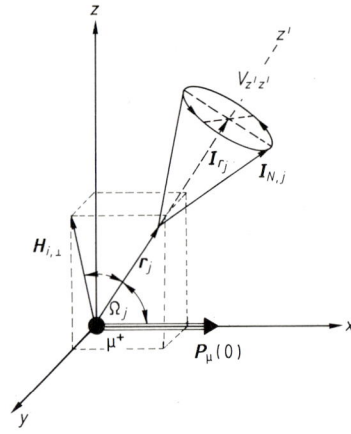

Figure 2.23 Precession of nuclear spin $I_{N,j}$ around the EFG, $V_{z'z'}$ directed radially away from the μ^+. The static component, I_{r_j}, produces a field $H_{i,\perp}$ transverse to the initial μ^+ polarisation $P(0) = P_x(0)\hat{x}_0$ ($\hat{x}_0 =$ unit vector).

$$M_2^{ZF,Q,I=\text{integer}} = \sum \gamma_\mu^2 (\overline{H_{x,j}^2} + \overline{H_{y,j}^2})$$

$$= (\hbar \gamma_\mu \gamma_N)^2 \sum_{j=1}^{N} \frac{1}{r_j^6} \left\langle I_{r_j,\perp} - \frac{3(\mathbf{I}_{r_j} \cdot \mathbf{r}_j) r_{j,\perp}}{r_j^2} \right\rangle$$

$$= \tfrac{1}{3} I(I+1)(\hbar \gamma_\mu \gamma_N)^2 \sum_{j=1}^{N} \frac{4}{r_j^6} \sin^2 \Omega_j \qquad (2.140)$$

where

$$|I_{r_j,\perp}| = |I_{r_j,x} + I_{r_j,y}| = I_{r_j} \sin \Omega_j$$

$$|r_{j,\perp}| = |r_{x,j} + r_{y,j}| = r_j \sin \Omega_j \qquad (2.141)$$

and the powder or polycrystalline average reads

$$M_{2,\text{polyc}}^{ZF,Q,I=\text{integer}} = \tfrac{1}{3} I(I+1) \tfrac{8}{3} (\hbar \gamma_\mu \gamma_N)^2 \sum_{j=1}^{N} \frac{1}{r_j^6}. \qquad (2.142)$$

Comparing this with equation (2.123), it is found that

$$M_{2,\text{polyc}}^{ZF,Q,I=\text{integer}} / M_{2,\text{polyc}}^{VV} = 10/3.$$

As before we have

$$\Delta^{Q,I=\text{integer}} = \tfrac{1}{2} M_{2,\text{polyc}}^{ZF,Q,I=\text{integer}} = \tfrac{1}{3} I(I+1) \tfrac{4}{3} (\hbar \gamma_\mu \gamma_N)^2 \sum_{j=1}^{N} \frac{1}{r_j^6}. \qquad (2.144)$$

For odd half-integer spins, one has to resort to a fully quantum mechanical calculation. This can, for instance, be done by splitting the full dipole–dipole Hamiltonian into a secular term with respect to the quadrupole interaction, that is, a term that commutes with $\mathcal{H}_Q$, and a non-secular term. As before, the second moment is calculated from the trace of the double commutator, equation (2.117). However, while the full trace of the double commutator containing the secular part of $\mathcal{H}_{\text{dip}}$ is calculated, the double commutator containing the non-secular part is evaluated only for the nuclear spin states, $m = \pm \tfrac{1}{2}$, which are degenerate with the μ^+ spin states. The full and partial traces are most conveniently calculated on the basis of spin states quantised with respect to the direction of the electric field gradient. This implies, again, that the quadrupole coupling is by far the most dominating interaction. The more general case has been discussed by Meier (1984).

For an arbitrarily directed electric field gradient at each nucleus, one obtains the following formula

$$M_2^{ZF,Q,I=\text{odd}\tfrac{1}{2}} = \tfrac{1}{3} I(I+1)(\hbar \gamma_\mu \gamma_N)^2 \sum_{j=1}^{N} \frac{1}{r_j^6} \left(\sin^2 \gamma_j (1 - 3 \cos^2 \theta_j)^2 \right.$$

$$\left. + \frac{3}{4} \frac{(I+\tfrac{1}{2})}{I(I+1)} \times [2 + 3 \sin^2 \Omega_j - \sin^2 \gamma_j (1 - 3 \cos^2 \theta_j)^2] \right) \qquad (2.145)$$

Note that the first term in the curly brackets not containing the spin factor, represents the integer spin solution. The angles occurring in the formula are defined as follows: γ_i = angle between $P_\mu(0)$ and the direction of the electric field gradient (EFG) at the jth nucleus, θ_j = angle between r_j and the direction of the EFG at the jth nucleus, Ω_j = angle between $P_\mu(0)$ and r_j. For radially directed EFGs, $\theta_j = 0$, $\gamma_j = \Omega_j$ and the formula reduces to (Grebinnik et al 1978, Hayano et al 1979)

$$M_2^{\text{ZF},Q,I=\text{odd}\frac{1}{2}} = \tfrac{1}{3}I(I+1)(\hbar\gamma_\mu\gamma_N)^2 \sum_{j=1}^{N} \frac{1}{r_j^6}\left(4\sin^2\Omega_j + \frac{3}{4}\frac{I+\tfrac{1}{2}}{I(I+1)}(2-\sin^2\Omega_j)\right)$$

(2.146)

and the powder average

$$M_{2,\text{polyc}}^{\text{ZF},Q,I=\text{odd}\frac{1}{2}} = \tfrac{1}{3}I(I+1)(\hbar\gamma_\mu\gamma_N)^2 \frac{8}{3}\left[1 + \frac{3}{8}\frac{I+\tfrac{1}{2}}{I(I+1)}\right]\sum_{j=1}^{N}\frac{1}{r_j^6}$$ (2.147)

and

$$\Delta^{Q,I=\text{odd}\frac{1}{2}} = \tfrac{1}{2}M_{2,\text{polyc}}^{\text{ZF},Q,I=\text{odd}}.$$ (2.148)

Note that for $I = \tfrac{1}{2}$, equations (2.145) to (2.147) correctly reduce to equations (2.135) and (2.136), since in that case the quadrupole interaction vanishes. The second moments (2.135), (2.140), (2.145)) do not contain the field spread in the $P(0)$ direction. The second moments associated with this direction can also be calculated. In the case of (I = integer) for instance, one obtains

$$M_{2,\parallel}^{\text{ZF},Q,I=\text{integer}} = \tfrac{1}{3}I(I+1)(\hbar\gamma_\mu\gamma_N)^2 \sum \frac{4}{r_j^6}\cos^2\Omega_j.$$ (2.149)

For a cubic or polycrystalline metal with $\langle\cos^2\Omega_j\rangle = \tfrac{1}{3}$ one finds

$$M_{2,\parallel}^{\text{ZF},Q,I=\text{integer}} = \tfrac{1}{2}M_{2,\perp}^{\text{ZF},Q,I=\text{integer}}$$ (2.150)

or, in general,

$$M_{2,x}^{\text{ZF}} = M_{2,y}^{\text{ZF}} = M_{2,z}^{\text{ZF}}.$$ (2.151)

For non-cubic single crystal samples, equation (2.150) is not fulfilled, and the derivation of the Kubo–Toyabe formula is not possible in closed analytical form. A numerical calculation of the zero field relaxation function for $M_{2,x}^{\text{ZF}} = M_{2,y}^{\text{ZF}} = \tfrac{1}{8}M_{2,z}^{\text{ZF}}$ has been presented by Petzinger and Wei (1984).

The effect of a transverse field B, which is small in comparison with the quadrupole interaction but larger than the internal field components ($B > H_i$), on the second moment can now be deduced by the following geometrical argument (Grebinnik et al 1978). In the case of a zero field, the second moment, $M_2^{\text{ZF},Q}$, was determined by both the dipolar field components in the x and the y direction, these being perpendicular to $P(0) = P_z(0)\hat{z}_0$

$$M_2^{ZF,Q} = M_2^{ZF,Q}(H_{i,x}^2) + M_2^{ZF,Q}(H_{i,y}^2) \qquad (2.152)$$

and the powder average, given by

$$M_{2,\text{polyc}}^{ZF,Q} = 2M_{2,\text{polyc}}^{ZF,Q}(H_{i,x}^2) = 2M_{2,\text{polyc}}^{ZF,Q}(H_{i,y}^2). \qquad (2.153)$$

If a small field, B_0, is now applied along, say, the x axis, the total field will read:

$$|B_{\text{tot}}| = [(B_0 + H_{i,x})^2 + H_{i,y}^2]^{1/2} = (B_0 + H_{i,x})\left[1 + \left(\frac{H_{i,y}}{B_0 + H_{i,x}}\right)^2\right]^{1/2}$$

$$\simeq B_0 + H_{i,x} \qquad (2.154)$$

This means the field spread relevant for the relaxation in a transverse field is determined solely by the spread of the dipolar field components in the x direction. Hence the second moment, $M_{2,\text{polyc}}^{Q,B_0}$, appropriate for the transverse field relaxation function is given by:

$$M_{2,\text{polyc}}^{Q,B_0} = \tfrac{1}{2} M_{2,\text{polyc}}^{ZF,Q} = \Delta^Q. \qquad (2.155)$$

If the internal field spread is described by an isotropic Lorentzian distribution, as in dilute magnetic systems (Walstedt and Walker 1974), i.e.

$$f(H_i) = \pi^{-1}[a/(a^2 + H_i^2)] \qquad (2.156)$$

where a is the half width at half maximum, then the Kubo–Toyabe relaxation function assumes the form (Kubo 1981)

$$P_z(t) = \tfrac{1}{3} + \tfrac{2}{3}(1 - at)\exp(-at). \qquad (2.157)$$

Recently the validity of the random local field approximation in deriving the zero field Kubo–Toyabe function has been challenged by Celio and Meier (1983, 1984) and Meier (1984). These authors considered the full dynamics of the coupled muon nuclear spin system by explicitly evaluating the expression

$$P(t) = \text{Tr}\{\rho \exp[(i/\hbar)\mathcal{H}_{\text{dip}}t]\sigma_\mu \exp[-(i/\hbar)\mathcal{H}_{\text{dip}}t]\} \qquad (2.158)$$

with

$$\mathcal{H}_{\text{dip}} = \sum_{j=1}^{N} \mathcal{H}_{\text{dip},j}.$$

where $\mathcal{H}_{\text{dip},j}$ is given by equation (2.97). The calculation involves the determination of the $2(2I+1)^N$ eigenvalues and eigenstates of the Hamiltonian $\mathcal{H}_{\text{dip}}$, and the calculation of the trace in the basis of these eigenvectors. Such a calculation is only feasible if the number of nuclei is restricted to a few (4–6) nearest μ^+ neighbours. The main difference to the classical Kubo–Toyabe function is an oscillatory behaviour in the long term, instead of a simple recovery to $\tfrac{1}{3}$. Since the μ^+ in fact is coupled to a large number of nuclei, it can be expected that for large t its initial polarisation will be shared by all these nuclei, leading finally to an almost total

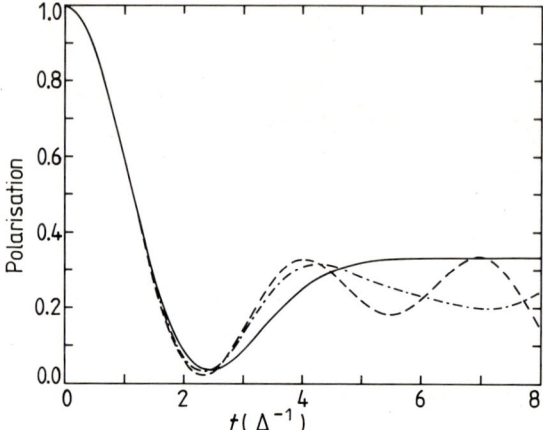

Figure 2.24 Time dependence of the muon polarisation in zero field. The chain curve/broken curve has been calculated for a μ^+ at an octahedral/tetrahedral site, interacting with $N = 6/4$ nuclei, with $I = \frac{1}{2}$. The full curve shows the Kubo–Toyabe function. All are FCC structures. The time unit (Δ^{-1}) is chosen in such a way that all the curves coincide initially (Celio and Meier 1984).

depolarisation of the μ^+. As an example, figure 2.24 shows the zero field relaxation function for a μ^+ at a tetrahedral and an octahedral site, with respectively 4 and 6 nearest nuclei, with $I = \frac{1}{2}$. Petzinger and Wei (1984) have shown that the occurrence of an oscillatory behaviour of $P_z(t)$ at long times can be traced back to a non-gaussian field distribution. This arises in the exact treatment of the dipolar coupling of muon and nuclear spins.

2.4 Total Magnetic Field at the μ^+

The presence of a magnetic field at the μ^+ site, perpendicular to its initial polarisation, will lead to a Larmor precession as discussed in §2.1.3. By determining the Larmor frequency from a transverse field μSR histogram, it will be possible to determine the local field, B_μ. It is, therefore, important to have a principal understanding of the possible contributions to the average total magnetic field at the μ^+ in the interior of the target sample. Aside from an external field B_{ext}, internal fields will result from magnetic dipole moments and contact hyperfine fields which will be discussed in chapter 4. In order to account for the internal field contributions it is convenient to introduce the so-called Lorentz sphere around the probe particle, which separates the sample into two regions (figure 2.25). Inside a Lorentz sphere of suitable diameter, the

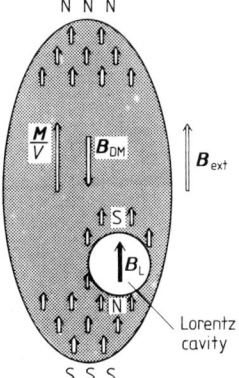

Figure 2.25 Illustration of the macroscopic magnetic field inside an empty sphere, the Lorentz sphere or cavity, in a magnetised ellipsoidal sample with the applied field parallel to the large axis of the ellipsoid. In the continuum approximation all magnetic dipoles outside the cavity will produce a field inside the cavity given by $B_{DM} + B_L = -(N - 4\pi/3)M$, where M is the sample magnetisation, and N is the demagnetisation factor. The arrows indicate the static component of the individual magnetic moments induced by the external field.

sources of magnetic fields are treated individually (microscopically), and outside the Lorentz sphere continuum theory is applied. Most generally the total magnetic field at the μ^+ can be expressed as follows.

$$B_\mu = B_{ext} - B_{DM} + B_L + B_c + B_{dip} \qquad (2.159)$$

where B_{ext} is the externally applied field, B_{DM} is the demagnetisation field stemming from the magnetisation of the finite sample, B_L is the so-called Lorentz field which is produced by the empty Lorentz sphere inside a magnetised environment, B_c is a contact hyperfine field at the μ^+ (including diamagnetic effects) and B_{dip} is the net field from magnetic dipoles inside the Lorentz sphere. The demagnetisation field and the Lorentz field depend on the sample magnetisation, M,

$$B_{DM} = -NM \qquad (2.160)$$

$$B_L = \frac{4\pi}{3} M \qquad (2.161)$$

where N is the shape dependent demagnetisation factor. Equation (2.160) only strictly corrects for an ellipsoidal sample if the magnetisation is directed along one of the principal ellipsoid axes. If the magnetisation is induced by the external field one has

$$M = \chi_{tot} B_{ext} \qquad (2.162)$$

where χ_{tot} is the bulk magnetic susceptibility. In ferromagnetic samples there will always be a Lorentz field which involves the domain or saturation magnetisation (domain size $\geqslant$ Lorentz sphere). The demagnetisation field, on the other hand, will depend on the net magnetisation of the whole sample.

The dipole field B_{dip} is simply the superposition of all fields, b_i, from the magnetic moments at position r_i, inside the Lorentz sphere or cavity

$$B_{dip} = \sum_{r \in \Omega^L} b_i(r_\mu - r_i). \qquad (2.163)$$

References

Abragam A 1970 *The principles of nuclear magnetism* (Oxford: Clarendon) p 103
Brewer J H, Fleming D G and Percival P W 1982 *Fourier, Hadamard and Hilbert transformations in chemistry* ed. A G Marshall (New York: Plenum) pp 345–85
Brown J A, Heffner R H, Leon M, Dodds S A, Vanderwater D A and Estle T L 1979 *Phys. Rev. Lett.* **43** 1751
Burkhard P, Roduner E, Hochmann F and Fischer H 1984 *J. Phys. Chem.* **88** 773
Camani M, Gygax F N, Klempt E, Rüegg W, Schenck A, Schulze R and Wolf H 1978 *Phys. Lett.* **77B** 326
Celio M and Meier P F 1983 *Phys. Rev. B* **27** 1908
——1984 *Hyperfine Interactions* **17–19** 435
Christiansen J, Mahnke H E, Recknagel E, Riegel D, Schatz G, Weber G and Witthuhn W 1970 *Phys. Rev. C* **1** 613
Coffin T, Garwin R L, Penman S, Ledermann L M and Sachs A M 1958 *Phys. Rev.* **109** 973
Fuller H 1976 *J. Phys. Chem. Ref. Data* **5** 835
Garner D M 1979 *Diss.* University of British Columbia
Goldhaber M, Grodzins L and Sunyar A W 1957 *Phys. Rev.* **106** 826
——1958 *Phys. Rev.* **109** 1015
Grebinnik V G, Gurevich I I, Zhukov V A, Ivanter I G, Klimov A I, Maiorov V N, Manych A P, Nikolskii B A, Pirogov A V, Ponomarev A N, Selivanov V I and Suetin V A 1978 *Zh. Eksp. Teor. Fiz.* **75** (Engl. transl. 1980 *Sov. Phys.–JETP* **48** 1002)
Hagedorn R 1964 *Relativistic Kinematics* (New York: Benjamin)
Hartmann O 1977 *Phys. Rev. Lett.* **39** 832
Hayano R S, Uemura Y J, Imazato J, Nishida N, Yamazaki T and Kubo R 1979 *Phys. Rev. B* **20** 850
Holzschuh E, Kündig W and Patterson B D 1981 *Helv. Phys. Acta* **54** 552
Kitaoka Y, Takigawa M, Yasuoka H, Itoh M, Takagi S, Kuno Y, Nishiyama K, Hayano R S, Uemura Y J, Imazato J, Nakayama H, Nagamine K and Yamazaki T 1982 *Hyperfine Interactions* **12** 51
Klempt E, Schulze R, Wolf H, Camani M, Gygax F N, Rüegg W, Schenck A and Schilling H 1982 *Phys. Rev. D* **25** 652
Kubo R and Toyabe T 1967 *Magnetic Resonance and Relaxation* ed. R Blinc (Amsterdam: North Holland)

References

Kubo R 1981 *Hyperfine Interactions* **8** 731
Matthias E 1967 *Hyperfine Interactions* ed. A J Freeman and R B Frankel (New York: Academic)
Meier P F 1984 *Hyperfine Interactions* **17–19** 427
Nagamine K and Yamazaki T (ed.) 1981 *UT-MSL Newsletter No* 1 (University of Tokyo, Meson Science Laboratory)
——1982 *UT-MSL Newsletter No* 2 (University of Tokyo, Meson Science Laboratory)
Nakayama H, Imazato J, Hayano R S, Nishiyama K, Kuno Y, Fukuchi K, Sasaki H and Kataoka Y 1981 *UT-MSL Newsletter No* 1 (University of Tokyo, Meson Science Laboratory)
Okun L B 1965 *Weak Interactions of Elementary Particles* (Oxford: Pergamon) p 53
Petzinger K G and Wei S H 1984 *Hyperfine Interactions* **17–19** 441
Pfifer A E, Bowen T and Kendall K R 1976 *Nucl. Instrum. Methods* **135** 39
Schenck A 1970 *Phys. Lett.* **32A** 19
——1976 *Nuclear and Particle Physics at Intermediate Energies* ed. J B Warren (New York: Plenum) p 159
Schenck A, Williams D L, Brewer J H, Crowe K M and Johnson R F 1972 *Chem. Phys. Lett* **12** 544
Slichter C P 1978 *Principles of Magnetic Resonance* (Berlin: Springer)
Trower W P 1966 *Berkeley Radiation Laboratory Report* UCRL 2426
Van Vleck J H 1948 *Phys. Rev.* **74** 1168
Walstedt R E and Walker L R 1974 *Phys. Rev.* B **9** 4857
Yamazaki T 1979 *Hyperfine Interactions* **6** 115

3

Muon Spin Rotation in Metals: Localisation, Diffusion, Trapping

3.1 Small Polaron State of μ^+ in Metals—a Variation of the Hydrogen in Metals Problem

We begin this chapter by discussing some basic properties of a μ^+ implanted in an ideal metal matrix. These concern its site of localisation and the related static effects which can be understood on the basis of a small polaron picture of the μ^+. One important feature is self-trapping at an interstitial site. The possible resulting modes of motion (which are generally quite rapid) will be discussed in the next section, as well as trapping or freezing effects caused by lattice imperfections. Here we will be concerned with the static properties of the self-trapped state and their experimental implications. We have to realise at this point, that, in this context the μ^+ is nothing but a light isotope of the proton. Like the proton it carries a single elementary charge and introduces the same strongly disturbing Coulomb potential into the metal lattice. The state of the μ^+ in a metal can therefore be discussed on the basis of models developed for hydrogen in metals.

Discussion of more sophisticated calculations of the hydrogen electronic structure in metals will be delayed until §3.3.1. We will consider here, in a more phenomenological way, the interaction of the μ^+ with the surrounding host atoms, following the description of Kehr (1978) and Sugimoto and Fukai (1980).

The host atoms and the μ^+ are bound elastically to each other, the interaction potential of the μ^+ and the host atoms being described by a sum over the two-body interaction potentials

$$V = \sum_m V(R^m + u^m - r_\mu). \tag{3.1}$$

R^m is the position of the mth host atom in the pure lattice, and u^m is the shift in position due to the presence of the μ^+. The displacements are determined by the condition that the total energy assumes a minimum. It follows then that

$$u^m = (D^{mn})F^n. \tag{3.2}$$

$(D)^{mn}$ is the static phonon Green's function in the harmonic approximation

$$D_{ij}^{mn} = \frac{1}{N} \sum_{k \neq 0, \lambda} \frac{e_i^{\lambda}(k) e_j^{\lambda}(k)}{M \omega_{k\lambda}^2} \exp i k(R^m - R^n) \qquad (i, j = x, y, z) \qquad (3.3)$$

with M = mass of metal atom; $\omega_{k\lambda}$ are the harmonic phonon frequencies with wave vector k and polarisation index λ and $e^{\lambda}(k)$ is the polarisation vector for the phonon mode k, λ. F^n is the force (called the Kanzaki force) that the μ^+ exerts on the nth metal atom, and is given by

$$F^m = -\int dr_{\mu} |\psi_{\mu}(r_{\mu})|^2 \operatorname{grad}_u V(R^m + u^m - r_{\mu}) \qquad (3.4)$$

where $\psi_{\mu}(r_{\mu})$ is the wavefunction of the μ^+ in the potential V.

In effect, the presence of the μ^+ in the metal lattice leads to a distortion of the near neighbour environment or the build-up of a short range strain field. It can be pictured as a dressing of the μ^+ by local static phonons. This new entity, the phonon-dressed μ^+, is called a small polaron. The small polaron is described here in a form equally applicable to large polarons. However, it is important to recall that there is a distinctive difference between the two systems (see e.g. Jones and March 1973). The large polaron carries a long range strain field (a few lattice constants) and is characterised by a weak coupling to phonons, while the small polaron carries a short range strain field (typically of the order of a lattice constant) and involves a very strong phonon coupling. This has important consequences in particular for the transport properties of the two entities. The large polaron can be pictured as a wide band quasi-particle and is therefore quite distinct from the small polaron which assumes the features of a 'localised' state. Its motional modes are described in §3.2.

The lattice relaxation around the μ^+ leads to an energy gain of

$$\Delta U = -\tfrac{1}{2} u^m (\phi^{mn}) u^n \qquad (3.5)$$

where $(\phi)^{mn}$ is derived from the lattice potential function $\phi(x)$ ($x = \{R^m + u^m\}_{m=1...N}$)

$$\phi_{ij}^{mn} = \frac{\partial^2 \phi(x)}{\partial u_i^m \, \partial u_j^n}. \qquad (3.6)$$

ϕ_{ij}^{mn} and the static lattice Green's function D_{ij}^{mn} are related by the expression

$$\phi_{\sigma\mu}^{sm} D_{\mu\nu}^{mn} = -\delta_{sn} \delta_{\sigma\nu}. \qquad (3.7)$$

This energy gain is shown schematically in figure 3.1 (Kehr 1978). The broken curve represents the potential with respect to a lattice coordinate u for a hydrogen free lattice. Insertion of hydrogen, without allowing for relaxation, would lower this curve by U_{el}, the electronic contribution to the binding energy (this contribution could also be positive, if hydrogen is expelled from the lattice). If relaxation to a new equilibrium position $\hat{u}$ is now allowed a further energy gain of ΔU is obtained. Of interest is also the energy difference

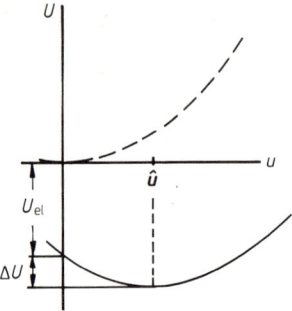

Figure 3.1 Potential energy U of a metal–hydrogen system as a function of one representative lattice coordinate u. The broken curve indicates the potential energy without hydrogen in an harmonic approximation. U is a quadratic function of u. The full curve indicates the potential energy after inserting one hydrogen atom. U_{el} is an electronic contribution to the binding energy of the hydrogen atom. The parabola is shifted such that its slope at $u=0$ gives the force $-F$ acting on a representative lattice atom. The lattice relaxes to a new equilibrium configuration $\hat{u}$, thereby lowering the energy by ΔU (Kehr 1978).

δU between a proton/μ^+ at some site 'b' next to site 'a' of the same type and the proton/μ^+ at site 'a', around which the lattice is fixed in the relaxed configuration

$$\delta U_{ba} = \tfrac{1}{2}(^a u^m - {}^b u^m)(\phi^{mn})(^a u^n - {}^b u^n). \tag{3.8}$$

This expression is always positive, which means the μ^+ has created its own trap at site 'a' by means of the lattice relaxation. It is customary to speak in this respect of a self-trapped state. This does not mean that the μ^+ has to be localised necessarily at this site. Rather, the self-trapped, small polaronic state may be described as a narrow band involving the possibility of coherent motion (see next section). However, the type of interstitial lattice site involved is of course determined by considering which site assignment leads to a minimal total energy, $U_{el} + \Delta U$.

The small polaron state of the μ^+ in a metal is thus characterised by the type of interstice occupied, and the lattice distortion formed around this site. The possible interstitial sites in FCC, BCC and HCP crystal structures are pictured in figures 3.2, 3.3 and 3.4.

The lattice site of the μ^+ can be determined in those metals in which second moment measurements are possible. This requires metals in which the nuclei carry spin and magnetic moments, and give rise to a distribution of internal fields over the available lattice sites. For single crystals in a strong transverse external field the second moment of this internal field distribution has been discussed in §2.3.1 and is once more quoted here:

$$M_2 = 2\sigma^2 = \tfrac{1}{3}I(I+1)\gamma_\mu^2 \gamma_N^2 \hbar^2 \sum \frac{(1-3\cos^2\theta_i)}{r_i^6}. \tag{3.9}$$

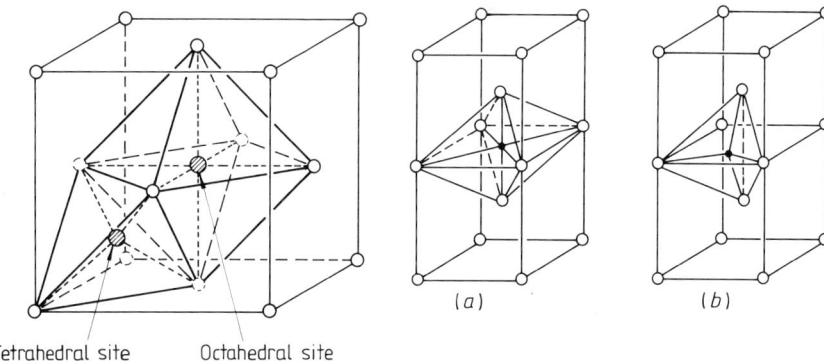

Tetrahedral site Octahedral site
Figure 3.2

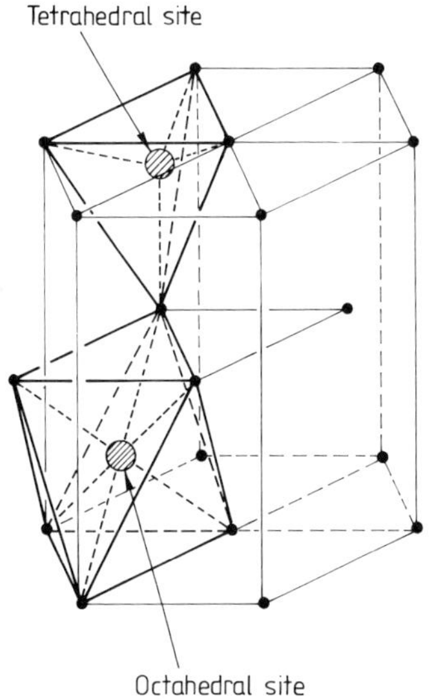

Tetrahedral site

Octahedral site
Figure 3.4

Figure 3.2 FCC crystal structure with tetrahedral and octahedral sites indicated. There are two tetrahedral and one octahedral interstitial sites per atom.

Figure 3.3 BCC crystal structure with (a) octahedral (three per atom) and (b) tetrahedral (six per atom) sites indicated.

Figure 3.4 HCP crystal structure with tetrahedral (two per atom) and octahedral (one per atom) sites indicated.

Table 3.1 gives a list of σ values calculated with equation (3.9) for various metals, assuming no lattice relaxation. A lattice dilatation around the μ^+ should lead to a decreased second moment, whilst a lattice contraction leads to an increased one relative to the rigid lattice value. So far, second moment data in the strong field Van Vleck limit are only available from Cu (Camani *et al* 1977, Schilling *et al* 1982), Al (Hartmann *et al* 1978b, Kehr *et al* 1982), V (Schilling *et al* 1982) and Nb (Birnbaum *et al* 1978, Schilling *et al* 1982, Gygax *et al* 1984a). In the case of Ta it was impossible to reach the Van Vleck limit because of the strong quadrupole interaction, previously discussed in §2.3.2 (Schilling *et al* 1978). Some of the experimental results are also listed in table 3.1. Only the results from Cu and Al allow a straightforward interpretation. The change of M_2 with orientation is only compatible with specific site

Table 3.1 Compilation of theoretical and experimental $\sigma = (\frac{1}{2}M_2)^{1/2}$ in the Van Vleck limit.

$B_{ext}\parallel$		σ_{calc} for a rigid lattice (μs^{-1}); values in parentheses from nearest neighbours only		σ_{exp} (μs^{-1})		Reference
		Octahedral	Tetrahedral			
Cu	$\langle 100 \rangle$	0.308 (0.302)	0.077 (0.0)	0.265		Schilling et al (1982)
	$\langle 110 \rangle$	0.165 (0.151)	0.279 (0.268)	0.15	$30 \leqslant T \leqslant 80$ K	
	$\langle 111 \rangle$	0.067 (0.0)	0.319 (0.310)	0.07		
Al	$\langle 100 \rangle$	0.31	0.08	0.09		Hartmann et al (1978)
	$\langle 110 \rangle$	0.16	0.28		$T = 15$ K	Kehr et al (1982)
	$\langle 111 \rangle$	0.07	0.32	0.24		
V	$\langle 100 \rangle$	0.616 (0.590)	0.449 (0.430)	0.21		Schilling et al (1982)
	$\langle 110 \rangle$	0.367 (0.295)	0.386 (0.363)	0.17	$T \leqslant 27$ K	
	$\langle 111 \rangle$	0.231 (0.0)	0.362 (0.338)	0.16		
Nb	$\langle 100 \rangle$	0.562 (0.538)	0.409 (0.392)	0.28		Gygax et al (1984a)
	$\langle 110 \rangle$	0.335 (0.269)	0.351 (0.331)	0.195	$T = 14$ K	
	$\langle 111 \rangle$	0.211 (0.0)	0.330 (0.308)	0.195		
Ta	$\langle 100 \rangle$	0.219 (0.209)	0.159 (0.153)			Schilling et al (1982)
	$\langle 110 \rangle$	0.131 (0.105)	0.136 (0.128)	0.107†	$T = 30$ K	
	$\langle 111 \rangle$	0.082 (0.0)	0.128 (0.120)			

Within the experimental column, additional values at $T = 40$ mK for Al: 0.22, 0.09; and for Nb: 0.29, 0.25, 0.24 at $T = 37$ K.

† Quadrupole interaction dominant, Van Vleck limit was not reached.

assignments, such as an octahedral one in Cu, a tetrahedral site in Al at 15 K and an octahedral site assignment at 40 mK in Al.

Consulting table 3.1 it is seen that the experimental value for $B\|\langle 111\rangle$ in Cu ($B\|$ is the external field parallel to the respective directions) is almost unchanged (slightly larger in fact) from the rigid lattice prediction, while $B\|\langle 100\rangle$ and $B\|\langle 110\rangle$ are definitely smaller than the rigid lattice predictions. These findings are nicely explained by noting that the nearest Cu neighbours do not contribute to the second moment for the $\langle 111\rangle$ axis parallel to B orientation. It then follows that only the nearest Cu neighbours are dilated outwardly from the μ^+ with $\delta r/r \simeq 5\%$, while the second nearest neighbours are unchanged in position or even slightly contracted towards the μ^+.

The nearest neighbour displacements can be used to calculate the force constant between the μ^+ and a nearest neighbour Cu atom via equation (3.2) in the so called Kanzaki formalism (Kanzaki 1957). Equation (3.2) can thus be rewritten in the following form:

$$u_i^m = \sum_{j,k} V_{ij}(-k)\Gamma_j(-k)\exp i k \cdot R^n. \quad (3.10)$$

The sum has to be taken over all k vectors in the first Brillouin zone, $V_{ij}(-k)$ is the host metal force constant matrix, and $\Gamma_j(-k)$ is the Fourier transform of F_j^m (k is the reciprocal lattice vector). V_{ij} has been calculated for an FCC lattice by Miller and Heald (1975) on the basis of nearest and second nearest neighbour host atom interactions. Including only nearest neighbour μ^+ lattice interactions, and assigning the μ^+ to an octahedral interstitial position, $\Gamma_j(-k)$ is given by the expression (Miller and Heald 1976):

$$\Gamma_j(-k) = -2ig \sin(k_j a/2) \quad (3.11)$$

where g is now the μ^+ nearest neighbour Cu atom force constant and a the lattice constant. Equation (3.10) has been evaluated by Schilling et al (1982). Equating u_x^1 by the measured nearest Cu neighbour dilatation (i.e. $u_x^1 = 0.05a/2$), one obtains $g = 3.39$ eV a^{-1}. This value is in good agreement with a direct model calculation of g from equation (3.4) by Teichler (1978) for a μ^+ at an octahedral site in Cu, $g = 2.0$ eV a^{-1}. Using linear elasticity theory (Wagner and Horner 1974) the force constant g can be used to calculate the total volume expansion ΔV due to a μ^+:

$$\Delta V = 3ga/(C_{11} + 2C_{12}). \quad (3.12)$$

C_{11} and C_{12} are elastic stiffness constants. Using $g = 3.39$ eV a^{-1} one obtains

$$\Delta V_{\mu^+} = 3.91 \text{ Å}^3.$$

for μ^+ in Cu. For protons in metals it is known (Peisl 1978) that the total volume expansion for many metals and alloys is pretty constant, namely

$$\Delta V_p = 2.9 \text{ Å}^3.$$

The difference between ΔV_p and ΔV_{μ^+} is attributed to a larger zero point vibration amplitude of the μ^+ (Schilling *et al* 1982).

The results in Al at 15 K indicate a dilatation of the μ^+ nearest neighbour atoms of $\delta r/r \simeq 10\%$. As will be discussed in §3.8.1 localisation of the μ^+, which is a prerequisite to measuring σ or M_2, is only observed in the presence of small amounts of certain impurities. At the lowest temperatures a transition to an octahedral site occupation is indicated (see figure 3.8). In view of these observations, it is questionable whether the intrinsic self-trapped small polaron state is observed.

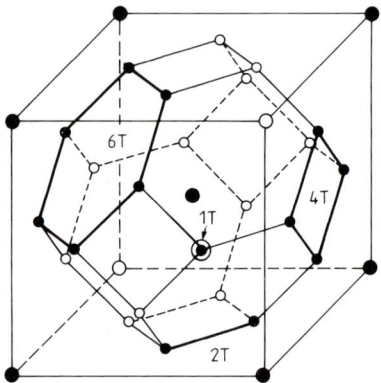

Figure 3.5 Tunnelling configurations 1T, 2T, 4T and 6T, involving one, two, four and six tetrahedral sites in the BCC lattice.

With reference to the results in V, Nb and Ta, trapping near impurities is always indicated (see also §3.8.2). The intrinsic self-trapped state does not seem to be observed anywhere. The high field results in Nb and V have been interpreted as caused by a mixed occupation of tetrahedral and octahedral sites (Nb: tetra/octahedral $\simeq 2.0$, V: tetra/octahedral $\simeq 4.4$) involving average nearest neighbour displacements $\delta R/R$ of $+31\%$ (V) and $+23\%$ (Nb), respectively (Schilling *et al* 1982). Compared with the corresponding number for hydrogen in these metals (Johnston and Scholl 1980, Peisl 1978), these numbers appear unphysically high. In any case, a preponderance towards a tetrahedral interstitial site occupation is indicated in agreement with the fact that hydrogen is always found at the tetrahedral interstices in BCC metals.

To explain the reduced experimental values for the second moment, it was suggested that the μ^+ might actually be self-trapped in an extended tunnelling configuration involving 2, 4, 6 etc tetrahedral interstices (see figure 3.5) (Lankford *et al* 1978). Assuming a δ function distribution of the μ^+ over these sites, the second moment follows from the generalisation of equation (2.119) (see also equation (3.38)) and reads

$$M_2 = \tfrac{1}{3}\hbar^2 \gamma_N^2 \gamma_\mu^2 I(I+1) \sum_{i,m,n} \frac{(1 - 3\cos^2 \theta_{i,m})(1 - 3\cos^2 \theta_{i,n})}{r_{i,m}^3 r_{i,n}^3}. \quad (3.13)$$

The sum extends over all host nuclear sites i, and over all μ^+ sites involved in the tunnelling configuration. Table 3.2, taken from Lankford et al (1978), contains a collection of M_2 values for the rigid Nb lattice with various tunnelling configurations involving tetrahedral interstices.

Table 3.2 Calculated depolarisation rates (in $10^{-6} \, s^{-1}$) in the Van Vleck limit for muons occupying interstitial sites in Nb. 1T is the tetrahedral site and 1O the octahedral site. The next four columns are for tetrahedral site multiplets: 2T—the near-neighbour pair; 4T—the square array in (100) planes; 5T—a tetrahedrally coordinated 'star'; 6T—the hexagon-shaped array, parallel to (111) planes (Lankford et al 1978, van der Wal ETH, Zürich).

z axis	1T	1O	2T	4T	5T	6T
⟨100⟩	4.038	5.552	3.546	3.270	2.665	2.129
⟨110⟩	3.472	3.310	2.975	2.474	2.143	2.010
⟨111⟩	3.261	2.083	2.758	2.144	1.937	1.969

The self-trapped state for μ^+ and protons in Nb has been investigated theoretically by Sugimoto and Fukai (1980) by numerically solving the appropriate Schrödinger equation and expressing the μ^+ nearest neighbour interaction potential by a double Born–Mayer potential. The parameters in this potential were adjusted so as to reproduce the experimentally known force dipole tensor **P** and the first two excitation energies of protons in the local potential as determined from inelastic scattering of neutrons. The total energy is given by the sum

$$E_{\text{tot}} = E_\alpha(\{u_i^m\}) + E_L(\{u_i^m\}) \quad (3.14)$$

where $E_\alpha(\{u_i^m\})$ is the lattice relaxation energy equation (3.5), and $E_L(\{u_i^m\})$ is an energy eigenvalue of the Schrödinger equation. The atomic displacements are determined by requiring self-consistency among the solutions of the Schrödinger equation and equations (3.1), (3.2) and (3.4). Various tunnelling configurations were used as input configurations as well as a simple tetrahedral site assignment. For protons the lowest total energy was obtained for a single tetrahedral (T) site occupancy. For μ^+ interestingly, the lowest energy was achieved for a 4T tunnelling confinguration (see table 3.3). The spread of the wavefunctions for μ^+ and p in the 4T state are shown in figure 3.6. In contrast to the proton, the μ^+ wavefunction is peaked at the octahedral position with a relatively large RMS spread. This situation is more properly described as an octahedral interstitial site occupany. Table 3.3 also lists the

Table 3.3 Properties of the self-trapped state of positive muons in Nb (Sugimoto and Fukai 1980).

Configuration	Self-trapping energy E (eV)	Displacements ($a/200$)				P tensor Tr P (eV)	$\dfrac{3(A-B)}{\text{Tr P}}$
		first neighbour		second neighbour			
		u_1	u_2	u_3	u_4		
1T	−0.938	8.6	3.1	−0.3	−0.2	12.90	−0.25
4T	−0.981	14.4		0.8		11.18	0.93
6T	−0.697	7.1		3.0	0.0	15.16	0.19

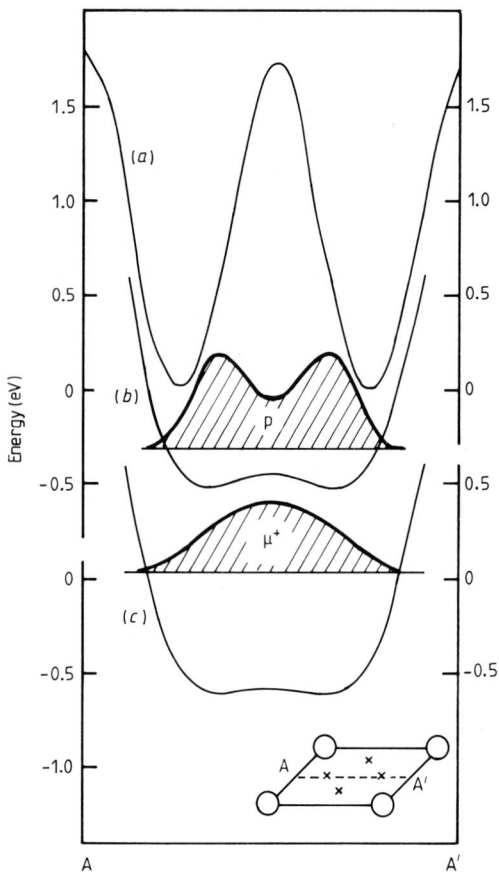

Figure 3.6 Potential profiles and groundstate wavefunctions of a proton and a positive muon in the 4T configuration in Nb. Variation is along the line AA′ as shown in the inset. (a) Potential profile in the undistorted lattice, (b) with proton, and (c) with muon (Sugimoto and Fukai 1980).

displacements of the nearest neighbours atoms. Scaling the second moment for an octahedral interstitial position (table 3.1) by this increase in distance, one predicts $\sigma\langle 100\rangle = 0.39\ \mu s^{-1}$, $\sigma\langle 110\rangle = 0.23\ \mu s^{-1}$, $\sigma\langle 111\rangle = 0.14\ \mu s$. Comparing these numbers with the experimental values in table 3.1, one finds an obvious disagreement. An octahedral site occupany is also ruled out by the field dependence of the second moment, as discussed below.

Information on the site occupation of the μ^+ can also be obtained in ferro- and antiferromagnetic single crystal metals with BCC and HCP crystal structure due to the nonvanishing dipole fields at the interstitial position. This will be discussed in §4.3.3.

Another static property of the μ^+ state in a metal, which is evidenced by the second moment behaviour, concerns the electric field gradient (EFG) exerted by the electron screened μ^+ on its neighbours. As discussed in §2.3.2, such EFGs will only become manifest if the host nuclei possess spin $I > \frac{1}{2}$, and a non-zero spectroscopic quadrupole moment. Characteristic modifications of the second moment values will then result (Hartmann 1977). This modification can be supressed in sufficiently strong applied fields when the Zeeman interaction energy is much larger than the quadrupole interaction energy (Van Vleck limit). The effect of the EFGs have so far been observed in Cu (Camani et al

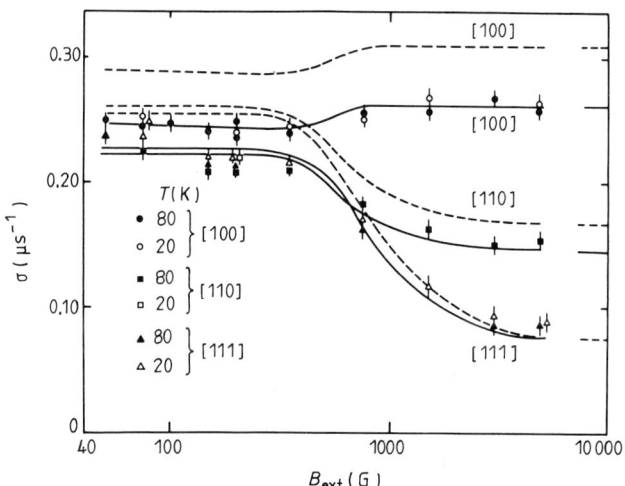

Figure 3.7 Field and orientation dependence of the relaxation constant σ in Cu at 20 and 80 K. The broken curves are the predictions of the Hartmann theory for an octahedral site assuming a rigid lattice and the EFGs directed radially away from the μ^+. The full curves represent a fit of the Hartmann theory to the data, yielding a nearest neighbour outward displacement of $\delta r/r = 5\%$ and an EFG at the nearest neighbour Cu nuclei of $V_{z'z'} = 0.30\ \text{Å}^{-3}$ (Camani et al 1977).

1977, Hartmann et al 1981), Al (Hartmann et al 1980, Hartmann 1981), Nb (Borghini et al 1978, Gygax et al 1983), V (Schilling et al 1982) and Ta (Schilling et al 1978, 1982). The field dependent σ in Cu and Al is displayed in figures 3.7 and 3.8. From the analysis of $\sigma(B)$, one obtains an EFG of $V_{z'z'} = 0.30 \pm 0.15$ Å^{-3} in Cu (the large error is mainly due to the uncertainty in the Cu nuclear quadrupole moments) and $V_{z'z'} = 0.4$ Å^{-3} and $V_{z'z'} = 0.3$ Å^{-3} in Al at 15 K and 40 mK respectively, corresponding to tetrahderal and octahedral site occupation, with similar errors to Cu.

The field dependence of σ in Nb cannot be reproduced purely by the Hartmann theory on the basis of an EFG directed radially away from the μ^+. It appears likely that other contributions to the total EFG at the Nb nuclei may arise from the local lattice distortion which destroys the cubic symmetry, and from impurity atoms close to where the μ^+ has become trapped (see §3.8.2). A theoretical calculation of the μ^+ induced EFG in Cu has been published by Jena et al (1978). All the results are discussed in detail by Schenck (1982). This leads to a diffusion rate of $\Gamma \simeq v_{th}/\lambda$, where λ is the muon mean free path

3.2 Classification of Possible Motional Modes of μ^+ in Metals

Before we discuss the various techniques and approaches on how information on μ^+ diffusion can be obtained from μSR spectra, it is necessary to have a principal understanding of the possible modes, or patterns of motion, that the μ^+ might display in a metal.

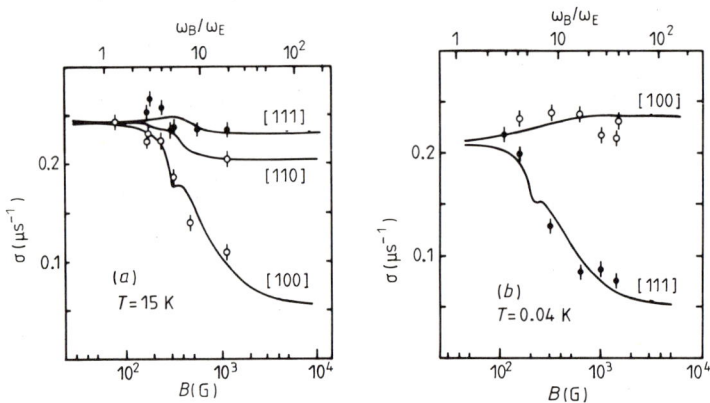

Figure 3.8 Field and orientation dependence of the relaxation parameter σ at (a) 15 K and (b) 40 mK in an AlMn (57 ppm) sample. The data at 15 K are consistent with a tetrahedral interstitial site while the data at 40 mK imply an octahedral interstitial site. The full curves represent fits of the Hartmann model to the data (Kehr et al 1982).

total jump-rate, is then $\Gamma = \tau_s^{-1}$. We also introduce a transition probability w_{if} for a particular jump from interstitial site i (initial) to interstitial site f (final). If there are z nearest neighbour interstitial sites f to interestitial site i, we have

$$\Gamma = \tau_s^{-1} = zw_{if}.$$

In metals of cubic symmetry the jump-rate Γ is related to the macroscopic diffusion constant D by the expression (following from Einstein's relation)

$$D = \frac{zd^2}{6} w_{if} = \frac{1}{\tau_s} \frac{d^2}{6} \qquad (3.15)$$

where d is the jump distance from site i to site f. Some values for d in terms of the lattice constant, a, are listed in table 3.4.

Table 3.4 Compilation of jump distances, d, between nearest neighbour interstitial sites in FCC and BCC metals, where a is the lattice constant.

Crystal structure	Tetrahedral–tetrahedral	Octahedral–octahedral
FCC	$\frac{1}{2}a$	$2^{-1/2}a$
BCC	$2^{-3/2}a$	$\frac{1}{2}a$

Since, as emphasised earlier, the μ^+ can be regarded as a light isotope of the proton, it seems reasonable to expect diffusional patterns similar to those of hydrogen in metals. However, the μ^+ is quite distinct from the proton in two respects. One is fundamental, and concerns the appreciable mass difference between the two particles ($m_\mu \simeq m_p/9$). As a result, the vibrational properties of the two particles in some interstitial potential will become quite different, introducing large isotope effects wherever a quantum mechanical description becomes necessary.

The second distinction is related to the μ^+ being a transient probe, and thus tells us only about the effects which happen during its lifetime of 2.2 μs after it has slowed down. It is therefore not entirely out of the question that the μ^+ might not have reached its lowest energy state in the lattice yet. In particular, that it has not yet dropped into the self-trapped or small polaron-like state discussed in §3.1. Indeed, Emin (1981) and Browne and Stoneham (1981) have proposed that there could be a time delay to self-trapping. Emin (1981) discusses the occurrence of a barrier between a non-polaronic state and the self-trapped small polaron-like state, while Browne and Stoneham (1981) consider the possible occupation of metastable states by the μ^+. In each case it is considered highly probable that the presence of impurities and other lattice imperfections could catalyse a fast transition to the lowest energy self-trapped configuration. Before that happens the muon may diffuse freely through the lattice with a velocity determined by its thermal energy, $v_{th} = (3k_B T/m_\mu)^{1/2}$. This leads to a diffusion rate of $\Gamma \simeq v_{th}/\lambda$, where λ is the muon free mean path between scattering events causing the diffusion constant D to be roughly $\Gamma\lambda^2 = v_{th}^2/\Gamma$ (Browne and Stoneham 1981).

A variety of motional modes has been considered in the literature for the μ^+ now in its self-trapped or small polaron-like state. Because of their quantum nature these modes are referred to as quantum diffusion. Starting at the lowest temperatures with virtually no phonons present, the small polaron, which is described by a narrow band, may move by coherent propagation, similar to a conduction electron. The basic mechanism is tunnelling from one site to the next, with unchanged phonon numbers (Flynn and Stoneham 1970, Kagan and Klinger 1974) in the initial and final states. Compared to the propagation in a rigid lattice, the small polaron coherent propagation is slowed down considerably because of the 'dressing' of the particle by a strain field. This implies, that the local displacements of host lattice atoms in a tunnelling event must also be shifted (translated) to the next site. This 'dressing' may be described in terms of an increased effective mass of the moving particle, with a correspondingly decreased tunnelling matrix element, or decreased bandwidth. A particular feature of the small polaron bandwidth is its relatively weak temperature dependence (Kehr 1975).

The velocity of propagation in the band state is given by the group velocity of the wave-packet representing the particle

$$v_g = \frac{1}{\hbar}|\nabla_k E(k)| \simeq \frac{\hbar k}{m_{eff}(T)} \simeq \frac{\hbar k_{max}}{m_{eff}(T)} \qquad (3.16)$$

where $E(k)$ is the energy at the momentum k and $m_{eff}(T)$ is the temperature dependent effective small polaron mass. For practical purposes, k is represented by its value at the zone boundary: $k_{max} = 1/a$, where a is the lattice constant. The effective mass can be approximated by

$$m_{eff}(T) = \frac{\hbar^2}{d^2 E(k)/dk^2} \simeq \frac{\hbar^2 k_{max}^2}{J_{eff}} \qquad (3.17)$$

where J_{eff} is an effective bandwidth or tunnelling matrix element, which is given by the product of a rigid lattice tunnelling matrix element, J_0, and the so-called phonon overlap or band narrowing factor, $\exp -S(T)$. This factor is formally analogous to a Debye–Waller factor (Kehr 1978). $S(T)$ is weakly temperature dependent and using an isotropic Debye model, can be approximated as $S(0) \simeq \frac{5}{4}\delta U_{ba}/2\hbar\omega_D$. The rate of diffusion Γ_{coh} is then given by

$$\Gamma_{\text{coh}} = \frac{v_g}{d} \simeq \frac{J_{\text{eff}}}{\hbar k_{\text{max}} d} \simeq \frac{J_{\text{eff}}}{\hbar} \tag{3.18}$$

where d is the distance between neighbouring sites, approximated very roughly by the lattice constant a. In a BCC metal, J_{eff} may be of the order of 1 meV (Teichler 1981a,b) which yields as a rough estimate: $\Gamma_{\text{coh}} \simeq 10^{12}$ s^{-1}.

In reality the ideal band propagation is limited by the finite lifetime of band-states, due to dynamical destruction (Kagan and Klinger 1974). This is the result of phonon induced (intra-site) scattering between different band-states which destroys the coherent nature of band propagation. According to Kagan and Klinger (1974), the diffusion rate should display the following temperature dependence for $T \ll \theta_D$ (θ_D = Debye temperature)

$$\Gamma_{\text{coh}} = \frac{1}{\hbar^2} \frac{J_{\text{eff}}^2}{\Omega(T)} \simeq \frac{1}{\hbar^2} \frac{J_{\text{eff}}^2}{B\omega_D} \left(\frac{\theta_D}{T}\right)^9 \tag{3.19}$$

where $\Omega(T)$ represents the intra-site scattering rate, and $B \simeq 10^7$ $(\omega_D/\omega_p)^4$ (ω_D is the Debye frequency, $\omega_p \hbar$ is the characteristic gap for the intra-well energy level of the proton (or μ^+)).

A similar formula was derived by Fujii (1979a,b) for the special case of a BCC lattice

$$\Gamma_{\text{coh}} = \frac{1}{\hbar^2} \frac{J_{\text{eff}}^2}{\Omega(T)} g_D \tag{3.20}$$

where g_D is a function that varies from 1 for $J_{\text{eff}}^2/\Omega(T) = 0$, to 0.50 (0.44) as $J_{\text{eff}}^2/\Omega(T)$ increases for octahedral,(tetrahedral) interstitial sites. For low temperatures, $\Omega(T)$ is given by:

$$\Omega(T) = \Omega(\theta_D) 5.6! (T/\theta_D)^7. \tag{3.21}$$

The equations (3.19) and (3.20), in which the effective tunnelling matrix element appears in quadratic form in contrast with equation (3.18), reflect the fact that the motion of the particle is no longer that of a freely evolving wavepacket, but is characterised by the random destruction of its evolution momentarily localising the particle.

This motional mode, which derives from the band picture of the small polaron is in competition with a purely stochastic process called phonon-assisted tunnelling, or incoherent tunnelling. It will dominate the diffusional motion at increased temperature and may be pictured as follows: because of the local lattice deformation, the energy level in the potential well occupied by the proton or muon is lowered appreciably as compared with the corresponding unoccupied level in a neighbour's potential well. Therefore a direct tunnelling event, by which the bare particle (p, μ^+) is shifted beneath the potential barrier into the next site, is not possible. However, such an event will become possible, if by way of lattice vibrations the two neighbouring potentials are momentarily distorted so as to produce degeneracy between the occupied and the unoccupied levels (coincidence configuration). The transition rate is then given by a product of a tunnelling matrix element, and the probability for the required local lattice configuration. The theory of small polaron motion by phonon-assisted tunnelling was first developed by Holstein (1959) and was later extended to the motion of hydrogen in metals by Flynn and Stoneham (1970) and Kagan and Klinger (1974). In the wake of the μ^+ diffusion studies, the theory has been further refined and extended by Teichler (1977), Fujii and Uemura (1978), McMullen and Bergerson (1978), Emin et al (1979), Fujii (1979a), Teichler and Seeger (1981), Teichler (1981a, 1981b) and Petzinger (1982). Without going into details here, we will merely list some of the main results.

Phonon-assisted or incoherent tunnelling motion may be characterised by the number of phonons involved in the jump event. At low temperatures, ($T \ll \theta_D$), one- and two-phonon processes have to be considered, and these are in competition with coherent diffusion.

One-phonon processes are only possible under certain conditions (Teichler and Seeger 1981, Teichler 1981a,b). One such condition is, that the orientation of the strain field associated with the μ^+/p (i.e. the small polaron), must be different in the initial and the final site. This can actually be fulfilled in BCC metals, where nearest neighbour tetrahedral or octahedral sites possess differently oriented axes of tetragonal symmetry (see figure 3.9). One-phonon processes also become possible in the presence of lattice strain, induced, for example, by impurities. This will lead to level shifts, and hence to a finite distribution of the energy difference between the initial and final levels occupied by the μ^+/p. The incoherent diffusion rate, Γ_h, induced by one-phonon-assisted tunnelling, is characterised by a linear temperature dependence (see also e.g. Sussman 1971):

$$\Gamma_h \simeq J_{\text{eff}}^2 T. \qquad (3.22)$$

Here and in equations (3.23) and (3.24) J_{eff} is the effective tunnelling matrix element introduced before.

Two-phonon processes are generally to be expected at low temperatures. Using an isotropic Debye model and considering only one interstitial site of a given type per unit cell (as in FCC metals) Flynn and Stoneham (1970) derived

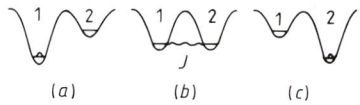

Figure 3.9 Phonon assisted tunnelling. (a) The μ^+ is self-trapped in site No 1, (b) thermal fluctuations have brought the levels in sites No 1 and No 2 to the same height and tunnelling is possible, (c) the μ^+ is finally self-trapped in site No 2.

the following expression for the transition probability, w_{if}:

$$w_{if} \simeq 57\,000 \, \frac{\pi}{\hbar} \, \frac{E_a^2}{(\hbar\omega_D)^3} \, J_{\text{eff}}^2 \left(\frac{T}{\theta_D}\right)^7. \tag{3.23}$$

E_a is the activation energy, and is explained shortly. If there is more than one interstitial site of a given type per unit cell (as in BCC metals), Fujii (1979) derived a modified expression, also within the isotropic Debye approximation

$$w_{if} = 48\pi^3 \, \frac{E_a^2}{\omega_D^3} \, J_{\text{eff}}^2 \left(\frac{T}{\theta_D}\right)^3. \tag{3.24}$$

The modified power occurs because now the two sites are connected by tunnelling, they are translationally inequivalent.

At higher temperatures (some fraction of θ_D), more and more phonons are becoming involved, and the diffusion is described as a multiphonon-assisted tunnelling mechanism. For temperatures well above θ_D the diffusion rate is given by an expression of the Arrhenius type

$$w_{if} = \frac{J_0^2}{2\hbar} \left(\frac{\pi}{E_a k_B T}\right)^{1/2} \exp\left(-\frac{E_a}{k_B T}\right). \tag{3.25}$$

Here J_0 is the rigid lattice tunnelling matrix element introduced before. The activation energy, E_a, appearing in equation (3.25) is not the potential barrier height, E_0, but is related to the lattice relaxation energy δU, discussed in §3.1

$$E_a = \tfrac{1}{4}\delta U. \tag{3.26}$$

Equation (3.25) can be interpreted as follows: $(1/T)^{1/2} \exp(-E/k_B T)$ is a measure of the probability that thermal fluctuations lead to a degeneracy of the energy levels at the occupied interstitial site, and at an unoccupied adjacent interstitial site (coincidence event). This is a prerequisite for a tunnelling event to occur, and J_0^2 is the tunnelling probability in that case.

The classical 'jumping over the barrier situation' can be described by an Arrhenius expression of the same type as equation (3.25) i.e. (Vineyard 1957)

$$\Gamma_{\text{class}} = 1/\tau_{\text{class}} = v_0 \exp(-E_0/k_B T). \tag{3.27}$$

where v_0 is the vibrational frequency of μ^+/p in the potential well. Upon comparing the two the pre-exponential factor and the activation energy E_a of the tunnelling case are much smaller than v_0 and E_0. Experimental verification of these properties would be clear evidence for the quantum nature of the diffusion process.

Multiphonon-assisted tunnelling at intermediate temperatures, around $\theta_D/2$, in FCC metals has been examined by Teichler (1977). He obtained the following expression for the diffusion rate:

$$\Gamma_h = \frac{1}{\tau_h} = \frac{J_0^2}{\hbar} \left(\frac{\pi}{4E_a k_B T_0 h_1(T/T_0)}\right)^{1/2} \exp\left(-\frac{E_a}{k_B T_0 h_2(T/T_0)}\right). \tag{3.28}$$

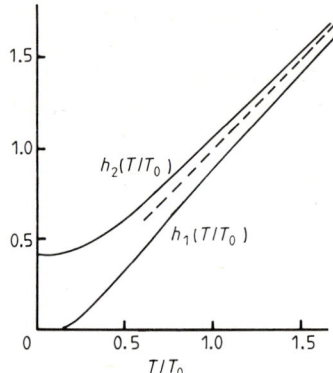

Figure 3.10 Plot of the universal functions, $h_1(T/T_0)$, $h_2(T/T_0)$ appearing in equation (3.28) (Teichler 1977).

Here $T_0 = \theta_D/2$ and $h_1(T/T_0), h_2(T/T_0)$ are universal functions valid for all FCC metals. These functions are shown in figure 3.10.

The intermediate temperature range has also been treated by Fujii (1979a,b) for BCC metals. Numerical results are shown in figure 3.11 for various ratios of $E_a/\hbar\omega_D$. The high and low temperature limits of the curves correspond to equations (3.25) and (3.24) respectively.

The transition from coherent (band-like) to incoherent (stochastic) motion has recently been treated by Petzinger (1982) in much detail for a simple cubic

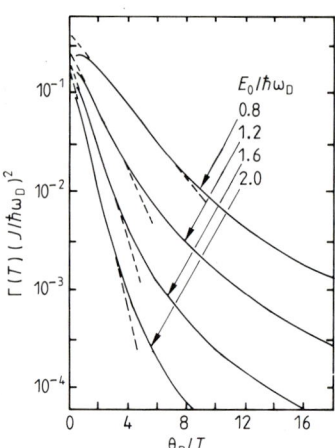

Figure 3.11 Temperature dependence of calculated hopping rates $\Gamma = \tau_h^{-1}$ for various ratios $E_a/\hbar\omega_D$ in BCC metals (ω_D = Debye frequency, θ_D = Debye temperature). The hopping rate is expressed in units of $(J_0/\hbar\omega_D)^2$ (Fujii 1979).

lattice. Neglecting the dynamical destruction of the ideal band-like propagation, Petzinger finds that the transition temperature, T^*, when the motion changes from predominantly coherent to predominantly incoherent, is rather independent of the assumed lattice distortion energy, E_a, and the rigid band width, J_0, and assumes values in the range: $0.3 < T^*/\theta_D < 0.7$. Including dynamical destruction (which is a result of higher order i.e. quadratic phonon, interactions) has the effect of appreciably lowering the transition temperature T^*. Numerical results appropriate for μ^+ in Cu are shown in figure 3.12. Nevertheless, transition to the coherent motion, and even to the ideal band propagation, should occur at temperatures easily accessible to experiment.

In all the treatments listed so far, the rigid lattice tunnelling matrix element is supposed to be independent of the lattice atom coordinates (this is called the Condon approximation). Treatments beyond the Condon approximation have been discussed by Flynn and Stoneham (1970) for FCC metals, by Emin *et*

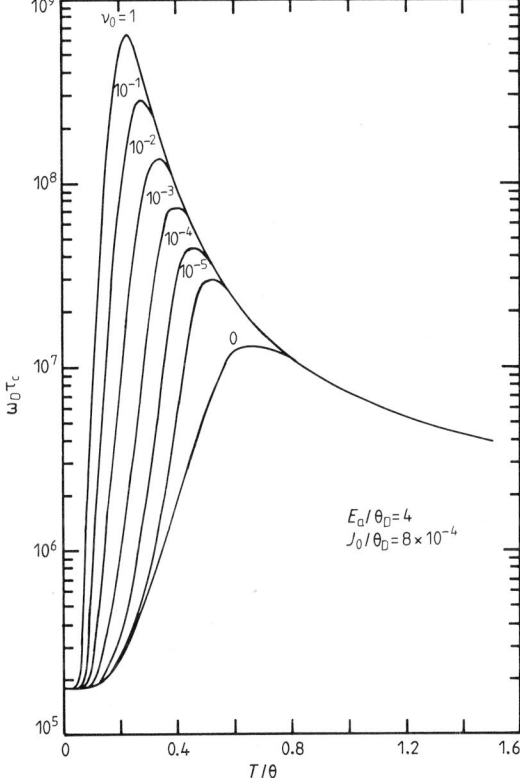

Figure 3.12 The mean residence time, τ_c, as a function of temperature for $J_0/\theta_D = 8 \times 10^{-4}$ and $E_a/\theta_D = 4$, corresponding roughly to μ^+ in Cu. The various curves correspond to different values of the two-phonon scattering rate at the Debye temperature (Petzinger 1982).

al (1979) for BCC metals and more generally by Teichler (1981a,b). One result of Teichler (1981a) is the occurrence of one phonon processes at low temperatures. Flynn and Stoneham find that the activation energy in expression equation (3.25) has to be replaced by $E_a + E_s$, where E_a is related to the lattice relaxation energy as before, and E_s is a lattice activation energy, related to the displacement of the host atoms which hinder the motion of the interstitial. In addition, the pre-exponential factor becomes temperature independent and is given by $\frac{1}{4}\hbar^{-1} J_0^2 (E_a E_s)^{-1/2}$.

So far we have only discussed types of long-range diffusion. Other forms of motion may be important as well. Of course, there is vibrational motion of the particle in its interstitial potential well, and this probably shows strong features of anharmonicity. Excitation to higher vibrational levels may also speed up the transition to a neighbouring interstitial site, thus helping in promoting long range diffusion. Another term of local motion concerns formation of an extended tunnelling state over an array of equivalent interstitial sites, e.g. over the four planar tetrahedral sites in a BCC lattice, as shown in figure 3.5 (Stoneham 1972, Buckholz *et al* 1973, Birnbaum and Flynn 1976).

So far, motional processes have mainly been discussed with reference to an ideal lattice. In reality, a metal sample will never be free of imperfections, and it can be expected that these will have their effect on the diffusional properties of the μ^+ or proton. In particular, it is expected that the coherent motion may be inhibited to some extent, since any imperfections will act as scattering centres, limiting the lifetime of a specific band-state (see e.g. Richter 1983).

Finally, lattice imperfections (substitutional and interstitial impurities, vacancies and other defects) may act as trapping centres by providing deeper potential wells than regular interstitial sites. This will slow down the overall diffusion rate, and may even immobilise the μ^+/p totally for temperatures below a certain critical temperature, where escape from the trap becomes impossible. As will be seen in §3.8, trapping at impurities is an important feature in all μ^+ diffusion studies.

It has also been suggested that the proposed tunnelling states over equivalent interstitial sites may actually form around impurities (Birnbaum *et al* 1978, Morkel *et al* 1978, Wipf *et al* 1983).

3.3 Modification of Static Dipolar Relaxation Function in Presence of Diffusion

3.3.1 *Motional narrowing in transverse fields*

In §2.3 we considered the interaction of the μ^+ spin, fixed in space, with many surrounding nuclear spins in the presence of a strong external magnetic field.

As a result, the exact magnetic field was shown to vary from one μ^+ site to another (i.e. the interstitial sites), and each implanted μ^+ therefore precessed with a slightly different frequency, leading to an overall decrease of the spin polarisation of the ensemble by dephasing. The relaxation function was described by a Gaussian function:

$$G_x(t) = \exp(-\sigma^2 t^2) \qquad (3.29)$$

with $\sigma^2 = \frac{1}{2}M_2$, where M_2 is the second moment of the dipolar field distribution (in frequency units).

We now allow the μ^+ to move through the host lattice or, more precisely, to move randomly from one interstitial site to another of a given type. The interaction between the system of nuclear spins and the μ^+ spin will become time-dependent, the corresponding Hamiltonian (equation (2.97)), a stochastic function of time. If the motion is fast enough, the μ^+ will average over the distribution of fields from site to site, that is, it will experience a narrower field distribution and the depolarisation rate due to dephasing will consequently decrease. The counter phenomenon in CW NMR experiments is known as motional line narrowing. It has been treated, for example, in Abragam's classical book on NMR in the framework of a Gaussian–Markovian theory (Abragam 1970).

The result (Abragam p 432) can be carried over directly to μSR. One then obtains the modified transverse field relaxation

$$P_x(t) = G_x(t) = \exp\left(-\int_0^t (t-\tau)G_{zz}(\tau)\,d\tau\right). \qquad (3.30)$$

This applies if the jump-frequency of the μ^+ is smaller than the nuclear Larmor frequency. $G_{zz}(\tau)$ is the correlation function of the z component of the nuclear dipole fields at the μ^+

$$G_{zz}(\tau) = \langle B_z(t)B_z(t-\tau)\rangle. \qquad (3.31)$$

Equation (3.30) only describes the effect stemming from dephasing. In addition the moving μ^+ experiences time dependent fields at its site which will give rise to spin-flip transitions (non-secular terms). This is particularly important in the so-called extreme motional narrowing limit for very rapidly moving spins at high temperatures. This additional dynamic contribution has been treated extensively by many authors (see e.g. Eisenstadt and Redfield 1963, Wolf and Jung 1975, Torrey 1953, Sholl 1974, 1975), and with special reference to μ^+ depolarisation by McMullen and Zaremba (1978) and Hayano et al (1979). However, the resulting relaxation rates (also manifest in a longitudinal field as a spin-lattice relaxation rate, T_1^{-1}), are too small in non-magnetic metals to be observable by μSR. The situation is different in certain magnetic materials and these will be discussed in §3.6. Other corrections to equation (3.30) have been discussed in the elaborate papers of Baryshevsky and Kuten (1976) and McMullen and Zaremba (1978). The alterations to equation (3.30) are, however, so minor in the time range accessible to μSR, that, for all practical purposes, equation (3.30) provides a very good approximation to the true relaxation function.

The main task we are left with, therefore, is the evaluation of the correlation function, $G_{zz}(\tau)$. Since we are mainly interested in the diffusion of the μ^+, it would be helpful if one could separate the effects on the correlation function, $G_{zz}(\tau)$ of μ^+ diffusion on the one hand, and the nuclear spin system on the other. Following McMullen and Zaremba (1978) this can be done with the assumption that the nuclear spins are only negligibly influenced by the presence of the μ^+, that is, it is assumed that the nuclear spin system behaves as if the μ^+ were absent. According to McMullen and Zaremba (1978) the correlation function $G_{zz}(\tau) = \langle B_z(t) B_z(t-\tau) \rangle$ may then be written as

$$G_{zz}(t'-t'') = \tfrac{1}{2} \int d^3r' \int d^3r'' [G_s(\mathbf{r}', t'|\mathbf{r}'', t'') F_{zz}(\mathbf{r}', t'|\mathbf{r}'', t'') \\ + G_s(\mathbf{r}'', t''|\mathbf{r}', t') F_{zz}(\mathbf{r}'', t''|\mathbf{r}', t')] \qquad (3.32)$$

with

$$G_s(\mathbf{r}', t'|\mathbf{r}'', t'') = \langle \delta[\mathbf{r}' - \mathbf{r}_m(t')] \delta[\mathbf{r}'' - \mathbf{r}_m(t'')] \rangle \qquad (3.33)$$

and

$$F_{zz}(\mathbf{r}', t'|\mathbf{r}'', t'') = \langle B_z(\mathbf{r}', t') B_z(\mathbf{r}'', t'') \rangle. \qquad (3.34)$$

$\mathbf{r}_m$ is the μ^+ position operator, $G_s(\mathbf{r}', t'|\mathbf{r}'', t'')$ is called the μ^+ self-diffusion function and $F_{zz}(\mathbf{r}', t'|\mathbf{r}'', t'')$ is called the field correlation function. Equation (3.32) neglects, in particular, the coupling of the nuclear quadrupole moments to the electric field gradients set up by the μ^+, whose effect was discussed in §2.3.2. If this coupling is present, the factorisation in equation (3.32) is no longer possible. We will return to this point later.

As the μ^+ can only be found at sites unoccupied by the host ion cores (e.g. at interstitial sites) and not everywhere in the lattice, it is convenient to express equation (3.32) in terms of a site representation. (McMullen and Zaremba 1978):

$$G_{zz}(\tau) = \tfrac{1}{2} \sum_{l,m} [G_s(l, \tau|m) F_{zz}(l, \tau|m) + G_s(l, -\tau|m) F_{zz}(l, -\tau|m)] \qquad (3.35)$$

with

$$G_s(l, \tau|m) = \langle n_l(t) n_m(t-\tau) \rangle \qquad (3.36)$$

and

$$F_{zz}(l, \tau|m) = \int d^3r \int d^3r' |\phi_l(r)|^2 F_{zz}(r, t|r', t-\tau) |\phi_m(r')|^2. \qquad (3.37)$$

$G_s(l, \tau|m)$ is now the correlation function for the particle occupation number, $n_j(t)$, at sites l and m at times t and $(t-\tau)$. $\phi_l(r)$ is the μ^+ wave function centred at site l.

When the precessional motion of the nuclear spins around the external field is neglected, this eliminates the time dependence of the field correlation function, $F_{zz}(l, \tau/m)$, which is then given by:

$$F_{zz}(l|m) = (16\pi/15)(\hbar\gamma_N)^2 I(I+1) \sum_{i,m} J_{i,l}^{(0)} J_{i,m}^{(0)*} \qquad (3.38)$$

with

$$J_{i,l}^{(0)} = \int d^3r \frac{|\phi_l(r)|^2 Y_2^{(0)}(\theta_i, \phi_i)}{|r - R_i|^3}. \qquad (3.39)$$

The index, i, refers to the position of nuclear spins, and l and m refer to the μ^+ sites. If $m = l$, $F_{zz}(l|l) = M_2$. Actually, the term involving $F_{zz}(l|l) = M_2$ will dominate the sum in equation (3.35), since $F_{zz}(l|m)$ will decrease with increasing site separation, $R_l - R_m$. In practice, it again suffices to only retain this term, and equation (3.35) simplifies to:

$$G_{zz}(\tau) = M_2 \sum_l G_s(l, \tau|l). \qquad (3.40)$$

The self-diffusion function has to be evaluated on the basis of some diffusional motion model. In the preceding section, various models were discussed which led to predictions for the transition rate, w_{ij} from a given site i to site j. Assuming a random walk picture, one can write down a semiclassical rate equation for the site occupation number, $n(t)$

$$\dot{n}_j(t) = \sum_i [w_{ji} n_i(t) - w_{ij} n_j(t)]. \qquad (3.41)$$

The first term in the brackets describes the diffusion towards site j, and the second, the diffusion away from site j. The solution of this differential equation is the self-diffusion function $G_s(j, \tau|i) = \langle n_j(\tau) n_i(0) \rangle$. If the sites form a Bravais lattice, and if one is only interested in the long time behaviour, one finds (McMullen and Zaremba 1978):

$$G_s(j, \tau|i) \simeq \frac{1}{N} \left(\frac{1}{4\pi D\tau}\right)^{3/2} \exp\left(\frac{-|R_j - R_i|^2}{4D\tau}\right). \qquad (3.42)$$

$$D = \frac{1}{6} \sum_n w_{n0} R_{n0}^2 \qquad (3.43)$$

is the diffusion constant and this reduces to equation (3.15), if only nearest neighbour site jumps are allowed. N is the total number of possible sites. The on-site ($R_j = R_i$) occupation probability decreases as $\tau^{-3/2}$. If the diffusing particle were banned from returning to its initial site, the on-site occupation probability would have a long term behaviour of (McMullen and Zaremba 1978):

$$G_s(j, \tau|j) \simeq N^{-1} \exp(\tau/\tau_c). \qquad (3.44)$$

This simple exponential form is usually used, in practice, for data analysis. As McMullen and Zaremba (1978) have pointed out, equation (3.44) is a good approximation to the full solution of equation (3.41) in the case of a μ^+ diffusion rate which is not too slow, and τ_c is a good measure of the mean time of residence, τ_s, at some given lattice site (see figure 3.13). Both McMullen and Zaremba (1978) and Beryshevsky and Kuten (1976) have also calculated the correlation function on the basis of coherent motion in very narrow bands (the tight-binding approximation). In the narrow-band limit the motion can be described as that of a classical particle, and the correlation function again follows from the solution of equation (3.41).

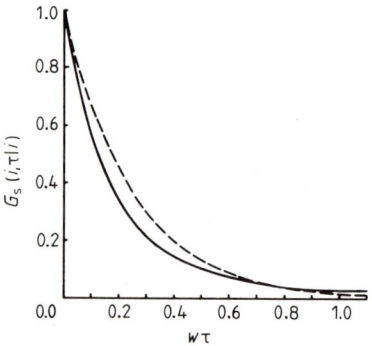

Figure 3.13 Comparison of the exact solution for $G_s(i, \tau|i)$ (full curve) with the exponential approximation equation (3.30) (broken curve). The time constant τ_c, in equation (3.44), is defined as the mean time of stay at a particular site, and is related to the transition rate, w by $\tau_c = 0.246/w$ (McMullen and Zaremba 1978).

We now insert equation (3.44) into equation (3.40) and the latter into equation (3.30) and arrive at

$$G_x(t) = \exp\left(-M_2 \int_0^t (t-\tau) \exp(-t/\tau_c) \, d\tau\right)$$
$$= \exp\{-M_2\tau_c^2[\exp(-t/\tau_c) - 1 + t/\tau_c]\}. \tag{3.45}$$

This is the relaxation function that has been used in analysing all the transverse field μSR data on μ^+ diffusion in metals. In the limit of very slow diffusion, this function reduces correctly to the static depolarisation function equation (3.29). In the limit of very fast diffusion, and for times $t \gg \tau_c$ equation (3.45) reduces to the exponential expression:

$$G_x(t) \rightarrow \exp(-M_2\tau_c t) \tag{3.46}$$

with the transverse relaxation time

$$1/T_2 = M_2 \tau_c. \qquad (3.47)$$

As mentioned before this expression would have to be supplemented by additional terms, caused by spin-flip transitions in the extreme motional narrowing limit.

Additionally equation (3.46) is exact in the limit $t \gg \omega_0^{-1} \gg \tau_c$ if M_2 is replaced by an effective field spread (McMullen and Zaremba 1978, Hayano et al 1979)

$$\langle M_2 \rangle_{\text{MN}} = \tfrac{1}{6}\gamma_\mu^2 \gamma_N^2 I(I+1) \sum_i (3\cos^2\theta_i + 7) R_{ij}^{-6} \qquad (3.48)$$

where the subscript, MN, refers to motional narrowing. This is a result of including the field fluctuations perpendicular to the external field. ω_0 is the average precession frequency determined by the external field. For a crystal with cubic symmetry, like for example Cu, $\langle M_2 \rangle_{\text{MN}}$ will become isotropic, that is, the relaxation rate will no longer depend on the orientation of the crystal with respect to the external magnetic field. However, a more in-depth analysis shows that this effect will not become important until $\omega_0 \tau_c \lesssim 1$, at which point the relaxation rate has dropped to such a small value that it would be very hard to detect experimentally.

So far the correlation function, equation (3.44), has been calculated by ignoring the possibility that the nuclear spin system may be subject to an electric quadrupole interaction. It appears, however, that equation (3.45) is again a very good approximation to the actual relaxation function. This is suggested by a comparison of τ_c values obtained in Cu in external fields ranging from zero to 0.45 T, in which the quadrupole interaction in the latter case is almost completely quenched (see §2.3.2). The comparison is shown in figure 3.14 where τ_c data are plotted semi-logarithmically against reciprocal temperature (Arrhenius plot). The values at 20 mT and 0.45 T were extracted from the μSR histogram by fitting equation (3.45) to the data. M_2 was thereby taken from low temperature data (20–80 K), where the μ^+ appears to be frozen at the octahedral interstitial site (Schilling et al 1982). The zero field values (Emmerich et al 1982) were extracted from the μSR histogram on the basis of a strong collision model which will be discussed in the next section. As can be seen, all three sets of τ_c values agree nicely with each other, implying that the mean time of residence of the μ^+ at its interstitial site can reliably be determined.

3.3.2 Effect of diffusion on zero field relaxation function

The static Kubo–Toyabe function in zero external field, $P_z(t) = g_z(t)$, was derived in §2.3.3. Its main characteristic feature is the polarisation recovery to $\tfrac{1}{3}$ in the direction of the initial polarisation vector

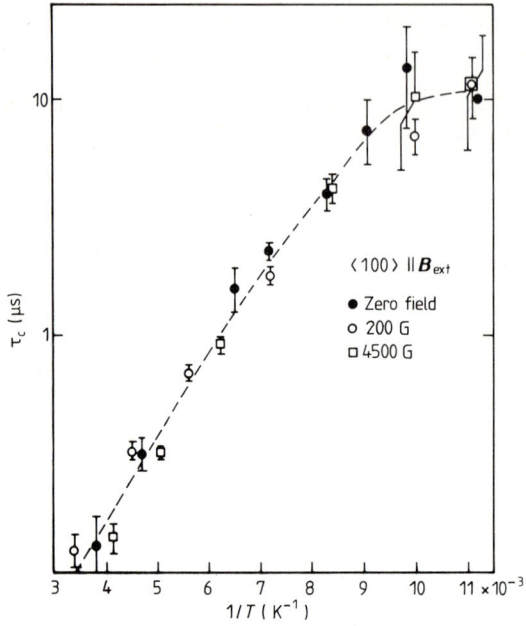

Figure 3.14 Arrhenius plot of correlation times τ_c in Cu from measurements in zero field and in fields of 20 mT and 450 mT (Schilling et al 1982, Emmerich et al unpublished).

$$g_z(t) = \tfrac{1}{3} + \tfrac{2}{3}(1 + \Delta^2 t^2)\exp(-\tfrac{1}{2}\Delta^2 t^2). \tag{3.49}$$

The effect of the local fields' stochastic time dependence on the zero field relaxation function was first treated by Kubo and Toyabe (1966), (see also Kubo 1981) and later with special reference to zero field μSR by Hayano et al (1979). We will follow here the latter treatment, which is based on the so-called strong collision model (Kehr et al 1978, Hayano et al 1979). It closely reflects a particular feature of random jump diffusion, in that it is assumed that the correlation between the local fields before and after a jump is completely lost. In other words, the diffusing particle experiences a sudden, rather than a gradual, change in local field in jumping from one site to another. In the gradual situation the process can be described as a Gaussian–Markovian process, while the strong collision model suggests a Markovian, but not a Gaussian process (Kubo 1981).

The strong collision assumption implies that the time evolution of the μ^+ polarisation follows the static Kubo–Toyabe function, $g_z(t)$, until it undergoes a jump, at which instant the evolution starts all over again, with, however, a reduced amplitude. The relaxation function $G_z(t)$ at some time t is therefore a superposition of contributions from muons that did not jump up to time t, and from those that performed one jump, two jumps etc until t:

$$G_z(t) = \sum_{n=0}^{\infty} g_z^{(n)}(t). \tag{3.50}$$

n indicates the number of jumps. If we define v to be the jump rate, the fraction of muons that will not have jumped will have decreased by e^{-vt} by time t, i.e.

$$g_z^{(0)}(t) = \exp(-vt) g_z(t). \tag{3.51}$$

The μ^+ that have jumped just once at $t' \leq t$ will contribute

$$g_z^{(1)}(t) = v \int_0^t dt' \exp[-v(t-t')] g_z(t-t') \exp[-vt'] g_z(t') \tag{3.52}$$

where $v \, dt'$ is the jumping probability at t' in the interval dt, and the integration extends over all jump times $t' \leq t$. Likewise we can write

$$g_z^{(2)}(t) = v^2 \int_{t''}^t \int_{t'}^{t''} dt'' \, dt' \exp[-v(t-t'')] g_z(t-t'') \exp[-v(t''-t')]$$

$$\times g_z(t''-t') \exp[-vt'] g_z(t') \tag{3.53}$$

where t' indicates the time of the first jump, and t'' the time of the second. Higher order terms can be derived in the same way, leading to the general formula

$$g_z^{(n)}(t) = v^n \int_{t^{(n)}}^t \ldots \int_0^{t''} dt^{(n)} \ldots dt' g_z(t-t^{(n)}) \ldots g_z(t') \exp -vt. \tag{3.54}$$

In order to perform the sum in equation (3.50), we make use of the Laplace transformation technique by introducing:

$$f_z^{(n)}(s) = \int_0^{\infty} g_z^{(n)}(t) \exp[-st] \, dt. \tag{3.55}$$

Inserting equation (3.54) into equation (3.55) one finds:

$$f_z^{(n)}(s) = v^n \left(\int_0^{\infty} dt \exp[-(\gamma+s)t] g_z(t) \right)^{n+1} = v^n [f_z(s)]^{n+1} \tag{3.56}$$

The Laplace transform of the sum, equation (3.50), is then:

$$F_z(s) = \int_0^{\infty} dt \, G_z(t) \exp -st$$

$$= \sum_n v^n (f_z(s))^{n+1} = \frac{f_z(s)}{1-vf_z(s)} \tag{3.57}$$

where we have made use of the fact that $vf_z(s) < 1$. In the case of zero field, $f_z(s)$ can be expressed as

$$f_z(s) = \frac{1}{3s} + \frac{2s}{3\Delta^2} \left(1 - s \int_0^{\infty} \exp(-\tfrac{1}{2}\Delta^2 t^2 - st) \, dt \right). \tag{3.58}$$

Finally the relaxation function $G_z(t)$ is numerically obtained by an inverted Laplace transformation of equation (3.57). $G_z(t)$ is not expressable in closed form. Some numerical results for $G_z(t)$ for various ratios of Δ/v are shown in figure 3.15(a). As a limiting case the static zero field Kubo–Toyabe function is indicated. With a rising hopping rate v, the polarisation recovery to $\frac{1}{3}$ is inhibited first, without much affecting the fast decrease of the relaxation function at earlier times. It is interesting to compare the zero field relaxation function, with the transverse field motional narrowing function equation (3.45) in the region of $\Delta/v \gtrsim 2$ (see figure 3.15(b)). While the motional narrowing curves are not very different from one another, the zero field

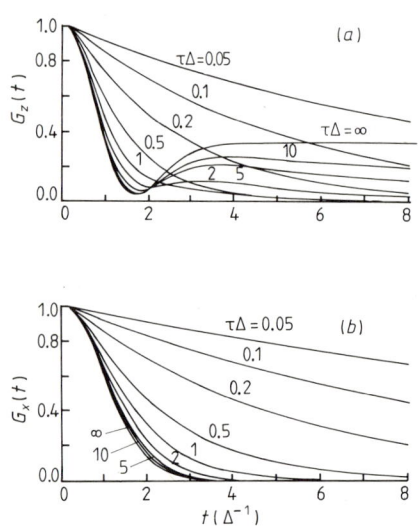

Figure 3.15 (a) Zero field relaxation function for various hopping rates, $v = \tau^{-1}$; (b) transverse field relaxation function equation (3.58) for the same $\tau \Delta$. The time is measured in units of Δ^{-1}.

relaxation functions vary significantly with Δ/v in the time region beyond $t/\Delta \simeq 2.5$. It thus appears that the zero field technique is particularly powerful for monitoring slow fluctuations, which cannot be distinguished by the transverse field technique from the static limit. For larger jump rates ($\Delta/v \lesssim 1$), both techniques seem to be of equal value, as can be seen from figure 3.15.

It should be mentioned that the zero field relaxation function calculated by Kubo and Toyabe (1966) on the basis of a Markovian–Gaussian process is rather similar to the curve which results from the strong collision model, the latter decaying somewhat slower than the former. The inverse of the jump rate v, introduced in equation (3.51), finds its counterpart in the Markovian–Gaussian process in a correlation time τ_c, which characterises the decay of the field correlation function $\langle B(t)B(0)\rangle \simeq \exp(-t/\tau_c)$. This was also the basic assumption in deriving the motional narrowing function equation (3.45). The motional narrowing function and the strong collision 'zero field relaxation function' are not therefore derived on exactly the same physical basis. One should be aware of this when comparing results from zero and transverse field measurements.

3.3.3 Effect of trapping on transverse field relaxation function

In the preceding two sections it was assumed that the μ^+ only diffuse among one type of interstitial site, involving a unique residence time τ_s and a unique fieldspread $\sigma^2 = \frac{1}{2}M_2$. This ideal intrinsic diffusion may, however, be influenced by imperfections of the crystal, leading to trapping of the μ^+ at certain sites in the crystal, which are associated with different τ_s and M_2 values. In order to have an effect on the relaxation function, the residence time in the trap must be longer than the residence time at some undisturbed interstitial site, i.e. the μ^+ must be more strongly bound to the trap than to the site of self-trapping. As mentioned previously various types of defects may act as trapping centres: atomic impurities, lattice strain, vacancies, vacancy-clusters etc. In effect, what generally would be expected most is a whole hierarchy of traps of various depths, concentrations and associated field spreads, M_2.

As before the external field will be directed along the z axis and the initial μ^+ polarisation, in the x direction: $\boldsymbol{P}(0) = P_x(0)\hat{x}_0$. We are interested in the time evolution of the μ^+ polarisation in the x direction: $P_x(t) = G_x(t)$. The effect of trapping on the transverse field relaxation function has been calculated on the basis of various approximations. One way is to start, in analogy with the treatment of positron trapping in metals (Frank and Seeger 1974), with a set of coupled differential equations (rate equations) on the occupation probability of the μ^+ at the various sites. Let us denote by $n_0(t)$ the number of untrapped μ^+, and the number of trapped μ^+ at a defect of type m at time t by $n_m(t)$. Let us further denote by σ_m, the capture probability by a type m trap per unit trap concentration, by c_m the concentration of traps of type m and by $v_m = 1/\tau_m$ the escape rate out of a type m trap. If we have j different types of traps, one obtains a set of $(j+1)$ coupled differential equations (Herlach 1981, Kossler et al 1978)

$$\frac{dn_0(t)}{dt} = -\left(\sum_{m=1}^{j} \sigma_m c_m\right) n_0(t) + \sum_{m=1}^{j} v_m n_m(t)$$

$$\frac{dn_m(t)}{dt} = \sigma_m c_m n_0(t) - v_m n_m(t). \tag{3.59}$$

Note that a direct transfer between traps is excluded. The $n_m(t)$ are normalised such that $\sum_{m=0}^{j} n_m(t) = 1$. A particularly simple case is given, if one assumes that once a μ^+ has reached a trap, it will stay there for the rest of its lifetime ($v_m = 0$). To simplify the treatment further, we will also assume that only one type of trap is present ($j = 1$). The solution of the system (3.59) is then easily obtained and reads:

$$n_0(t) = n_0(0) \exp(-\sigma_1 c_1 t)$$
$$n_1(t) = n_1(0) + n_0(0)[1 - \exp(-\sigma_1 c_1 t)] \quad (3.60)$$

with the initial condition $n_0(0) \gg n_1(0)$ (i.e. thermalisation at random sites). We now define a relaxation function $g_x^f(t)$ associated with the free μ^+ and a relaxation function $g_x^t(t)$ associated with the trapped μ^+. The total relaxation function is then obtained by again invoking a strong collision argument (i.e. no correlation between the fluctuating fields at the normal and the trapping sites)

$$G_x(t) = n_0(t)g_x^f(t) + n_1(0)g_x^t(t) + \int_0^t dt' g_x^t(t-t')g_x^f(t')n_0(t'). \quad (3.61)$$

The relaxation function for the trapped μ^+ is the static relaxation function of the Gaussian form already discussed

$$g_x^t = \exp(-\sigma_t^2 t^2) \quad (3.62)$$

where σ_t^2 describes the field spread experienced by the μ^+ at the trapping sites. $g_x^f(t)$ can be chosen to be of the exponential form (3.46), since the intrinsic diffusion must be rapid enough to let the μ^+ reach a trap, usually of low concentration, during its lifetime

$$g_x^f = \exp(-t/T_2). \quad (3.63)$$

With these assumptions $G_x(t)$ can be explicitly expressed as (Herlach 1981):

$$G_x(t) = \exp[-(\sigma_1 c_1 + T_2^{-1})t]$$
$$\times \{1 + (\pi^{1/2}\sigma_1 c_1/2\sigma_t) \exp(\alpha^2)[\text{erf}(\alpha) - \text{erf}(\alpha - \sigma_t t)]\}$$

where $\alpha = (\sigma_1 c_1 + T_2^{-1})\sigma_t^{-1}$.

The capture probability, σ_1, is related to the macroscopic diffusion constant D by:

$$\sigma_1 = 4\pi r_0 D/\Omega_a \quad (3.65)$$

with r_0 = capture radius of the trap and Ω_a = atomic volume (Seeger 1975). Experimental observation of the μ^+ diffusion towards permanent traps opens up, therefore, the interesting possibility of measuring the intrinsic macroscopic diffusion constant directly, rather than a correlation or residence time τ_s. This in turn would allow us to check the relationship of the macroscopic diffusion constant and the microscopic correlation time, τ_s.

If repeated trap release and capture have to be included, the situation becomes more complicated. On the basis of the relaxation function treatment given by McMullen and Zaremba (1978) and by applying perturbation theory, Petzinger et al (1979) derived an expression for the relaxation function in the presence of traps:

$$G_x(t) = \exp\left(-\sum_i M_{2,i} \int_0^t dt_1 \int_0^{t_1} dt_2 \frac{1}{N}\sum_j G_{ij}(t_2)G_{ii}(t_1 - t_2)\right). \quad (3.66)$$

$M_{2,i}$ is the second moment associated with site i, which could be an undisturbed interstitial site or a trapping site. $G_{ij}(t) = \langle n_i(t)n_j(0) \rangle$ is the probability that a transition from site i to site j has been made in time t and $G_{ii}(t) = \langle n_i(t)n_i(0) \rangle$ is the auto-correlation function of site i. These functions correspond to the self-diffusion function $G_s(r', t'|r'', t'')$ defined in equation (3.33). The sums over i and j extend over all possible μ^+ sites. N is the total number of sites. In the derivation of equation (3.66), nonsecular terms were neglected, it was assumed that the μ^+ hop stochastically from site to site and that the initial distribution of the μ^+ over available sites is random, i.e. not in thermal equilibrium. $N^{-1}\sum_i G_{ij}(t)$ can be interpreted as the occupation probability of site j at time t. The sum over j can be regrouped in such a way that it consists of one term involving all undisturbed sites, and m terms corresponding to m different types of trapping sites (Kossler et al 1978). One then obtains

$$G_x(t) = \exp\left(-\sum_{k=0}^m M_{2,k} \int_0^t \int_0^{t_1} f_k(t_2)g_k(t_1 - t_2)\,dt_1\,dt_2\right). \quad (3.67)$$

$f_k(t)$ is now the occupation probability of the sites of type k with $k=0$ corresponding to undisturbed interstitial sites and $k=1,\ldots,m$ corresponding to m types of trapping sites. The auto-correlation function $g_k(t)$ for any site of type k is again approximated by (see equation (3.44))

$$g_k(t) = \exp(-v_k t) \quad (3.68)$$

where v_k is the escape rate from a trap of type k. The $f_k(t)$ can be identified with $n_k(t)$ and the solutions of the coupled rate equations (3.41). The nice feature of this approach is that it can be easily adapted to any number of different types of trapping sites.

A Markovian theory of spin depolarisation for μ^+ diffusion in crystals with traps has been developed by Kehr et al (1978). It is based on a random walk picture and the strong collision assumption, that is, complete loss of any correlation after each hop. This theory was worked out for a two-state situation, i.e. μ^+ are either freely diffusing or are trapped in just one kind of a trap. The following functions are introduced: $P_0(t) = g_x^f(t)$, the relaxation function of μ^+ in the untrapped state; $P_1(t) = g_x^t(t)$ the relaxation function in the trapped state, given as before by the static Gaussian function $P_1(t) = \exp(-\sigma_t^2 t^2)$. Further, $Q_{ij}(t)$ denotes the average polarisation of a μ^+ now in state i ($i=0, 1$) at time t but which was in state j ($j=0, 1$) at $t=0$. Intermediate state changes are included. The probability that the μ^+ is found in state i at time t after having been placed into this state at $t=0$ is expressed by the exponential function $\psi_i = \exp(-t/\tau_i)$. τ_i therefore denotes the average time spent in state i (i.e. $\tau_0^{-1} = c_1 \sigma_1$). By following the treatment in §3.3.2, we can write

$$Q_{00}(t) = \psi_0(t)P_0(t) + \frac{1}{\tau_0^2} \int_0^t \int_0^{t'} dt' \, dt'' \psi_0(t-t')P_0(t-t')$$
$$\times \psi_1(t'-t'')P_1(t'-t'')\psi_0(t'')P_0(t'') + \cdots$$

$$Q_{10}(t) = \frac{1}{\tau_1} \int_0^t \psi_1(t-t')P_1(t-t')\psi_0(t')P_0(t') + \ldots \qquad (3.69)$$

and likewise for $Q_{11}(t)$ and $Q_{01}(t)$. $Q_{00}(t)$ thus consists of a series of terms representing the μ^+ that was never trapped and the μ^+ that was trapped once, twice etc. $Q_{10}(t)$, which always involves a change of state, consists of a series of terms representing the μ^+ that experienced one change of state, three changes of states etc. The infinite series can be expressed as the coupled integral equations (Kehr et al 1978)

$$Q_{00}(t) = \psi_0(t)P_0(t) + \tau_0^{-1} \int_0^t dt' \psi_0(t-t')P_0(t-t')Q_{10}(t)$$

$$Q_{10}(t) = \tau_1^{-1} \int_0^t dt' \psi_1(t-t')P_1(t-t')Q_{00}(t)$$

$$Q_{11}(t) = \psi_1(t)P_1(t) + \tau_1^{-1} \int_0^t dt' \psi_1(t-t')P_1(t-t')Q_{01}(t')$$

$$Q_{01}(t) = \tau_0^{-1} \int_0^t dt' \psi_0(t-t')P_0(t-t')Q_{11}(t'). \qquad (3.70)$$

The total relaxation function for μ^+ diffusing freely as well as being repeatedly captured and released into and out of traps is then given by

$$G_x(t) = (1-c_1)[Q_{00}(t) + Q_{10}(t)] + c_1[Q_{01}(t) + Q_{11}(t)] \qquad (3.71)$$

where c_1 is the concentration of traps. c_1 is usually very small and it suffices to consider only the first term. The integral equations can be solved by applying Laplace transformation techniques. $Q(t)$ is then finally obtained by an inversion of the Laplace transformed $\tilde{Q}_{ij}(s)$.

A very simple approximate treatment of the relaxation function for the two-state case has been postulated by Borghini et al (1978) and discussed in more detail by Kehr et al (1978). The transitions between the two states can be compared with two chemical phases between which atoms perform chemical exchange reactions. Zimmerman (1957) has calculated expressions for the NMR line shape in just such a situation from a Markovian theory of transitions between the two states. In analogy to this treatment, one can define the following pair of coupled differential equations, where $Q_i(t)$ ($i=0, 1$) is the polarisation function in the state i and $\lambda_i(t)$ is the depolarisation rate in state i at time t. The τ_i have the same meaning as before

$$dQ_0(t)/dt = -[\lambda_0(t) + \tau_0^{-1}]Q_0(t) + \tau_1^{-1}Q_1(t)$$
$$dQ_1(t)/dt = -[\lambda_1(t) + \tau_1^{-1}]Q_1(t) + \tau_0^{-1}Q_0(t). \quad (3.72)$$

The infinitesimal change of polarisation of the μ^+ in state i is either caused by depolarisation, or loss to the other state, or transfer of polarisation from the other state. The depolarisation rate λ_i in state i is obtained from the undisturbed relaxation function $P_i(t)$ in the state i, assuming no transfer events:

$$\lambda_i = dP_i(t)/dt. \quad (3.73)$$

In the treatment of Borghini *et al* (1978) and Kehr *et al* (1978) it is assumed that in both states $P_i(t)$ can be represented by the motional narrowing function equation (3.45)

$$P_i(t) = \exp\{-M_{2,i}\tau_{c,i}^2[\exp(-t/\tau_{c,i}) - 1 - t/\tau_{c,i}]\}. \quad (3.74)$$

The $M_{2,i}$ are the second moments of the dipolar field spreads in the undisturbed and the trapping sites, $i = 0, 1$, respectively. The $\tau_{c,i}$ are the corresponding correlation times and these are identified with the mean time of stay at a given site. It is obvious that $\tau_{c,1} = \tau_1$, but note that $\tau_{c,0} \neq \tau_0$. The pair of coupled differential equations can be solved numerically with the initial conditions $Q_1(0) = 1$, and $Q_0(0) = 0$ (i.e. initially no μ^+ are in a trap). The total relaxation function is then given by:

$$G_x(t) = Q(t) = Q_1(t) + Q_0(t). \quad (3.75)$$

The time constants (correlation times), τ_0, $\tau_{c,0}$ and τ_1, are temperature dependent and can be generally approximated by the Arrhenius expressions (see §3.2.1):

$$\tau_0 = \tau_0^0 \exp(-E_0/k_B T)$$
$$\tau_{c,0} = \tau_{c,0}^0 \exp(-E_{c,0}/k_B T) \quad (3.76)$$
$$\tau_1 = \tau_1^0 \exp(-E_1/k_B T).$$

Remembering that $\tau_0^{-1} = c_1(4\pi r_0 D/\Omega_a) \propto \tau_{c,0}^{-1}$ (equation (3.65)) we can identify E_0 with the effective activation energy $E_{c,0}$ for an intrinsic diffusion event. By adopting appropriate values for describing experimental results for $E_{c,0}$, E_1, τ_0^0, $\tau_{c,0}^0$ and σ_i in Nb (Borghini *et al* 1978). Kehr *et al* (1978) have calculated the relaxation function both from the solution of the integral equations (3.70) and the solution of the differential equations (3.72).

To foster a convenient comparison of the two calculations, a damping parameter, λ, is defined which is the reciprocal of the time at which $Q(t)$ has declined to $1/e$. Figure 3.16 shows λ as a function of temperature as calculated by two approaches. As can be seen, the two approaches yield quite similar results justifying the use of the approximate differential equations.

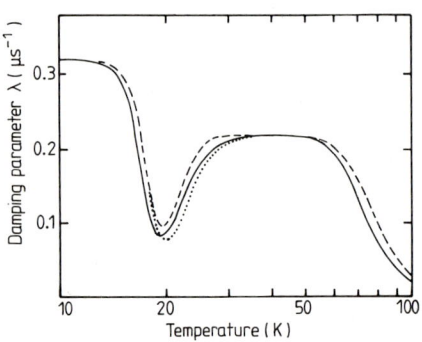

Figure 3.16 Comparison of the damping parameter, $\lambda(T)$, obtained from the solution of the integral equations (3.70) (full curve), the solution of the approximate differential equations (3.72) (broken curve), and the Zimmerman–Brittin theory (dotted curve). The result of the Zimmerman–Brittin theory coincides with that of the differential equations where no separate dotted curve appears (Kehr et al 1978).

It is interesting to compare the various treatments with one another. It is easily seen that in the limit $\tau_1 \to \infty$ (no release from a trap), the relaxation functions $G_x(t)$ defined by (3.71) and (3.75) reduces exactly to the $G_x(t)$ defined by (3.61). It is also easily verified that (3.67) reduces to (3.70) and (3.75) in the limit $\tau_0 \to 0$ and $\tau_1 \to \infty$. If no trapping occurs at all ($\tau_0 \to \infty$), all three expressions reduce to the usual motional narrowing function. In the other extremes, $\tau_{c,0} \to 0$, $\tau_0 \to 0$ and $\tau_1 \to 0$ (rapid capture and release events), equilibrium occupation between the two states is achieved. It follows from equation (3.59) that in this limit:

$$n_0(t) = (\tau_1^{-1}/\tau_0^{-1})n_1(t) = \text{constant} \tag{3.77}$$

with the normalisation $n_0(t) + n_1(t) = 1$. Equation (3.67) then reduces to:

$$G_x(t) = \exp\left(\frac{-M_{2,0}\tau_{c,0}t(\tau_0/\tau_1)}{1+(\tau_0/\tau_1)}\right) \exp\left(\frac{-M_{2,1}\tau_2 t}{1+(\tau_0/\tau_1)}\right). \tag{3.78}$$

The same expression is also obtained from the approximate differential equations (Kehr et al 1978). We see that all approaches describe the limiting cases in mutual agreement.

3.3.4 Effect of trapping on zero field (Kubo–Toyabe) relaxation function

In principle, all considerations from the previous section may be applied as well to zero field relaxation. We will limit ourselves here to the simple case of permanent trapping once a μ^+ has reached a trapping site. As emphasised by Petzinger (1980), zero field measurements should allow us to distinguish whether the μ^+ are moving towards a trap or are escaping from the trap. If one assumes that the intrinsic diffusion is sufficiently fast so as to inhibit any depolarisation in the free state, one should always observe an asymptotic value of $\frac{1}{3}$ for the residual polarisation. This occurs because all muons will eventually experience the static field spread associated with the trapping sites. In contrast, if there are releases from the traps taking place, the relaxation function will tend to zero with time, as was discussed in §3.3.2.

In deriving the zero field relaxation function for permanent trapping after capture into a trap we follow exactly the same treatment as in the transverse field case. In analogy to equation (3.61) we write

$$G_z(t) = n_0(t) + \int_0^{t'} dn_0(t') g_{z,t}^{KT}(t'-t). \tag{3.79}$$

$n_0(t)$ is again the occupation probability of the free state and $g_{z,t}^{KT}(t)$ is the static Kubo–Toyabe relaxation function associated with the trapping sites. Note that we have approximated the dynamic Kubo–Toyabe relaxation function in the free state by 1. As before $n_0(t) = \exp(t/\tau_0)$ where the mean residence time in the free state, τ_0, is related to the macroscopic diffusion constant D by (equation (3.65)):

$$\tau_0^{-1} = c\sigma = c(4\pi r_0 D/\Omega_a). \tag{3.80}$$

The parameters have the same meaning as in §3.3.3. Inserting the appropriate expressions into equation (3.79) we obtain

$$G_z(t) = \tfrac{1}{3} + \tfrac{2}{3}\exp(-t/\tau_0) + \tfrac{2}{3}\tau_0 \int_0^t \exp -t'/\tau_0$$
$$\times \{1 - \Delta^2(t-t')^2 \exp[-\tfrac{1}{2}\Delta^2(t-t')^2]\}\, dt'. \tag{3.81}$$

The same expression has also been derived by Petzinger (1980) in a somewhat different way. It can be evaluated further to yield:

$$G_z(t) = \tfrac{1}{3} + \tfrac{2}{3}\exp(-t/\tau_0)[1-(\tau_0\Delta)^{-2}]$$
$$+ \tfrac{2}{3}\exp(-\tfrac{1}{2}\Delta^2 t^2)[\tau_0^{-1} + (\tau_0\Delta)^{-2}]$$
$$- \tfrac{2}{3}\exp(-t/\tau_0)(\tau_0\Delta)^{-3}(\pi/2)^{1/2}\exp[\tfrac{1}{2}(\tau_0\Delta)^{-2}]$$
$$\times \{\text{erf}\,2^{-1/2}[\Delta t - (\tau_0\Delta)^{-1}] + \text{erf}\,(2^{1/2}\tau_0\Delta)^{-1}\}. \tag{3.82}$$

In the limit $(\tau_0\Delta)^{-1} \gg 1$, the static Kubo–Toyabe relaxation function is recovered (all μ^+ are quickly placed into traps). In the other limit of slow capturing into traps, that is $(\tau_0 \Delta)^{-1} \ll 1$ one obtains

$$G_z = \tfrac{1}{3} + \tfrac{2}{3} \exp -t/\tau_0. \tag{3.83}$$

As pointed out before, an asymptotic limit of $\tfrac{1}{3}$ is always reached.

3.4 Diffusion and Relaxation in Presence of Dilute Magnetic Impurities

Many metals do not possess nuclei with a magnetic moment and spin, or contain only a small fraction of isotopes carrying spin and magnetic moment. In many of these metals it would be highly interesting to perform μ^+ diffusion studies, for example in Pd, Pt, Au, Ag, the alkaline earth metals and others. However, because of the absence of an intrinsic inhomogeneous field distribution, observation of diffusion by monitoring the spin relaxation function is not possible. A way out would be to introduce artificially an intrinsic inhomogeneous field distribution by alloying the metals with a dilute amount of paramagnetic impurities. As stated in §2.3.1, a random and static distribution of dilute magnetic impurities leads to an exponential relaxation function for fixed μ^+. We now want to discuss the changes in the relaxation rates brought about by the motion of the μ^+. The theoretical description of the depolarisation of diffusing μ^+ by paramagnetic impurities has been worked out by Schillaci *et al* (1981). The theory finds its analogue in the treatment of relaxation in super-ionic conductors doped with paramagnetic impurities, which has been formulated by Richards (1978).

Two interaction mechanisms between the μ^+ and the magnetic impurities have to be considered:

(i) long range interactions which are of dipolar (and RKKY (Ruderman–Kittel–Kasuya–Yosida)) origin and

(ii) short range interactions, which involve contact hyperfine fields at the μ^+ sites adjacent to the magnetic impurity.

The dipolar fields originating from the magnetic impurities will be felt by the μ^+ at any distance from the impurities, with the strength decreasing as r^{-3}, while the contact field is only felt, if the μ^+ are actually located next to an impurity.

The interactions' time dependence could now be one of two origins. It could result from either μ^+ diffusion (hopping rate $= \tau_h^{-1}$), or from the spin relaxation of the paramagnetic spins, whose rate is denoted by τ_e^{-1}.

The treatment of Schillaci *et al* (1981) is an improvement over the theory of Richards in that the long and the short range interactions are treated simultaneously, and the location of and transition between the possible μ^+ sites close to an impurity are more realistically represented. The change of polarisation is expressed by a system of coupled rate equations quite similar in spirit to the treatment of Kehr et al (1978) which lead to the system of differential equations (3.72). The different states for the μ^+ are now defined by its distance from the magnetic impurity. This is done by defining shells of sites with equal distance from the impurity. Each shell accommodates a certain number of sites, n_i, which of course rises with the shell radius. The occupation probability of the μ^+ on a certain shell is proportional to the number of sites on this shell. μ^+ diffusion is assumed to proceed in hops between nearest neighbour sites only. The relative transition probabilities between shells can then be calculated by purely geometrical considerations. Assume, for example, an FCC lattice with the impurity at a substitutional site and octahedral interstitial sites for the μ^+. The first shell would then consist of six sites, the second shell of eight and the third of 24 sites. A given site in the first shell has four nearest neighbour sites in its own shell, four nearest neighbour sites in the second and four nearest neighbour sites in the third shell. This exhausts all possible nearest neighbour sites. The relative transition probabilities from the first shell to the second and third shells are therefore $\frac{1}{3}$ each.

Initially the polarisation residing on each shell, $n_i p_i(0)$, is proportional to the number of sites on this shell, that is, uniform polarisation is assumed. Long range interactions have a depolarising effect on all shells, while depolarisation due to the contact field is restricted to the first shell. Loss of polarisation $n_i \, dp_i/dt$ in a given shell is therefore a result of transfer of μ^+ *to* other shells, and of *intra*-shell depolarisation due to the μ^+ magnetic impurity interactions. Polarisation can be gained by transfer of μ^+ *from* other shells. The change of polarisation in shell $1, 2, \ldots, N$ can then be written as

$$n_1 \frac{dp_1}{dt} = \frac{1}{\tau_h} \sum_{j \neq 1} R_{j1} n_j p_j - \frac{1}{\tau_h} \sum_{k \neq 1} R_{1k} n_1 p_1 - C n_1 p_1 \frac{1}{r_1^6} - Q n_1 p_1$$

$$n_2 \frac{dp_2}{dt} = \frac{1}{\tau_h} \sum_{j \neq 2} R_{j2} n_j p_j - \frac{1}{\tau_h} \sum_{k \neq 2} R_{2k} n_2 p_2 - C n_2 p_2 \frac{1}{r_2^6}$$

$$\vdots$$

$$n_N \frac{dp_N}{dt} = \frac{1}{\tau_h} \sum_{j \neq N} R_{jN} n_N p_N - \frac{1}{\tau_h} \sum_{k \neq N} R_{Nk} n_N p_N - C n_N p_N \frac{1}{r_N^6}. \qquad (3.84)$$

n_i is the number of sites per shell, R_{jk} is the relative transition probability between shells j and k and C/r_k^6 and Q are the intra-shell depolarisation rates. N is the number of the most distant shell included in the treatment. This number is determined by the mean volume per impurity. The total polarisation, $P(t)$ is then given by a sum over all shells:

$$P(t) = \sum_i n_i p_i(t) = \exp(-\lambda t) \qquad (3.85)$$

with λ the overall depolarisation rate. Next we need to know the intra-shell depolarisation rates C/r_i^6 and Q.

In each shell the μ^+ feels a uniform field in space and also, if the impurity spins are not relaxing, in time. Therefore no depolarisation of the μ^+ in a given shell should result. In reality, however, the impurity spins relax quite quickly and the μ^+ are subject to fluctuating magnetic fields. The relaxation time of the impurity spins, τ_e, is generally short enough to fulfil the exchange narrowing condition $\tau_e \Omega_i \ll 1$, where Ω_i is related to the dipole and contact field strength in shell i acting on the μ^+. Hence the μ^+ relaxation can be characterised by a transverse spin relaxation time, T_2, implying an exponential relaxation function (dynamical depolarisation). The appropriate formulae for T_2 can be taken directly from Abragam's book (1970, p 334) yielding the following expression for C and Q:

$$C = \tfrac{1}{15}(\hbar\gamma_\mu\gamma_s)^2 S(S+1)\tau_e$$

$$\times \left(4 + \frac{3}{1+\omega_\mu^2\tau_e^2} + \frac{6}{1+\omega_s^2\tau_e^2} + \frac{1}{1+(\omega_s+\omega_\mu)^2\tau_e^2} + \frac{6}{1+(\omega_s+\omega_\mu)^2\tau_e^2}\right), \quad (3.86)$$

and

$$Q = \tfrac{1}{3}S(S+1)A^2\tau_e\{1 + 1/[1+(\omega_s-\omega_\mu)^2\tau_e^2]\}. \quad (3.87)$$

$\omega_\mu = \gamma_\mu B = \mu^+$ Larmor frequency in the external field B, $\omega_s = \gamma_s B =$ impurity spin Larmor frequency in the external field and $A =$ contact hyperfine constant. Note that the same correlation time τ_e enters the expression caused by the long range dipole interaction, as well as the one for the short range contact interaction. τ_e can, in principle, be obtained independently from ESR measurements. Its temperature dependence is often approximated by the Korringa form (see e.g. Slichter 1978)

$$\tau_e^{-1} = \text{const } \gamma_s T. \quad (3.88)$$

The general solution of the coupled system of equations (3.84) can be obtained by many well known techniques, and some qualitative features can be directly derived. We will assume, hereby, that the contact interaction in the first shell will be the dominating process in relaxing the μ^+ spin. For low diffusion rates, most μ^+ will not reach the first shell during their lifetime and hence will not be subject to the contact interaction. Consequently the relaxation rate will turn out to be small. With an increasing diffusion rate, the μ^+ will start to reside at least once in the first shell. If the residence time in a first shell site, τ_h, is exceeding T_2 the μ^+ will be depolarised completely in this encounter. The depolarisation rate $\lambda(T)$ in this region is then simply proportional to the diffusion rate, i.e.

$$\lambda \simeq \tau_h^{-1} \quad (3.89)$$

and is independent of the nature and strength of the short range interaction (Brown et al 1981). Since τ_h^{-1} rises with temperature, so does $\lambda(T)$. At much higher temperatures, when τ_h becomes smaller than T_2, complete depolarisation is impossible in a single encounter. On the other hand, the μ^+ may now reside in a first shell site many times. The field fluctuations, felt by the μ^+ and leading to spin-flip transitions (dynamical depolarisation), are now not only governed by the electronic relaxation (correlation) time τ_e, but also by τ_h. These combine to give a new correlation time

$$\tau_c^{-1} = \tau_h^{-1} + \tau_e^{-1}. \tag{3.90}$$

In the extreme motional narrowing regime, application of equation (3.87) yields

$$\lambda(T) \simeq \tau_c \tag{3.91}$$

i.e. $\lambda(T)$ is now declining with τ_c and therefore also with rising temperature. In between, $\lambda(T)$ must show a maximum.

It is worth pointing out that τ_h^{-1} on the low temperature side is a kind of a macroscopic diffusion rate. This is because what is observed is the approach of a dilute magnetic impurity, which involves a large number of hops over many lattice constants. If the magnetic impurity could act as a trap, the higher temperature behaviour of τ_h should be distinct from that at low temperatures, since τ_h can now be interpreted as a trapping time.

3.5 Coherent Diffusion and Spin Relaxation in BCC Ferromagnets

So far we have considered systems in which the μ^+ are depolarised by random nuclear dipole fields, or fields from dilute magnetic impurities. The effect of diffusion was either to cause motional narrowing of the dipolar field spread or to induce relaxation by sufficiently close encounters with local magnetic moments. The situation is quite different in ferromagnetic metals, where the strong electronic magnetic moments are ordered (see §4.6.1). In FCC metals this leads to a cancellation of the dipole fields at the interstitial sites. In HCP metals the dipole fields do not cancel, but acquire a unique value at a given type of interstitial site. μ^+ that reside in these sites will therefore always experience the same field and no dephasing results. In BCC metals, on the other hand, one finds two magnetically inequivalent positions for a given type of interstitial site if the ordered moments point along the $\langle 100 \rangle$ or any equivalent axis. These have relative weights of 1:2 and dipole fields of $2H_{dip}$ and $-H_{dip}$, respectively (see §4.6.1). If the μ^+ were fixed for sufficiently long times at a given type of interstitial site, one should actually observe two quite distinct precession frequencies. If, on the other hand, the μ^+ perform hops at random times, motional averaging (see Abragam 1970, chapter 10) as well as the appearance of a longitudinal or spin–lattice relaxation rate, T_1 (see §3.6) will be observed. A new situation, however, arises if the μ^+ propagate in a band state with a mean free path longer than the dimensions of the unit cell. The μ^+ would now feel a periodic field with a well defined average, and no dephasing would result. Other states in the band may be occupied too and are associated with different average fields. If no transitions from one band state to another take place, the muon ensemble will be subject to a spread in effective fields just like in the static case in the presence of random nuclear dipole fields. If the μ^+ scatters between different states, something like motional averaging emerges, although in fact the μ^+ motion is slowed down. If the mean free path becomes of the order of the unit cell dimensions or smaller, the motion will acquire the character of random hopping.

What we have described here in words has been treated quantitatively by Fujii (1979b) on the basis of his quantum theory of diffusion mentioned in §3.2 (Fujii 1979a). It is thereby assumed that all time modulations are fast, and that one is always in the respective motional narrowing limits. It is further assumed that the bulk magnetisation points along the $\langle 100 \rangle$ axis (the easy axis) and that therefore no transverse field components ($\perp \langle 100 \rangle$ axis) exist. Hence, if the initial polarisation of the μ^+ is perpendicular to the $\langle 100 \rangle$ axis, we need only be concerned with the transverse relaxation, i.e. the damping of the precession signal.

The relaxation function is then given by

$$G_x(t) = \exp(-t/T_2^{\text{coh}}) \tag{3.92}$$

and the transverse relaxation time T_2^{coh} is calculated by Fujii (1979b)

$$T_2^{\text{coh}} = (2\gamma_\mu^2 H_{\text{dip}}^2)^{-1} \Omega(T)(1-g_M)g_M^{-1}. \tag{3.93}$$

$\Omega(T)$ is the scattering rate between different band states (intra-site scattering rate) which we have already introduced in §3.2, and g_M is a function which depends on the ratio $J_{\text{eff}}/\Omega(T)$, with J_{eff} the effective bandwidth (see equation (3.17). This function is shown in figure 3.17 for tetrahedral and octahedral interstitial sites. For small J_{eff}/Ω, g_M approaches 1 like

$$g_M = 1 - 12[J_{\text{eff}}^2/\Omega(T)^2] + \cdots \tag{3.94}$$

In the limit $J_{\text{eff}}/\Omega(T) \to 0$, i.e. rapid scattering between the band states, one thus obtains

$$T_2^{\text{coh}} = \frac{1}{2\gamma_\mu^2 H_{\text{dip}}^2} \Omega(T) \frac{12 J_{\text{eff}}^2}{\Omega(T)^2} = \frac{6 J_{\text{eff}}^2}{\gamma_\mu^2 H_{\text{dip}}^2 \Omega(T)}. \tag{3.95}$$

It is interesting now to look at the diffusion constant for band diffusion. With the help of equation (3.15) and equation (3.20), one obtains

$$D = \frac{d^2}{6} 8 \frac{J_{\text{eff}}^2}{\Omega(T)} g_D = \frac{d^2}{6} \Gamma_{\text{coh}} \tag{3.96}$$

with $g_D \to 1$ for $J_{\text{eff}}/\Omega(T) \to 0$. Consequently we find that

$$T_2^{\text{coh}} \simeq \Gamma_{\text{coh}}$$

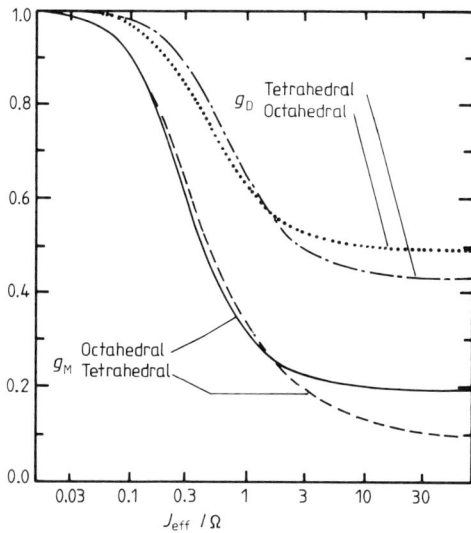

Figure 3.17 g_M and g_D as a function of $J_{\text{eff}}/\Omega(T)$ (Fujii 1979b).

that is, the relaxation time is proportional to the diffusion rate: the larger the diffusion rate the longer the relaxation time.

In the limit of very large $J_{\text{eff}}/\Omega(T)$ we have

$$1/T_2^{\text{coh}} \simeq \gamma_\mu^2 \langle H_p^2 \rangle_{\text{band}} \tau_b \qquad (3.97)$$

with $\langle H_p^2 \rangle_{\text{band}} = 2H_{\text{dip}}^2 g_M$ and $\tau_b^{-1} = \Omega(T)(1-g_M)$. The time constant, τ_b now has the meaning of some average lifetime of the band states and $\langle H_p^2 \rangle_{\text{band}}$ serves as a second moment of the field distribution associated with the band states. Compared with the diffusion rate, Γ_{coh}, which increases generally with $J_{\text{eff}}/\Omega(T)$, T_2^{coh} assumes a maximum around $J_{\text{eff}}/\Omega(T) = 1$, and starts to decrease at larger $J_{\text{eff}}/\Omega(T)$ with the relation

$$T_2^{\text{coh}} \simeq 1/\Gamma_{\text{coh}} \qquad (3.98)$$

that is, T_2^{coh} now decreases with increasing diffusion rate as is shown in figure 3.18. This is a particular consequence of band diffusion in a periodic magnetic field array. If such an effect could be detected in, for example, BCC iron, it would be a strong indication for the presence of band diffusion.

So far, we have neglected any incoherent diffusion contribution, always in competition with coherent band diffusion, hence the formulae (3.92), (3.93) and (3.96) are still incomplete. The full formulae of Fujii (1979b) are listed below

$$T_2 = \frac{3}{4\gamma_\mu^2 H_{\text{dip}}^2}\left(8\Gamma_h + \tfrac{2}{3}\Omega(T)\frac{1-g_M}{g_M}\right) \qquad (3.99)$$

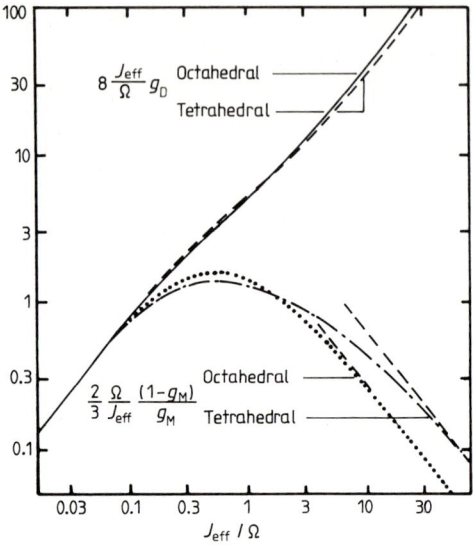

Figure 3.18 Comparison of T_2^{coh} (equation (3.93)) and the diffusion constant, D (equation (3.96)) as functions of J_{eff}/Ω (Fujii, 1979b).

$$D = \frac{d^2}{6}\left(8\Gamma_h + 8\frac{J_{eff}^2}{\Omega(T)}g_D\right) \qquad (3.100)$$

where Γ_h is the incoherent or hopping diffusion rate. Fujii's theory has been formulated with respect to ferromagnetic BCC metals, however, with proper care it should also be applicable to antiferromagnetic structures.

3.6 Longitudinal Relaxation in Magnetically Ordered Systems

As we will discuss later in §4.6.1, the ordered electronic moments in ferromagnetic, ferrimagnetic and antiferromagnetic systems produce, under certain symmetry conditions, non-vanishing strong dipole fields at the interstitial sites where the μ^+ may reside. Depending on the system these dipole fields may vary in strength and orientation from site to site. Given this situation, the randomly diffusing μ^+ will be exposed to fluctuating magnetic fields. As is well known from the theory of relaxation in NMR this will lead to spin-flip transitions which will relax the polarisation of a spin ensemble along a strong static field $\boldsymbol{B}_0$. This possibility of studying μ^+ diffusion in certain ferromagnetic hosts was first pointed out by Seeger (1978). Actually, only the fluctuating components transverse to $\boldsymbol{B}_0$ contribute to the relaxation mechanism. Let us denote the fluctuating field by $B_1(t)$ and define the z axis by the direction of B_0, i.e. $B_0 = B_0 \hat{z}_0$.

The interaction of the μ^+ with these fields is described by the Hamiltonian

$$\mathcal{H} = \gamma_\mu \hbar (I_z B_0 + B_1(t) \cdot I). \tag{3.101}$$

The initial polarisation of the μ^+ is assumed to be parallel to B_0 or the z axis. The relaxation rate in the z direction, the longitudinal or spin–lattice relaxation rate, caused by $B_1(t)$ can be calculated from the Redfield theory. Following Slichter (1978, pp 179ff). The relaxation rate is given by the expression

$$1/T_1 = \gamma_\mu^2 [k_{xx}(\omega_0) + k_{yy}(\omega_0)] \tag{3.102}$$

where

$$k_{qq'}(\omega) = \tfrac{1}{2} \int_{-\infty}^{+\infty} \langle B_{1,q}(t) B_{1,q'}(t+\tau) \rangle \exp(-i\omega_0 \tau) \, d\tau \tag{3.103}$$

and

$$q = x, y, z \qquad \omega_0 = \gamma_\mu B_0. \tag{3.104}$$

$\langle B_{1,q}(t) B_{1,q'}(t+\tau) \rangle = G_{qq'}(\tau)$ is the field correlation function, already introduced in §3.3.1. Assuming, as before, that

$$\langle B_{1,q}(t) B_{1,q'}(t+\tau) \rangle = \overline{B_{1,q}^2} \exp{-\tau/\tau_c} \tag{3.105}$$

we find

$$T_1^{-1} = \gamma_\mu^2 (\overline{B_{1,x}^2} + \overline{B_{1,y}^2}) \tau_c / [1 + (\omega_0 \tau_c)^2]. \tag{3.106}$$

It remains to determine the average squared values of the fluctuating transverse field components $B_{1,x}(t)$ and $B_{1,y}(t)$.

In FCC metals, like Ni, the cubic local symmetry at the possible interstitial sites leads to vanishing dipole fields from the ferromagnetically ordered moments, independent of the direction of the magnetisation (see §4.6.1). Hence μ^+ diffusion in ferromagnetic FCC metals will not produce any longitudinal relaxation.

In HCP metals, like Co and Gd, non-zero dipole fields are present at the interstitial sites, but those at any one type of interstitial site (tetrahedral, octahedral) are identical in direction and magnitude (§4.6.1). Excluding the possibility that diffusion may also take place between different types of interstitial sites, no fluctuating fields will be felt by the diffusing μ^+.

The situation is different in ferromagnetic BCC metals, like Fe. In the general case of arbitrary orientation of magnetisation or ionic magnetic moments, with respect to the crystal axes, we find that for each type of site, three magnetically inequivalent versions exist. This is associated with the fact that in BCC crystals the interstitial sites possess axes of tetragonal symmetry with orientation along $x' = \langle 100 \rangle$, $y' = \langle 010 \rangle$ and $z' = \langle 001 \rangle$ in the ratio 1:1:1.

In ferromagnetic iron in zero external field, the spontaneous magnetisation is either along the $\langle 100 \rangle$, the $\langle 010 \rangle$ or the $\langle 001 \rangle$ crystalline axis. For reasons of symmetry, B_{dip} at an interstitial site with its tetragonal symmetry axis parallel to the magnetisation is either parallel or antiparallel to the magnetisation direction. Let us denote this field by $B_{\text{dip}}^{\|}$. It can be calculated according to §4.6.1. If the magnetisation has some arbitrary direction with respect to the crystalline axis x', y', z' the dipolar field B_{dip} at an interstitial site with the tetragonal axis in $i = x'$, y', z' direction is given by the expressions (Seeger and Monachesi 1982).

$$\boldsymbol{B}_{\text{dip}}^{(x')} = \tfrac{1}{2} B_{\text{dip}}^{\|} (-2\beta_1, \beta_2, \beta_3)$$
$$\boldsymbol{B}_{\text{dip}}^{(y')} = \tfrac{1}{2} B_{\text{dip}}^{\|} (\beta_1, -2\beta_2, \beta_3) \qquad (3.107)$$
$$\boldsymbol{B}_{\text{dip}}^{(z')} = \tfrac{1}{2} B_{\text{dip}}^{\|} (\beta_1, \beta_2, -2\beta_3)$$

where β_1, β_2 and β_3 are the direction cosines of the magnetisation with respect to the axes x', y' and z'.

An arbitrary direction of the magnetisation can always be achieved by an external field strong enough in that direction to saturate the sample (see §4.6). We will always assume the initial polarisation of the μ^+ to be parallel to the magnetisation or the applied field $\boldsymbol{B}_0$ defining the z direction. The total field at the μ^+ is given by equation (2.159)

$$\boldsymbol{B}_\mu = \boldsymbol{B}_0 + [(4\pi/3) - N]\boldsymbol{M} + \boldsymbol{B}_{\text{dip}} + \boldsymbol{B}_{\text{hf}}. \qquad (3.108)$$

Above saturation $\boldsymbol{B}_0$, $\boldsymbol{M}$ and $\boldsymbol{B}_{\text{hf}}$ are collinear and the only transverse component, if present, must originate from $\boldsymbol{B}_{\text{dip}}$. The average squared transverse component is now calculated as follows

$$\overline{B_{\text{dip},\perp}^2} = \overline{B_{\text{dip},x}^2} + \overline{B_{\text{dip},y}^2} = \tfrac{1}{3} \sum_{i=1}^{3} (|\boldsymbol{B}_{\text{dip}}^{(i)}|^2 - \boldsymbol{B}_{\text{dip}}^{(i)} \hat{\boldsymbol{z}}_0) \qquad (3.109)$$

where i equals x', y' or z'. This equation will be evaluated for the experimentally relevant case: the magnetisation is confined to the (110) plane ($\beta_1 = \beta_2$). By rotating the magnetisation $\boldsymbol{M}$ in that plane we reach the three important orientations: $\boldsymbol{M} \| \langle 100 \rangle, \langle 110 \rangle$ and $\langle 111 \rangle$. We then arrive at

$$\overline{B_{\text{dip},\perp}^2} = (B_{\text{dip}}^{\|}/2)^2 \tfrac{3}{2} \sin^2 \theta (1 + 3\cos^2 \theta) \qquad (3.110)$$

where θ is the angle between the $\langle 100 \rangle$ axis and the magnetisation. Equation (3.106) then reads

$$T_1^{-1} = \Gamma_1 = \gamma_\mu^2 (B_{\text{dip}}^{\|}/2)^2 \tfrac{3}{2} \sin^2 \theta (1 + 3\cos^2 \theta) \frac{\tau_c}{1 + \omega_0^2 \tau_c^2} \qquad (3.111)$$

with

$$\omega_0 = \gamma_\mu \{ B_0 + [(4\pi/3) - N]M + B_{\text{hf}} \}. \qquad (3.112)$$

If M is along the $\langle 100 \rangle$ axis we find

$$\Gamma_1^{\langle 100 \rangle} = 0 \tag{3.113}$$

as is to be expected, since B_{dip} is collinear with M as previously pointed out. If M is along the $\langle 111 \rangle$ axis we have

$$\Gamma_1^{\langle 111 \rangle} = \tfrac{1}{2}(\gamma_\mu B_{\text{dip}}^\parallel)^2 \tau_c / (1 + \omega_0^2 \tau_c^2). \tag{3.114}$$

The longitudinal relaxation rate has been derived on the assumption that diffusion takes place only among sites of the same type. Seeger and Monachesi (1982) have treated the more general case of μ^+ diffusion between different types of sites. They have also derived an expression for the transverse relaxation rate $\Gamma_2 = 1/T_2$. Assuming again only diffusion between equivalent sites, one finds for example

$$\Gamma_2^{\langle 100 \rangle} = T_2^{-1} = \tfrac{1}{2}(\gamma_\mu B_{\text{dip}}^\parallel)^2 \tau_c \tag{3.115}$$

$$\Gamma_2^{\langle 111 \rangle} = \tfrac{1}{2}\Gamma_1^{\langle 111 \rangle}. \tag{3.116}$$

Experimentally these rates can be measured if the initial polarisation is perpendicular to the magnetisation when a relaxing precession signal can be observed.

Some important advantages in measuring longitudinal relaxation rates must be pointed out. According to equation (3.114) $\Gamma_1^{\langle 111 \rangle}$ will show a maximum at $\omega_0 \tau_c = 1$. This can be used for an absolute determination of τ_c at a given temperature by varying ω_0 and recording the value at which $\Gamma_1^{\langle 111 \rangle}$ reaches its maximum. ω_0 can be varied via the applied field (equation (3.112)). In the experiment one then determines the value of the applied field at which $\Gamma_1^{\langle 111 \rangle}$ assumes its maximum. To determine τ_c we need the corresponding value of ω_0 which can be calculated from equation (3.112) provided B_{hf} and $[(4\pi/3) - N]M$ are known, which is usually the case. No information on $B_{\text{dip}}^\parallel$ is required, which contrasts with the usual transverse field experiments in non-magnetic metals with nuclear dipole fields. In those, the relaxation rate in the motional narrowing limit is given by $1/T_2 = M_2 \tau_c$ (equation (3.47)) and moreover the determination of τ_c depends on a pre-knowledge of M_2. Once τ_c is known also $B_{\text{dip}}^\parallel$ can be determined. This information may be used to identify the interstice occupied by the μ^+.

In contrast to the transverse field experiments, the longitudinal relaxation rates are unaffected by trapping at impurities as long as repeated capture and release is excluded. Once a μ^+ is trapped it no longer experiences a fluctuating field, and the longitudinal relaxation rate disappears. In transverse field experiments, trapping will manifest itself as an enhanced relaxation rate due to dephasing in the resulting static field spread (see §3.3.3).

Finally if applicable this longitudinal field technique in magnetically ordered systems allows us to extend the measurement of diffusion rates to much higher values, since $B_{\text{dip}}^\parallel$ in equation (3.114) now originates from electronic

Muon spin rotation in metals

moments which are at least two orders of magnitude stronger than nuclear magnetic moments.

3.7 Change of Precession Frequency and Amplitude in the Presence of Trapping

If the Larmor frequency of the μ^+ at a trapping site is distinctly different from the frequency at an ordinary site, a new possibility for studying diffusion towards the trap is opened up. For simplicity let us assume that escape out of a trap is absent. The μ^+ will arrive at a trap at random times with random spin phases. Once in a trap it will start to precess with a different frequency. The superposition of all μ^+ that precess with the new frequency, but with random phases will lead to a more or less reduced precession amplitude. This depends on how fast the μ^+ has, on the average, reached the trap and on the difference between the two frequencies. Observation of the amplitude of the trap associated precession signal will therefore allow us to monitor the diffusion towards the trap. Of course, the trap associated precession frequency must be shifted enough from the ordinary site frequency to be studied separately. In analogy to equation (3.61), the total time dependent polarisation can be written in complex notation as follows

$$\tilde{P}(t) = \exp(-\lambda t)\exp(-i\omega_0 t)$$
$$+ \int_0^t \exp[-i\omega_1(t-t')]\exp(-t/T_2)\exp(-\omega_0 t')\,dn_0(t'). \quad (3.117)$$

ω_0 = Larmor frequency at ordinary sites, ω_1 = Larmor frequency at trapping sites, T_2 = transverse relaxation time of μ^+ at ordinary sites and $\lambda = T_2^{-1} + \tau_0^{-1}$. τ_0 is the average lifetime of the μ^+ in the free (untrapped) state and $n_0(t)$ is the occupation probability of μ^+ in the free state at time t. $dn_0(t')$ denotes the fraction of μ^+ that are transferred to the trapped state at time t' in the interval dt. It is assumed that all μ^+ are initially in the free state. $n_0(t)$ is again given by expression (3.60)

$$n_0(t) = \exp(-t/\tau_0) \quad (3.118)$$

and $dn_0(t) = -\tau_0^{-1} n_0(t)\,dt$. Equation (3.117) can now be evaluated and yields the result

$$\tilde{P}(t) = \exp(-i\omega_0 t)\exp(-\lambda t)$$
$$+ \frac{1}{\tau_0 \lambda}\frac{\exp[-i(\omega_1 t + \varphi)]}{[1+(\Delta\omega/\lambda)^2]^{1/2}}\{1-\exp[-(\lambda - i\Delta\omega)t]\}. \quad (3.119)$$

The phase φ follows from $\tan\varphi = \Delta\omega/\lambda$ and $\Delta\omega = \omega_1 - \omega_0$ and the real part of

$\tilde{P}(t)$ is given by

$$P(t) = \text{Re } \tilde{P}(t)$$

$$= \exp(-\lambda t) \cos \omega_0 t$$

$$+ \frac{1}{\tau_0 \lambda} \frac{\cos(\omega_1 t + \varphi)}{[1 + (\Delta\omega/\lambda)^2]^{1/2}} \{1 - \exp[-\lambda t] \cos \Delta\omega t\}. \tag{3.120}$$

For rapid capture into the traps, (τ_0 small and $t \gg \tau_0$) equation (3.120) reduces to

$$P(t) \simeq \cos(\omega_1 t + \varphi)/[1 + (\Delta\omega\tau_0)^2]^{1/2}. \tag{3.121}$$

By studying the amplitude and phase of the precession signal from the μ^+ in the traps it is thus possible to extract τ_0 and its temperature dependence (Möslang et al 1983). Equation (3.117) could be generalised so as to also include repeated capture and release events. For very rapid processes, thermal equilibrium will be established and the free and trapped states will show occupation probabilities of $\tau_0/(\tau_0 + \tau_1)$ and $\tau_1(\tau_0 + \tau_1)$, respectively. τ_1 is the lifetime in the trapped state. In that limit the precession is characterised by a single average precession frequency $\bar{\omega}$ with

$$\bar{\omega} = \frac{\tau_0}{\tau_0 + \tau_1} \omega_0 + \frac{\tau_1}{\tau_0 + \tau_1} \omega_1 \tag{3.122}$$

and

$$P(t) = \exp(-\Gamma t) \cos \bar{\omega} t \tag{3.123}$$

with $\Gamma = \langle \omega^2 \rangle \tau_c$, $\tau_c^{-1} = \tau_0^{-1} + \tau_1^{-1}$. $\langle \omega^2 \rangle$ serves as a second moment of the field spread over all the available μ^+ sites.

If the difference, $(\omega_0 - \omega_1)$, is small compared with $1/\tau_\mu$, the two signals at ω_0 and ω_1 cannot be resolved and the technique of measuring the amplitude of the signal at ω_1 fails. It is, however, still possible to measure some average frequency $\langle \omega \rangle$ (for example from the Fourier transform of equation (3.119)) which will be related to τ_0. Such an approach has recently been applied by Gygax et al (1983) in the analysis of a temperature induced change of the μ^+ Knight shift in Zn caused by trapping (see also Studer 1983).

3.8 Short Review of μ^+ Diffusion Results in Metals and a Comparison with Hydrogen Data

μ^+ diffusion was first observed in polycrystalline Cu in the pioneering measurement of Gurevich et al (1972). Since then μ^+ diffusion has been studied extensively in a number of other metals, notably Al, Nb, V, Ta, Au, Fe, Zn and also Be, Bi, Ga, In, Sb and Ni and Cr. See table 3.5. In some of the metals tried (Al, Nb, V) the amount of impurities acting as traps was very carefully controlled. Others had been exposed to electron and neutron irradiation (Al, Nb, Fe), or were subjected to mechanical and temperature treatments (Al). μ^+ diffusion has also been observed in some metal hydrides, namely in $LaNi_5H_x$ (Camani et al 1979, Gygax et al 1984b), ZrH_x and VH_x (Doyama et al 1981), β-PdH_x (Gygax et al 1984b) and β-NbH_x (Richter et al 1984). Early measurements in β-PdH_x (Kossler et al 1977) were restricted to temperatures below 77 K and did not indicate significant μ^+ mobility. In the following, some exemplary data are reviewed and will be compared to corresponding hydrogen data, if available. The metal hosts are grouped according to their crystal structure.

3.8.1 FCC metals: Cu, Al, Au, Ag

(1) Copper

As mentioned before, Cu was the first material in which μ^+ diffusion was successfully demonstrated (Gurevich et al 1972, Grebinnik et al 1975). Figure 3.19, taken from Grebinnik et al (1975), shows a plot of the relaxation rate λ versus temperature, which was obtained by using a single crystal Cu sample. λ is defined as the reciprocal time at which the relaxation function had decreased to $1/e$. On the low temperature side we see a plateau reflecting the depolarisation of immobile μ^+ due to the Cu nuclear dipole fields. This part of the data was already discussed in §3.1. Above 80 K the relaxation rate, λ, decreases with increasing temperature, signalling the onset of diffusion by motional averaging. The relaxation function is now described by equation (3.45). From a fit of this equation to the rate-against-time histograms, the correlation time τ_c can be extracted. Results for τ_c from the measurements by Schilling et al (1982) are displayed semilogarithmically against inverse temperature in figure 3.20. A similar figure (figure 3.14) was already shown in §3.3.1 which demonstrated the independence of the extracted τ_c values from the applied external magnetic field used in the experiments. According to §3.3.1, τ_c can be taken as a measure of the mean time of residence at a given site. τ_c^{-1} is thus the diffusion rate. Both the data of Grebinnik et al (1975) and Schilling et al (1982) show that the temperature dependence of the diffusion rate is not well described by a simple temperature activated (Arrhenius) process, which would correspond to the broken straight lines in figure 3.19. Instead a better fit is obtained by adopting the Teichler expression for the diffusion rate equation (3.28). The corresponding results for the activation energy, E_a, and the tunnelling matrix element, J_0, are listed in table 3.6, together with results for the pre-exponential factor, v_0, and the activation energy, E_0, by adopting the Arrhenius expression equation (3.27) for the diffusion rate.

Table 3.5 μ^+ diffusion studies in metals and their authors.

Metal	Crystal structure	References
Cu	FCC	Gurevich et al (1972), Grebinnik et al (1975, 1978b), Camani et al (1977), Gauster et al (1977), Hartmann et al (1980), Clawson et al (1982), Schilling et al (1982)
Al	FCC	Gauster et al (1977), Dorenburg et al (1978), Grebinnik et al (1978a), Hartmann et al (1978a), Kossler et al (1978, 1979), Brown et al (1979b), Metz (1980), Brewer et al (1981), Herlach (1981), Kehr et al (1981), Nakai et al (1981), Kehr et al (1982)
Au	FCC	Brown et al (1981), Schillaci et al (1981)
Ag	FCC	Schillaci et al (1983)
Ni	FCC	Graf et al (1979)
V	BCC	Fiory et al (1978), Grebinnik et al (1978b, 1981), Hartmann et al (1978a), Heffner et al (1979)
Nb	BCC	Birnbaum et al (1978), Borghini et al (1978), Grebinnik et al (1978b), Hartman et al (1978a), Lankford et al (1978), Brown et al (1979a), Boekema et al (1982), Hartmann et al (1983), Gygax et al (1984a)
Ta	BCC	Hartmann et al (1978a), Schilling et al (1978), Metz et al (1979)
Cr	BCC	Kossler et al (1977), Grebinnik et al (1978c), Weidinger et al (1981)
Fe	BCC	Grynszpan et al (1979), Yaouanc et al (1979), Graf et al (1980), Balzer et al (1981), Yagi et al (1981), Möslang et al (1983, 1984)
Be	HCP	Metz et al (1979)
Zn	HCP	Gygax et al (1983)
Ga	orthorhombic	Grebinnik et al (1977b)
In	tetragonal	Grebinnik et al (1977b, 1979)
Sb	rhombohedral	Grebinnik et al (1977b)
Bi	rhombohedral	Grebinnik et al (1977a), Barsov et al (1984), Gygax et al (1984c)

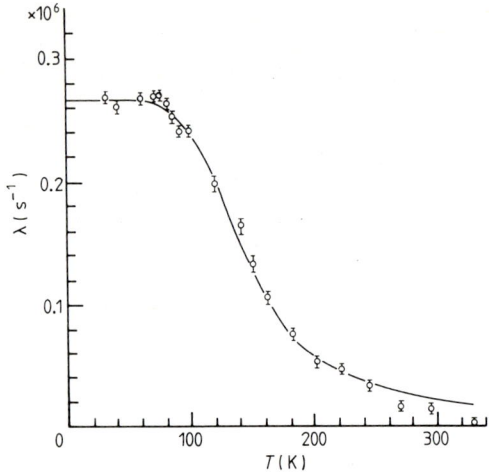

Figure 3.19 Temperature dependence of the relaxation rate, λ, of the transverse field μSR signal ($B = 6.2$ mT) in monocrystalline Cu (Grebinnik et al 1975).

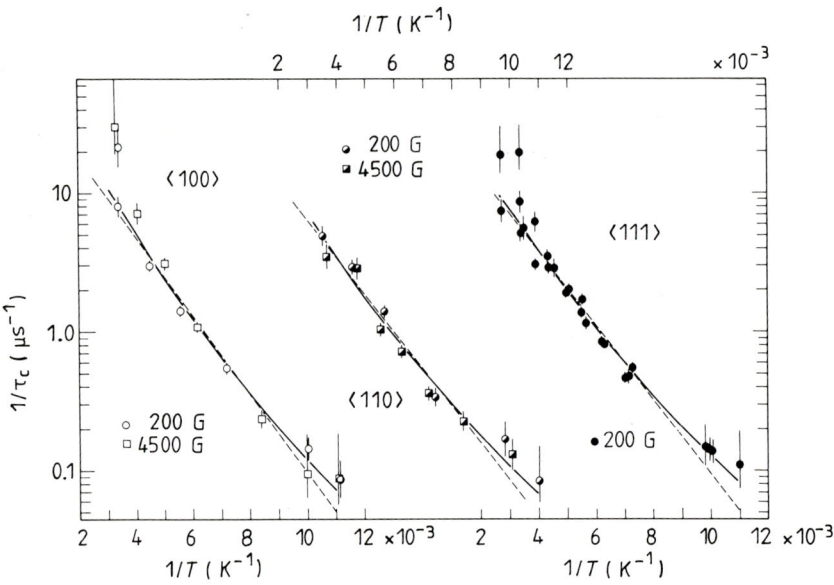

Figure 3.20 Semilogarithmic plot of the temperature dependence of the diffusion rate, τ_c^{-1}, from measurements in Cu (Schilling et al 1982)

Table 3.6 Compilation of diffusion parameters for μ^+ and protons in Cu, with the respective error bar values given in brackets.

	Temperature range (K)	Diffusion model	v_0 (s^{-1})	D_0 (cm^2 s^{-1})	E_0, E_a (meV)	J_0 (μeV)	Reference
μ^+	20–293	Arrhenius	$4.07\,(0.38)\times 10^7$	—	48.4 (1.5)	—	Grebinnik et al (1975)
	5–365	Arrhenius	$4.4\,(1.0)\times 10^7$	—	53.4 (3.6)	—	Schilling et al (1982)
	5–365	Teichler	—	$1.66\,(0.33)\times 10^{-8}$	79.2 (3.1)	18.3 (1.8)	Schilling et al (1982)
p	703–908	Arrhenius	1.06×10^{14}	—	423	—	Eichenauer (1965)
	723–1188	Arrhenius	1.09×10^{14}	11.3×10^{-3}	403	—	Katz et al (1971)

At this point it is interesting to compare the μ^+ results with results on hydrogen diffusion in Cu (Eichenauer et al 1965, Katz et al 1971), which are well reproduced by an Arrhenius expression. Results for the v_0 and E_0 of protons are also listed in table 3.6. It is seen that the pre-exponential factors and the activation energies are quite different displaying the ratios. $E_0(p)/E_0(\mu^+) = 8.66$ and $v_0(p)/v_0(\mu^+) = 3 \times 10^6$. According to classical rate theories (see e.g. Vineyard 1957), the activation energy should not display an isotopic mass dependence while the pre-exponential factors should behave like

$$v_0(p)/v_0(\mu^+) = (m_\mu/m_p)^{1/2} \simeq 2.9. \quad (3.124)$$

The pre-exponential factor for the μ^+ is reduced considerably in comparison to the hydrogen value. This led Grebinnik et al (1975) to the conclusion that μ^+ diffusion in Cu does not proceed by classical over-barrier jumping, but by under-barrier quantum tunnelling events.

In comparing the μ^+ and hydrogen diffusion data in Cu one has, of course, to bear in mind, that the data were taken in quite different temperature ranges, the μ^+ data below the Debye temperature ($\theta_D = 343$ K) and the proton data above. However, it is in the low temperature regime that quantum effects in diffusion are expected to show up. And it is this aspect of μ^+ diffusion that has attracted so much attention in recent years. Confirmation of the quantum nature of μ^+ diffusion in Cu below 300 K was finally indicated by the excellent agreement of the Teichler (1977) theory with the experiment.

According to the quantum theories of diffusion (see §3.2) the activation energy, E_a appearing in equation (3.28) is related to the lattice relaxation energy, or self-trapping energy, associated with the small polaron picture of the μ^+. According to Teichler (1977), the activation energy can be expressed as follows

$$E_a = 0.258 g^2/f \quad (3.125)$$

where g is the μ^+ nearest neighbour Cu force and f the elastic coupling constant between nearest neighbour Cu atoms. For a FCC lattice with lattice constant a, $f = aC_{44}$ (C_{44} = elastic stiffness constant). Inserting E_a from table 3.6 into equation (3.125) one obtains $g = 2.71$ (eV a^{-1}). In §3.1, the same force constant was deduced from the observed nearest neighbour dilatation by use of the Kanzaki formalism with the result $g = 3.39$ (eV a^{-1}) (Schilling et al 1982). The two values are sufficiently close to each other to claim that a consistent understanding of both the static and dynamic properties of the μ^+ in Cu has been reached: self-trapping below 80 K and phonon-assisted tunnelling of the small polaron structure above 80 K.

More recent measurements, however, have cast some doubt on this picture. Hartmann et al (1980, 1981) have found that below about 20 K the relaxation rate, $\lambda = \sigma$, starts to decrease with decreasing temperature (figure 3.21). The decrease is most drastic between about 5 K and 0.5 K and seems to flatten out at still lower temperatures (down to 30 mK). This behaviour could suggest,

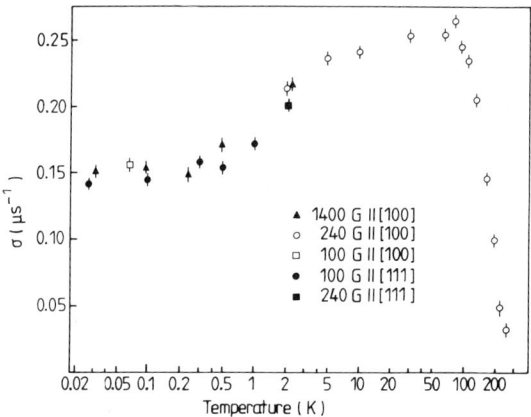

Figure 3.21 Low temperature behaviour of the gaussian relaxation constant, σ, in high purity Cu (Hartmann et al 1981).

analogous to the results in Nb (see below) that the plateau in the σ of Cu above 20 K is also due to trapping. Another interesting result is shown in figure 3.22 in which the field and orientation dependence at 2 K and 0.1 K are displayed. The data are still consistent with an octahedral site assignment and an unchanged electric field gradient (see §2.3.2). Recent zero field measurements (Clawson et al 1982, Kadono et al 1984) show a dynamically narrowed Kubo–Toyabe signal, clearly indicating that the μ^+ is diffusing at the lowest temperatures, approaching a jump-rate of 5×10^5 s^{-1}. The low temperature mobility can be suppressed by the presence of impurities, such as Fe and Si.

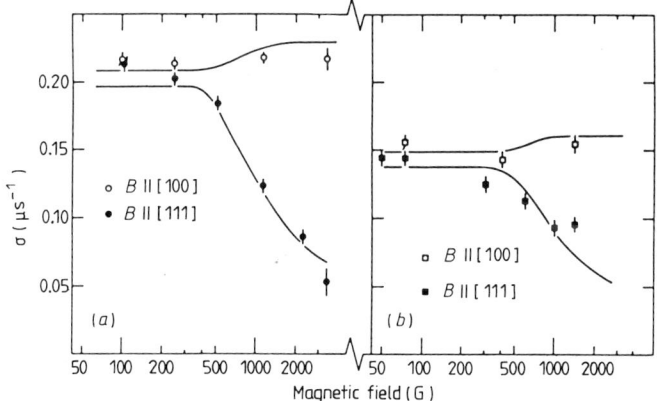

Figure 3.22 Magnetic field dependence of σ for fields parallel to crystal directions [100] and [111] at (a) 2 K and (b) 0.1 K in high purity Cu (Hartmann et al 1981).

(2) Aluminium

From the beginning, μSR experiments in pure or moderately pure Al were quite puzzling. No damping of the μSR signal could be observed, in complete contrast to the situation in Cu (Gauster *et al* 1977). This indicated a very fast μ^+ diffusion down to the lowest temperature applied at that time (100 mK) leading to pronounced motional narrowing. No self-trapping in particular seemed to be present. It was soon found that alloying Al with small amounts of Cu (Grebinnik *et al* 1977a, Kossler *et al* 1978), or deforming it mechanically (Kossler *et al* 1978), or irradiating Al with neutrons (Dorenburg *et al* 1978) produced a damping in the μSR signal. The temperature dependence of this looked generally complicated and was explained by trapping induced by the created defect structures.

The apparent fast diffusion at the lowest temperatures suggested it was of the nature of coherent band propagation. This hypothesis was tested by Hartmann *et al* (1978b) by alloying Al with small amounts (500, 1300 ppm) of Mn impurities, which are known to introduce long range strain fields. If these are of sufficient magnitude, one may expect the band propagation to be inhibited. It was indeed found that the μ^+ seemed to localise around 15 K, and that the field and orientation dependence of the relaxation rate was compatible with a tetrahedral site occupation (see figure 3.5). There was no indication that the μ^+ would come to rest in the immediate vicinity of a Mn impurity.

Subsequent measurements by Hartmann *et al* (1980) and Kehr *et al* (1981, 1982) in Al doped with considerably smaller quantities of Mn as well as Li and Ag and extended down to a temperature of 30 mK, showed that the situation was much more complex than previously thought. Some of the data are presented in figure 3.23. The diagram displays the relaxation parameter σ (the data were fitted with a gaussian damping function) as a function of temperature between 30 mK and 50 K. We see again a peak in σ around 15 K, the position of which is independent of the Mn impurity concentration. The peak height, however, depends strongly on the Mn concentration below 500 ppm. In the first experiments on 500 ppm and 1300 ppm AlMn samples (Hartmann *et al* 1978b) σ decreased slightly for temperatures below 15 K, giving the impression that the μ^+ were also predominantly localised for T < 15 K. In the samples with lower Mn concentrations, however, σ decreased drastically with decreasing temperature below 15 K down to about 2 K. At this point σ started to increase again with further decrease in temperature. Even in pure (6N) Al, a slight increase of σ is visible. The same behaviour is also found for samples of $AlLi_{75ppm}$ and $AlAg_{117ppm}$ (Kehr *et al* 1982). In contrast to Mn impurities, Li and Ag impurities in Al produce a strainfield which is either of negligible or of short range order only. The data around 15 K were analysed according to the two-state trapping model of Kehr *et al* (1978) (see §3.3.3), with the assumption that trapping by Mn impurities occurs in this temperature

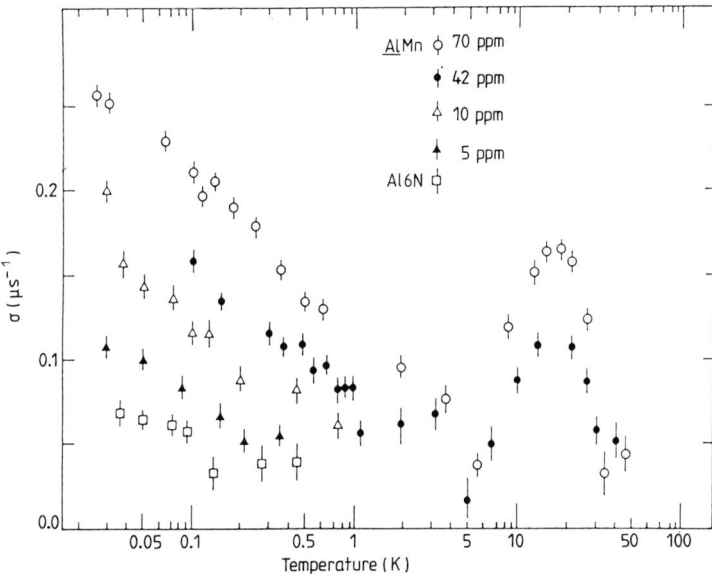

Figure 3.23 Temperature dependence of the gaussian damping parameter σ for Al and AlMn polycrystalline samples (Kehr et al 1982).

region whilst at higher temperatures, escape from the traps is observed. As discussed in § 3.3.3, the two-state model is characterised by two parameters, τ_0 and τ_1. τ_0 is the lifetime of the μ^+ in the untrapped state, or the average time needed to reach a trap and τ_1 is the residence time actually in the trap. It was found that τ_0^{-1} depended only weakly on temperature, and was proportional to the impurity concentration c_{Mn}:

$$\tau_0^{-1} \simeq c_{Mn} T^\beta \quad (3.126)$$

where $\beta = 0.89(3)$ for the AlMn system. The escape rate τ_1^{-1} was found to obey an Arrhenius relationship:

$$\tau_1^{-1} \simeq \exp(-E_0/k_B\tau) \quad (3.127)$$

with an activation energy of $E_0 = 120(6)$ K. The linear dependence of τ_0^{-1} on the Mn concentration is strong evidence for the applicability of the two-state trapping model.

The low temperature data were analysed using the motional narrowing relaxation function, equation (3.45) and yielding the correlation time τ_c. A double logarithmic plot of τ_c versus temperature is shown in figure 3.24. It is evident from the figure that the temperature dependence of τ_c follows a simple power law. In fact, the data can all be described accurately by the equation:

$$\tau_c^{-1}(s^{-1}) = (6.9 \pm 1.2) \times 10^5 \{[T(mK)]^{0.60(4)} [c_{Mn}(ppm)]^{-0.76(4)}\}. \quad (3.128)$$

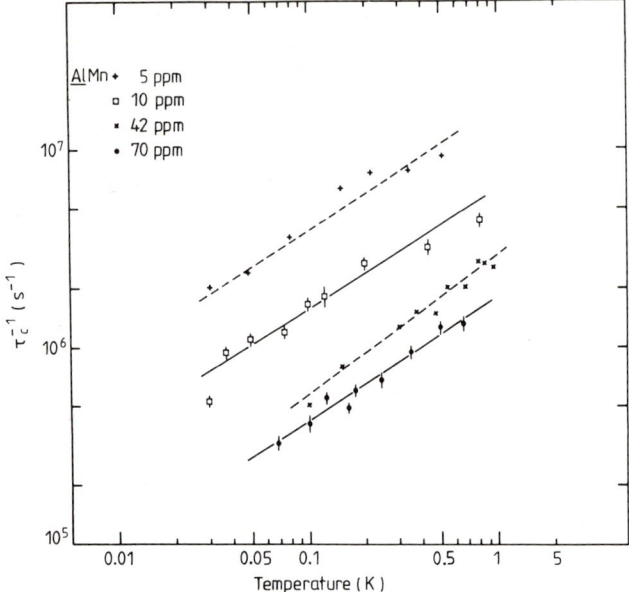

Figure 3.24 Double logarithmic plot of the temperature dependence of the correlation rate τ_c^{-1} for temperatures below 1 K in various AlMn samples. The correlation time, τ_c, was determined by fitting equation (3.45) to the data (Kehr *et al* 1982).

Additionally, the low temperature data were also analysed by the two-state model, assuming that the intrinsic diffusion rate increases with decreasing temperature leading eventually to trapping at an impurity. In this model, the diffusion would proceed by band propagation limited by the dynamical destruction of the band states as discussed in §3.2. According to equation (3.80), the lifetime τ_0 in the untrapped state is given by:

$$\tau_0^{-1} = 4\pi r_0 D n_t \tag{3.129}$$

where n_t is the number of trap sites per unit volume. The diffusion constant, D, for coherent diffusion may be expressed as (see §3.2)

$$D \simeq \tau^{-1}\lambda^2 = v^2 T^{-1} \tag{3.130}$$

where τ is the lifetime of the bandstate, λ the mean free path and v the velocity of the μ^+ which is an average over all occupied bandstates. Assuming that escape out of the trap sites at these low temperatures is impossible (i.e. $\tau_1^{-1} = 0$), the analysis yields a capture-rate of

$$\tau_0^{-1} = 3.0(6) \times 10^5 T^{-0.61(4)} c_{Mn}^{0.76(4)}. \tag{3.131}$$

All the exponent signs are simply reversed with practically the same absolute values when compared to equation (3.128).

Both methods of analysis lead to various interpretative difficulties. In the first case, the almost linear temperature dependence of τ_c^{-1} could be considered in terms of a one-phonon induced incoherent tunnelling mechanism as in §3.2. However, the observed concentration dependence appears incompatible with such a view (Kehr et al 1982). In the second case it is hard to understand the observed exponent in the temperature dependence. Combining equations (3.129), (3.130) and (3.131) and assuming that $v \simeq T^{1/2}$ (Kehr et al 1982) one obtains the following temperature and concentration dependence of the average lifetime of the bandstates

$$1/\tau \simeq c^{0.24} T^{1.6}. \tag{3.132}$$

This result is in complete contrast with the theory of Kagan and Klinger (1974) which predicts $\tau \simeq T^{-9}$ equation (3.19). Also, the residual concentration dependence is not explained. In summary, a consistent and quantitative understanding of the low temperature diffusion properties of the μ^+ in Al is still lacking.

Another feature of the low temperature data from the AlMn samples which is remarkable concerns the field and orientation dependence of the relaxation rate. At 15 K, as mentioned before, a tetrahedral site occupation is indicated, while at 40 mK an octahedral site occupation is clearly manifest (see figure 3.8). At intermediate temperatures the data point to an occupation of both types of sites (Kehr et al 1982). This behaviour is also not understood, although an octahedral interstitial site occupation is generally expected for FCC metals (see discussion in §3.1).

μ^+ diffusion in Al has also been studied under conditions in which trapping at vacancies was expected. There are two classes of experiments in which μ^+ trapping at vacancies has probably been observed. In the first, vacancies are thermally created by heating the sample and then cooling it rapidly to low temperatures. By this the vacancy concentration is conserved as they can no longer move and are frozen in. By careful control of this quenching procedure one can make sure that only mono-vacancies are formed and vacancy clustering is avoided (Nakai et al 1981). The first evidence for μ^+ trapping at vacancies in quenched Al came from an experiment by Brown et al (1979b). Further studies were reported by Nakai et al (1981). The latter study allowed the formation energy for vacancies in Al to be extracted with the result $E_a = 0.68 \pm 0.03$ eV, in excellent agreement with the normally accepted value of 0.69 eV (Hall et al 1974).

In the second class of experiments, the vacancies are created by irradiation, for example with neutrons (Dorenburg et al 1978) or electrons (Metz 1980, Herlach 1981). The disadvantage of this technique is that not only are vacancies created in general, but also other defect centres and clusters. However, such problems can be minimised by irradiating with suitable doses of electrons. Very careful measurements on electron irradiated Al have been made by Metz et al (Metz 1980, Herlach 1981). Results for the relaxation rate λ at various annealing temperatures are shown in figure 3.25. At low temperatures, the relaxation rate is very low and increases with rising temperature towards some saturation value. The saturation value coincides very well with the rigid lattice second moment value for a substitutional site. Other evidence for a vacancy trapping results from the dependence of the data on the annealing temperature. It is seen from figure 3.25, that the vacancy concentration starts to decrease for $T_a > 227$ K. This would imply faster diffusion rates on the average for reaching a vacancy during the μ^+ lifetime, and hence higher temperatures to reach saturation of the depolarisation rate. Annealing the sample at 350 K irreversibly removed the trapping centres.

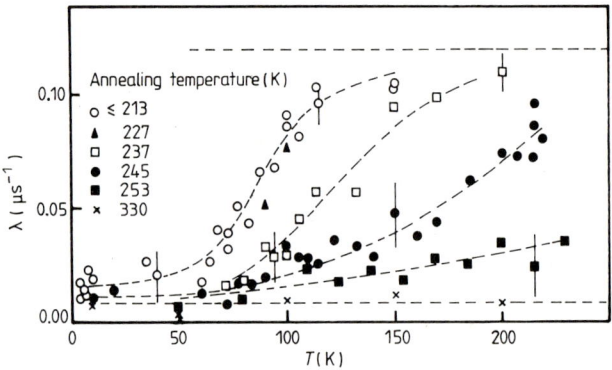

Figure 3.25 μ^+ depolarisation rate, λ, in electron irradiated Al. The different sets of data were taken after annealing the sample at the indicated temperatures. The upper horizontal dashed line indicates the value for the second moment at a vacancy site (Herlach 1981).

Both Metz (1980) and Nakai et al (1981) find rather similar activation energies for the intrinsic diffusion towards the vacancies, namely $E_a = (32.0 \pm 0.24)$ meV (Metz 1980), and $E_a = (39 \pm 15)$ meV (Nakai et al 1981) respectively. These values apply to the temperature range above 50 K. Through comparison with the low temperature results in AlMn, it is evident that a change in the diffusional mode somewhere between 15 and 50 K has to occur. At low temperatures the diffusion rate is high with a weak temperature dependence, and at higher temperatures (>50 K) a temperature activated diffusion mechanism sets in. Note that the activation energy in the latter case is larger than the activation energy observed in the escape process from the Mn related traps around 15 K ($E_a = 10$ meV).

(3) Noble metals: Au, Ag

Natural gold consists of only one isotope ($_{79}$Au^{197}) with spin $\frac{3}{2}$ and a magnetic moment, which is about 16 times smaller than the average nuclear magnetic moment of Cu. In addition the lattice constant of Au is about 30% larger than in Cu. Therefore the nuclear dipole fields at some interstitial site are much too weak to cause any measurable μ^+ depolarisation by dephasing. To be able to study μ^+ diffusion in Au, Brown et al (1981) alloyed Au with small amounts of magnetic Gd (350 ppm, 1200 ppm) and Er (300 ppm) as substitutional impurities. The magnetic interaction between the μ^+ and the magnetic impurities may lead to depolarisation, as discussed in §3.4. Long range diffusion which produces occasional encounters with a magnetic impurity can now be studied. Experimental depolarisation rates, λ, taken from Brown et al (1981) are displayed in figure 3.26. Below 75 K one finds very small relaxation times, indicating that the μ^+ are either immobile or diffusing so slowly that they will not experience a close encounter with a magnetic impurity.

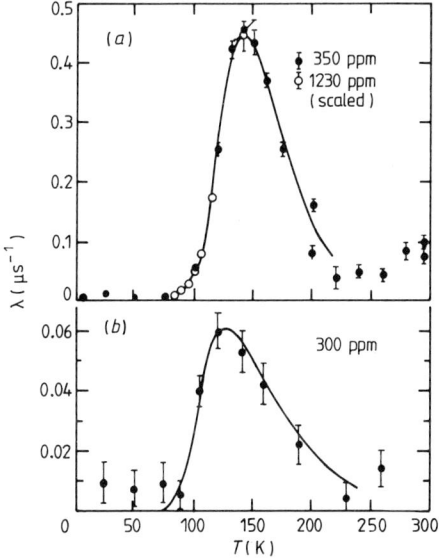

Figure 3.26 Temperature dependence of μ^+ depolarisation rate λ in (a) AuGd and (b) AuEr at 0.8 mT applied field (Brown et al 1981).

Above 75 K a strong increase of λ is observed, which reaches a maximum around 140 K, followed by a decline with further increases in temperature. The qualitative features of the data correspond well to what was described in §3.4. It is worthwhile pointing out that relaxation rates over the range of the peak in AuGd scale with the Gd concentration. This is taken as evidence that the Gd impurities do not act as trapping centres. The full curves in figure 3.26 are fits to the data of the model described in §3.4. The only parameters entering in are the activation energy, E_a, and the pre-exponential factor, v_0, characterising the intrinsic diffusion (a simple Arrhenius behaviour is assumed), the contact hyperfine field at a nearest neighbour site to a magnetic impurity, $H_c = (AS)/\gamma_\mu$, and a parameter b_s denoting the constant factor in equation (3.87) for the electronic relaxation time. In AuGd one finds $H_c = 24 \pm 10\,\text{kOe}$, $b_s = 6 \pm 4\,\text{Oe K}^{-1}$, and in AuEr $H_c = 7\,\text{kOe}$, $b_s = 33 \pm 11\,\text{OeK}^{-1}$. Results for E_a and v_0 are the same for both alloys and are shown in table 3.7, together with more recent results in Ag (Schillaci et al 1983) and corresponding results from hydrogen diffusion measurements.

Table 3.7 Compilation of μ^+ and proton diffusion parameters in Au and Ag hosts, and investigated temperature ranges, (ΔT).

	Host	E_a (K)	v_0 (s^{-1})	$v_0(\mu^+)/v_0(p)$	ΔT (K)	Reference
μ^+	Au	1350 ± 100	$10^{13.5 \pm 0.5}$	~ 8	85–230	Brown et al (1981)
p		2840	$10^{12.6}$		770–1200	Eichenauer et al (1962)
μ^+	Ag	3200 ± 200	$10^{13.5 \pm 0.5}$	~ 1	200–700	Schillaci et al (1983)
p		3620	$10^{13.8}$		950–1120	Katsuta et al (1979)

A direct comparison of the μ^+ and proton data is again questionable in view of the fact that the results were obtained in quite different temperature ranges. The μ^+ data stemmed from measurements around the Debye temperature, and the proton data from measurements above, typically, 3–4 times the Debye temperature. Nevertheless the magnitude of the pre-exponential factors indicate a more classical behaviour as compared, for example, to the situation in Cu. Indeed Schillaci et al (1983) conclude that incoherent tunnelling is not important for μ^+ diffusion in Ag and Au.

3.8.2 BCC metals: Nb, Fe

(1) Niobium

The study of μ^+ diffusion in Nb hosts has played and is still playing an important role in the development of a general understanding of μ^+ diffusion in metals. Measurements in Nb first showed that trapping is an important mechanism in μ^+ diffusion (Birnbaum et al 1978, Borghini et al 1978), even in samples with very low impurity concentrations. In fact, trapping at various impurities with different trap potentials seems to almost completely mask the intrinsic diffusion which appears fast, even at the lowest temperatures so far applied (Hartmann et al 1983). The same behaviour is also essentially observed in V (see table 3.5 for references).

Figure 3.27, taken from Borghini et al (1978) displays the relaxation rate λ as a function of temperature in Nb samples with different concentrations of interstitial N, O and C. The pattern observed in the sample with the lowest content (c) of impurities is very typical. Its qualitative features can be explained as follows: the low temperature plateau (which is not really a plateau) is caused by trapping near a substitutional Ta impurity (Hartmann et al 1983). Above about 15 K detrapping sets in (activation energy $\simeq 20$ meV) which produces motional narrowing. At still higher temperatures (>20 K) the μ^+ moves fast enough to reach the vicinity of an interstitial impurity (O, N) within its lifetime, where it becomes trapped again. This leads again to an increase of the relaxation rate and a second plateau is reached (static dipolar relaxation at the trapping sites). Above 60 K the μ^+ starts to escape out of the second type of trap and a second motional narrowing regime sets in (activation energy $\simeq 50$ meV). Depending on the interstitial impurity concentration, the

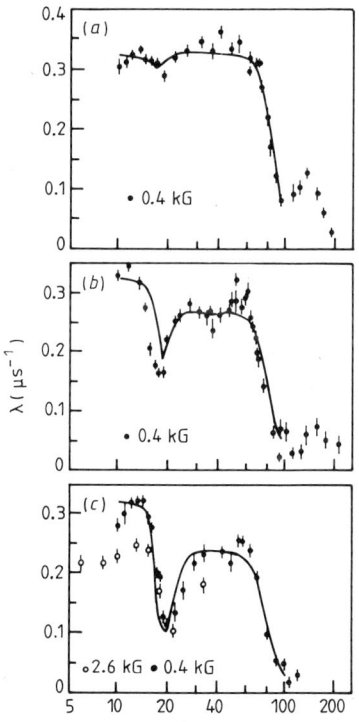

Figure 3.27 Temperature dependence of the μ^+ depolarisation rate λ in Nb, doped with different amounts of interstitial O and N and C impurity atoms (a) 3700 ppm, (b) $\lesssim 60$ ppm, (c) 10–20 ppm) (Borghini et al 1978).

dip in λ around 20 K will be more or less structured. Whether this model constitutes the only reasonable interpretation of the dip-structure in the relaxation rate is still a matter undergoing investigations (Gygax *et al* 1984a).

(2) Iron

Iron is an interesting system to study, in that it is possible to observe μ^+ diffusion in three different ways:

(i) by the usual transverse field technique (see §3.3.1),
(ii) by the longitudinal field technique (see §3.6),
(iii) by monitoring the change of the precession amplitude and frequency associated with trapping at a vacancy (see §3.7).

Fe has a BCC crystal structure, and therefore possesses for each type of interstitial site two magnetically inequivalent versions occurring in the ratio 1:2 and associated with different dipole fields (see §4.6). For the tetrahedral site one finds $B_{\text{dip}}^{\perp} = 2.6$ kG (tetragonal axis $\perp$ magnetisation direction, parallel to $\langle 100 \rangle$) and $B_{\text{dip}}^{\parallel} = -2B_{\text{dip}}^{\perp} = -5.2$ kG (tetragonal axis parallel to $\langle 100 \rangle$). For this distribution of fields over tetrahedral interstitial sites, one can define a second moment given by (Graf *et al* 1980)

$$M_2 = \gamma_\mu^2 (\tfrac{1}{3}(B_{\text{dip}}^{\parallel})^2 + \tfrac{2}{3}(B_{\text{dip}}^{\perp})^2) = \tfrac{1}{2}\gamma_\mu^2 (B_{\text{dip}}^{\parallel})^2. \tag{3.133}$$

For sufficiently fast and random jumps, the μ^+ transverse depolarisation rate, λ, will be given by the motional narrowing expression (see equations (3.46) and (3.47))

$$\lambda = M_2 \tau_c = \tfrac{1}{2}\gamma_\mu^2 (B_{\text{dip}}^{\parallel})^2 \tau_c. \tag{3.134}$$

Because M_2 is very much larger than an M_2 originating from nuclear dipole fields, one will be able to follow the diffusion to much higher rates. On the other hand, rates below several 10^7 s^{-1} will be hard to detect because the depolarisation rates would become too large. The longitudinal field technique discussed in §3.6 gives freedom from such a constraint. This technique has the further advantage that it will be insensitive to trapping, and only the intrinsic diffusion will be observed.

Figure 3.28 displays both the transverse relaxation data of Graf *et al* (1980) and the longitudinal field relaxation data of Yagi *et al* (1981). In the latter experiment, a strong field between 1.2T and 4.9T, applied along the $\langle 111 \rangle$ crystal axis (and that of the initial polarisation), is used to magnetise the Fe single crystal sample in that direction. As discussed in §3.6, the relaxation rate is then given by the expression

$$T_1^{-1} = \Gamma_1^{\langle 111 \rangle} = \tfrac{1}{2}(\gamma_\mu B_{\text{dip}}^{\parallel})^2 \tau_c / (1 + \tau_c^2 \omega_0^2). \tag{3.135}$$

Figure 3.28 shows that $\Gamma_1^{\langle 111 \rangle}$ is independent of the field (and ω_0) for temperatures above 20 K. This implies that $\omega_0 \tau_c < 1$ and that, consequently,

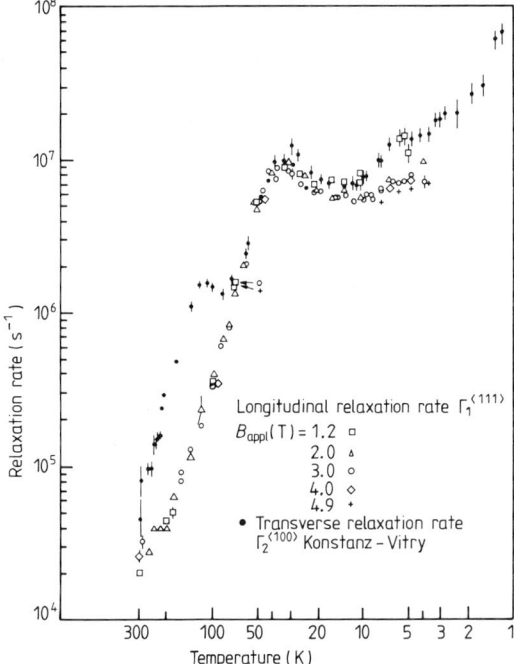

Figure 3.28 Temperature dependence of the longitudinal relaxation rate, $\Gamma_1^{\langle 111 \rangle}$ for various external fields along the $\langle 111 \rangle$ axis (Yagi et al 1981, 1984), plus the transverse relaxation rates of Graf et al (1980).

longitudinal and transverse relaxation rates are the same. This is indeed observed in the temperature range $20 < T < 70$ K, as can be seen from figure 3.28. Above 70 K the data clearly deviate, revealing trapping effects which show up only in the transverse field measurements. Below 20 K $\Gamma_1^{\langle 111 \rangle}$ starts to become field independant, implying that $\omega_0 \tau_c$ is no longer negligibly small compared to 1. In any case, the observed structure in $\Gamma_1^{\langle 111 \rangle}$ and T_2^{-1} below 40 K is a feature due to properties of the intrinsic diffusion and is not related to trapping effects. Yagi et al (1981) ascribe the structure to the change from a predominantly tetrahedral site occupation at temperatures below 20 K, to a mixed occupation of tetrahedral and octahedral sites above 20 K. On the other hand Graf et al (1980) have suggested that the observed structure could be explained by the theory of Fujii (1979a,b), which was outlined in §3.5. In this theory the minimum in T_2^{-1} at $\simeq 15$ K would signal the disappearance of coherent propagation and a slowdown of the total muon mobility until at around 35 K, incoherent tunnelling takes over as the dominant diffusion mechanism.

The longitudinal field data above 70 K are well fitted by an expression of the form of equation (3.25) with $E_a = 39 \pm 1$ meV.

Finally we mention a recent measurement by Möslang et al (1983) on μ^+ diffusion in electron irradiated Fe. Between 95 K and 220 K a new precession frequency has been observed which is only weakly dependent on temperature and displays almost no relaxation. Its value has been determined as $v_\mu \simeq 30$ MHz as compared to $v_\mu \simeq 50$ MHz in pure Fe. This new frequency is assigned to μ^+ trapped at mono-vacancies. The amplitude of this new signal, A_2, is temperature dependent and is displayed in figure 3.29 for two different irradiation doses. The temperature dependence of the amplitude can be

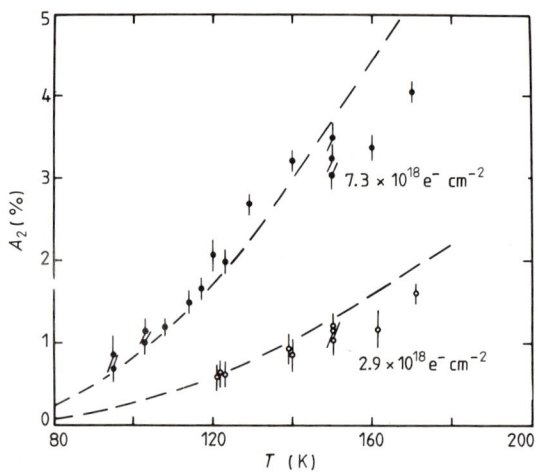

Figure 3.29 Temperature dependence of the 30 MHz amplitude in electron irradiated Fe. The two data sets correspond to different irradiation doses. The broken curves represent a fit of equation (3.120) to the data, assuming the diffusion coefficient to obey an Arrhenius law (Möslang et al 1983).

analysed on the basis of μ^+ diffusion towards the mono-vacancies, as was discussed in §3.7. By fitting equation (3.120) to the amplitude data, one can determine the temperature dependence of the average lifetime, τ_0 of the μ^+ in the free state. τ_0 is related to the macroscopic diffusion constant D by (see equation (3.65))

$$\tau_0^{-1} = c_v \sigma_v = c_v 4\pi r_v D/\Omega_a,$$

with c_v = concentration of vacancies (per atom), r_v = capture radius of the vacancy and Ω_a = atomic volume. Figure 3.30 displays the thus determined diffusion constants in an Arrhenius plot, yielding an activation energy of E_a = 38 meV. These data are in good agreement with the longitudinal field data of Yagi et al (1981) if one assumes that the diffusion proceeds by jumps over tetrahedral interstitial sites.

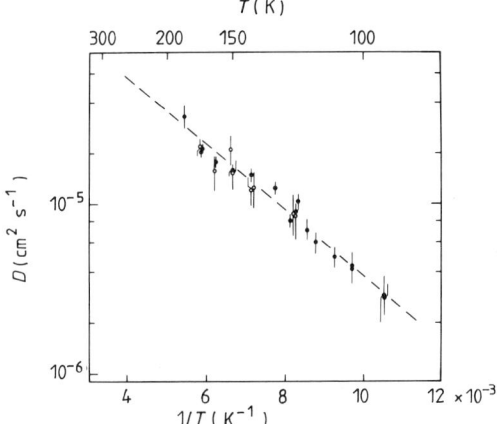

Figure 3.30 Arrhenius plot of the diffusion coefficient extracted from the two data sets for Fe in figure 3.29 (Möslang *et al* 1983), with $D_0 = 3.3 \times 10^{-4}$ cm^2 s^{-1} and $E_a = 38$ meV.

The disappearance of the new signal above 220 K is explained as either caused by migration of the vacancies or by annihilation from migrating interstitials. It appears that it will be possible by μSR studies to help in settling this controversial issue (see also Möslang *et al* 1984).

References

Balzer G, Graf H, Möslang T, Recknagel E, Weidinger A and Grynszpan R I 1981 *Hyperfine Interactions* **9** 589

Barsov S G, Getalov A L, Grebinnik V G, Gordeev V A, Gurevich I I, Kagan Yu M, Klimov A I, Kruglov S P, Kuzmin L A, Lazarev A B, Mikirtchyants S M, Nikolskii B A, Pirogov A V, Ponomarev A N, Selivanov V I, Shcherbakov G V, Suetin V A and Zhukov V A 1984 *Hyperfine Interactions* **17–19** 145

Baryshevsky V G and Kuten S A 1976 *Fiz. Tverd. Tela* **18** 2873 (Engl. Transl. 1976 *Sov. Phys.–Solid State* **18** 1677)

Birnbaum H K, Camani M, Fiory A T, Gygax F N, Kossler W J, Rüegg W, Schenck A and Schilling H 1978 *Phys. Rev.* A **17** 4143

Birnbaum H K and Flynn C P 1976 *Phys. Rev. Lett.* **37** 25

Boekema C, Heffner R H, Hutson R L, Leon M, Schillaci M E, Kossler W J, Numan M and Dodds S A 1982 *Phys. Rev.* B **26** 2341

Borghini M, Niinikoski T O, Soulie J C, Hartmann O, Karlsson E, Norlin L O, Pernestal K, Kehr K W, Richter D and Walker E 1978 *Phys. Rev. Lett.* **65A** 435

Brewer J H, Koster E, Schenck A, Schilling H and Williams D L 1981 *Hyperfine Interactions* **8** 671

Brown J A, Heffner R H, Hutson R C, Kohn S, Leon M, Olsen C E, Schillaci M E, Dodds S A, Estle T L, Vanderwater D A, Richards M and McMasters O O 1981 *Phys. Rev. Lett.* **47** 261

Brown J A, Heffner R H, Leon M, Parkin D M, Schillaci M E, Gauster W B, Fiory A T, Kossler W J, Birnbaum H K, Denison A B and Cooke D W 1979a *Hyperfine Interactions* **6** 233

Brown J A, Heffner R H, Leon M, Schillaci M E, Cooke D W and Gauster W B 1979b *Phys. Rev. Lett.* **43** 1513

Browne A M and Stoneham A M 1982 *J. Phys. C: Solid State Phys.* **15** 2709

Buchholz J, Völkl J and Alefeld G 1973 *Phys. Rev. Lett.* **30** 318

Camani M, Gygax F N, Rüegg W, Schenck A and Schilling H 1977 *Phys. Rev. Lett.* **39** 836

Camani M, Gygax F N, Rüegg W, Schenck A, Schilling H, Stucki F and Schlapbach L 1979 *Hyperfine Interactions* **6** 305

Clawson C W, Crowe K M, Kohn S E, Rosenblum S S, Huang C Y, Smith J L and Brewer J H 1982 *Physica* **109** B and **110** B 2164

Dorenburg K, Gladisch M, Herlach D, Mansel W, Metz H, Orth H, zu Putzlitz G, Seeger A, Wahl W and Wigand M 1978 *Z. Phys.* B **31** 165

Doyama M, Nakai R, Yamamoto R, Uemura Y J, Yamazaki T, Fukai Y and Suzuki T 1981 *Hyperfine Interactions* **8** 711

Eichenauer W and Liebscher D 1962 *Z. Naturf.* a **17** 355

Eichenauer W, Luesser W and Witte H W 1965 *Z. Metallkd.* **56** 287

Eisenstadt M and Redfield A G 1963 *Phys. Rev.* **132** 635

Emin D 1981 *Hyperfine Interactions* **8** 515

Emin D, Baskes M F and Wilson W D 1979 *Phys. Rev. Lett.* **42** 791

Emmerich K, Gygax F N, Pinkvos H, Schenck A, Schwink C and Studer W 1982 unpublished data

Fiory A T, Lynn K G, Parkin D M, Kossler W J, Lankford W F and Stronach C E 1978 *Phys. Rev. Lett.* **40** 968

Flynn C P and Stoneham A M 1970 *Phys. Rev.* B **1** 3966

Frank W and Seeger A 1974 *Appl. Phys.* **3** 61

Fujii S 1979a *J. Phys. Soc. Japan* **46** 1833

Fujii S 1979b *J. Phys. Soc. Japan* **46** 1843

Fujii S and Uemura Y 1978 *Solid State Commun.* **26** 761

Gauster W B, Heffner R H, Huang C Y, Hutson R L, Leon M, Parkin D M, Schillaci M E, Triftshaeuser W and Wampfer W R 1977 *Solid State Commun.* **24** 619

Graf H, Balzer G, Recknagel E, Weidinger A and Grynszpan R I 1980 *Phys. Rev. Lett.* **44** 1333

Graf H, Holzschuh E, Recknagel E, Weidinger A and Wichert T 1979 *Hyperfine Interactions* **6** 245

Grebinnik V G, Gurevich I I, Didyk A Yu, Zhukov V A, Manych A P, Melnikov E V, Nikolskii B A, Roganov V S, Selivanov V I and Suetin V A 1978a *Zh. Eksp. Teor. Fiz. Pis. Red.* **27** 33 (Engl. transl. 1978 *Sov. Phys–JETP Lett.* **27** 30)

Grebinnik V G, Gurevich I I, Klimov A I, Maiorov V N, Manych A P, Melnikov E V, Nikolskii B A, Piragov A V, Ponomarev A N, Selivanov V I, Suetin V A and Zhukov V A 1979 *Hyperfine Interactions* **6** 275

Grebinnik V G, Gurevich I I, Zhukov V A, Ivanter I G, Klimov A I, Maiorov V N, Manych A P, Nikolskii B A, Pirogov A V, Ponomarev A N, Selivanov V I and

Suetin V A 1978b *Zh. Eksp. Teor. Fiz.* **75** (Engl. transl. 1980 *Sov. Phys.–JETP* **48** 1002)
Grebinnik V G, Gurevich I I, Zhukov V A, Ivanter I G, Klimov A I, Manych A P, Melnikov E V, Nikolskii B A, Pirogov A V, Ponomarev A N, Selivanov V I, Suetin V A and Formichev S V 1978c *Zh. Eksp. Teor. Fiz. Pis. Red.* **28** 456 (Engl. transl. 1979 *Sov. Phys.–JETP Lett.* **28** 423)
Grebinnik V G, Gurevich I I, Zhukov V A, Klimov A I, Lozina L A, Maiorov V N, Manych A P, Nikolskii B A, Pirogov A V, Ponomarev A N, Selivanov V I and Suetin V A 1981 *Zh. Eksp. Teor. Fiz.* **80** 298 (Engl. transl. 1981 *Sov. Phys.–JETP* **53** 151)
Grebinnik V G, Gurevich I I, Zhukov V A, Klimov A J, Maiorov V N, Manych A P, Melnikov E V, Nikolskii B A, Pirogov A V, Ponomarev A N, Selivanov V I and Suetin V A 1977a *Zh. Eksp. Teor. Fiz.* **25** 322 (Engl. transl. 1977 *Sov. Phys.–JETP Lett.* **25** 299)
—— 1977b *Proc. Int. Symp. Meson Chemistry and Mesomolecular Processes in Matter (Dubna)*
Grebinnik V G, Gurevich I I, Zhukov V A, Manych A P, Maleshko E A, Muratova I A, Nikoliskii B A, Selivanov V I and Suetin V A 1975 *Zh. Eksp. Teor. Fiz.* **68** 1548 (Engl. transl. 1975 *Sov. Phys.–¹JETP* **41** 777)
Grynszpan R I, Nishida N, Nagamine K, Hayano R S, Yamazaki T, Brewer J H and Fleming D G 1979 *Solid State Commun.* **29** 143
Gurevich I I, Meleshko E A, Muratova I A, Nikolskii B A, Roganov V S, Selivanov V I and Sokolov B V 1972 *Phys. Lett.* **40A** 143
Gygax F N, Hintermann A, Rüegg W, Schenck A, Studer W, van der Wal A J, Brewer J H, Stucki F and Schlapbach L 1984b *Hyperfine Interactions* **17–19** 267
Gygax F N, Hintermann A, Rüegg W, Schenck A, Studer W, van der Wal A J and Kaplan N 1984a *Hyperfine Interactions* **17–19** 177
Gygax F N, Hintermann A, Rüegg W, Schenck A, Studer W, van der Wal A J and Wehr H 1983 *Phys. Rev. Lett* **51** 505
Gygax F N, Hintermann A, Rüegg W, Schenck A and van der Wal A J 1984c *SIN Newsletter No* 16 (Swiss Institute for Nuclear Research, Villigen)
Hall, T M, Goland A N and Snead C L 1974 *Phys. Rev.* B **10** 3062
Hartmann O 1977 *Phys. Rev. Lett.* **39** 832
Hartmann O 1981 *Hyperfine Interactions* **8** 525
Hartmann O, Karlsson E, Norlin L O, Niinikoski T O, Kehr K W, Richter D, Welter J M, Yaouanc A and Le Hericy J 1980 *Phys. Rev. Lett.* **44** 337
Hartmann O, Karlsson E, Norlin L O, Pernestal K, Borghini M, Niinikoski T and Walker E 1978a, *Hyperfine Interactions* **4** 824
Hartmann O, Karlsson E, Norlin L O, Richter D and Niinikoski T O 1978b *Phys. Rev. Lett.* **41** 1055
Hartmann O, Karlsson E, Wäppling R, Richter D, Hempelmann R, Schulze K, Patterson B, Holzschuh E, Kündig W and Cox S F J 1983 *Phys. Rev.* B **27** 1943
Hayano R S, Uemura Y J, Imazato J, Nishida N, Yamazaki T and Kubo R 1979 *Phys. Rev.* B **20** 850
Heffner R H, Brown J A, Hutson R L, Leon M, Parkin D M, Schillaci M E, Gauster W B, Carlson O N, Rehbein D K and Fiory A T 1979 *Hyperfine Interactions* **6** 237
Herlach D 1981 *Recent Developments in Condensed Matter Physics*, vol. 1 ed. J T

Devrese (New York: Plenum) p 93
Holstein T 1959 *Ann. Phys., NY* **8** 325, 343
Jena P, Das S G and Singwi K S *Phys. Rev. Lett.* **40** 269
Jones W and March N H 1973 *Theor. Solid State Phys.* vol. 2 pp 938-69 (London: Wiley)
Kadono R, Imazato J, Nishiyama K, Nagamine K, Yamazaki T, Richter D and Welter J M 1984 *Hyperfine Interactions* **17-19** 109
Kagan Y and Klinger M I 1974 *J. Phys. C: Solid State Phys.* **7** 2791
Kanzaki H 1957 *J. Chem. Phys. Solids* **2** 24
Katsuta H and McLellan R B 1979 *Scr. Metall.* **13** 65
Katz L, Guinan M and Borg R I 1971 *Phys. Rev. B* **4** 330
Kehr K W 1975 *Ber. der Kernforsch. Juelich Nr.* 1211
——1978 *Hydrogen in Metals I* ed. G Alefeld and J Völkl *Topics in Applied Physics* vol. 28 (Berlin: Springer)
Kehr K W, Honig G and Richter D 1978 *Z. Phys. B* **32** 49
Kehr K W, Richter D, Welter J M, Hartmann O, Karlsson E, Norlin L O, Niinikoski T O and Yaouanc A 1982 *Phys. Rev. B* **26** 567
Kehr K W, Richter D, Welter J M, Hartmann O, Norlin L O, Karlsson E, Niinikoski T O, Chappert J and Yaouanc A 1981 *Hyperfine Interactions* **8** 681
Kossler W J, Fiory A T, Lankford W F, Lindemuth J, Lynn K G, Mahajan S, Minnich R P, Petzinger K G and Stronach C E 1978 *Phys. Rev. Lett.* **41** 1558
Kossler W J, Fiory A T, Lankford W F, Lynn K G, Minnich R P and Stronach C E 1979 *Hyperfine Interactions* **6** 295
Kossler W J, Fiory A T, Murnick D E, Stronach C E and Lankford W F 1977 *Hyperfine Interactions* **3** 287
Kubo R 1981 *Hyperfine Interactions* **8** 731
Kubo R and Toyabe T 1966 *Magnetic Resonance and Relaxation* ed. R Blinc (Amsterdam: North-Holland) p 810
Lankford W F, Birnbaum H K, Fiory A T, Minnich R P, Lynn, K G, Stronach C E, Bieman L H, Kossler W J and Lindemuth J 1978 *Hyperfine Interactions* **4** 833
McMullen T and Bergerson B 1978 *Solid State Commun.* **28** 31
McMullen T and Zaremba E 1978 *Phys. Rev. B* **18** 3026
Metz H 1980 *Doctoral thesis* University of Stuttgart see also Herlach (1981)
Metz H, Orth H, zu Putlitz G, Seeger A, Teichler H, Vetter J, Wahl W, Wigand M, Dorenburg K, Gladisch M and Herlach D 1979 *Hyperfine Interactions* **6** 271
Miller K M and Heald P T 1975 *Phys. Status Solidi* b **67** 569
——1976 *Phys. Status Solidi* b **78** 391
Morkel C, Wipf H and Neumaier K 1978 *Phys. Rev. Lett.* **40** 947
Möslang A, Albert E, Recknagel E, Weidinger A and Moser P 1984 *Hyperfine Interactions* **17-19** 255
Möslang A, Graf H, Balzer G, Recknagel E, Weidinger A, Wichert Th and Grynszpan R I 1983 *Phys. Rev. B* **27** 2674
Nakai R, Doyama M, Yamamoto R, Uemura Y J, Yamazaki T and Brewer J H 1981 *Hyperfine Interactions* **8** 717
Peisl H 1978 *Hydrogen in Metals I*, ed. G Alefeld and J Völkl *Topics in Applied Physics* vol. 28 (Berlin: Springer)
Petzinger K G 1980 *Phys. Lett.* **75A** 225
——1981 *Hyperfine Interactions* **8** 639

——1982 *Phys. Rev.* B **26** 6530
Petzinger K G, Munjal R L and Zaremba E 1979 *Hyperfine Interactions* **6** 223
Richards P M 1978 *Phys. Rev.* B **18** 6358
Richter D 1983 *Neutron Scattering and Muon Spin Rotation, Springer Tracts in Modern Physics* vol. 101 (Berlin: Springer)
Richter D, Hempelmann R, Kehr K W, Hartmann O, Karlsson E and Norlin L O 1984 *Hyperfine Interactions* **17-19** 261
Schillaci M E, Boekema C, Heffner R H, Hutson R L, Leon M, Olsen C E, Dodds S A, McLaughlin D E and Richards P M 1983 *Electronic Structure and Properties of Hydrogen in Metals* P Jena and C B Satterthwaite (New York: Plenum)
Schillaci M E, Hutson R L, Heffner R H, Leon M, Dodds S A and Estle T L 1981 *Hyperfine Interactions* **8** 663
Schilling H, Camani M, Gygax F N, Rüegg W and Schenck A 1978 *Phys. Lett.* **67A** 231
——1982 *J. Phys. F: Met. Phys.* **12** 875
Seeger A 1975 *Phys. Lett.* **53A** 324
——1978 in *Hydrogen in Metals I* ed. G Alefeld and J Völkl *Topics in Applied Physics* vol. 28 (Berlin: Springer)
Seeger A and Monachesi P 1982 *Phil. Mag.* B **46** 283
Schenck A 1982 *Helv. Phys. Acta* **54** 471
Sholl C A 1974 *J. Phys. C: Solid State Phys.* **7** 3378
——1975 *J. Phys. C: Solid State Phys.* **8** 1737
Slichter C P 1978 *Principles of Magnetic Resonance* 2nd edn (Berlin: Springer)
Stoneham A M 1972 *J. Phys. F: Met. Phys.* **2** 417
Studer W 1983 *Doctoral thesis* Eidg. Tech. Hochsch. Zürich
Sugimoto H and Fukai Y 1980 *Phys. Rev.* B **22** 670
Süssman J A 1971 *Ann. Phys. NY* **6** 135
Teichler H 1977 *Phys. Lett.* **64A** 78
——1978 *Phys. Lett.* **67A** 313
——1981a *Hyperfine Interactions* **8** 505
——1981b *Phys. Status Solidi* b **104** 329
Teichler H and Seeger A 1981 *Phys. Lett.* **82A** 91
Torrey H C 1953 *Phys. Rev.* **92** 962
Vineyard G H 1957 *J. Chem. Phys.* **3** 121
Wagner H and Horner H 1974 *Adv. Phys.* **23** 587
Weidinger A, Balzer G, Graf H, Möslang T, Recknagel E, Wichert Th, Bigot J and Grynszpan R I 1981 *Hyperfine Interactions* **8** 534
Wipf H and Neumaier K 1983 *Electronic Structure and Properties of Hydrogen in Metals* ed. P Jena and C B Satterthwaite (New York: Plenum)
Wolf O and Jung P 1975 *Phys. Rev.* B **2** 3596
Yagi E, Bossy H, Döring K P, Gladisch M, Herlach D, Matsui H, Orth H, zu Putlitz G, Seeger A and Vetter J 1981 *Hyperfine Interactions* **8** 553
Yagi E, Flik G, Furderer K, Haas N, Herlach D, Major J, Seeger A, Jakobs W, Krause M, Krauth M, Mundinger H J and Orth H 1984 *Phys. Rev.* B **30** 441
Yaouanc A, Dufresne J F, Longobardi R, Pezzetti J P, Chappert J, Hartmann O, Karlsson E and Norlin L O 1979 *J. Phys. F: Met. Phys.* **9** 2157
Yaouanc A, Chappert J, Hartmann O, Norlin L O and Karlsson E 1981 *Hyperfine Interactions* **8** 667
Zimmermann J R 1957 *J. Phys. Chem.* **61** 1328

4

Hyperfine Fields at the μ^+ in Metals

4.1 Hyperfine Fields, Knight Shift and their Relation to Local Electronic Structure

4.1.1 Introduction

In §3.1 we discussed various static properties of the μ^+ state in the metal host lattice on the basis of a non-specific interaction potential, V (equation (3.1)), and by applying elasticity theory. This allowed us to relate the μ^+ nearest neighbour force constant, g, to the nearest neighbour displacements $\delta r/r$, although it did not allow the calculation of either directly. A step further was taken by Sugimoto and Fukai (1980) in assuming a model potential for V that was designed to reproduce certain known properties of H in Nb. This allowed the self-trapping site, or site configuration to be predicted for the μ^+ in Nb, and associate properties such as the lattice dilatation. The electronic structure around the μ^+ was not yet considered explicitly. A step in that direction was taken by Teichler (1978), who wrote the interaction potential between a μ^+ at position r_μ and a nearest neighbour host ion (i.e. Cu^+ for μ^+ in Cu) at position R_l as

$$V(r_\mu - R_l) = e^2 Z_{\text{eff}}(r_\mu - R_l)/(|r_\mu - R_l|) + \int d^3 r \, r \, \delta n(r - r_\mu) v_{\text{ps}}(r - R_l). \quad (4.1)$$

The first term represents the Coulomb energy of the μ^+ in the Chodorov potential (Burdick 1963) of the Cu^+ core. The effective charge, Z_{eff} becomes a function of the $\mu^+ - Cu^+$ distance. The second term represents the interaction energy between the conduction electron density, $\delta n(r - r_\mu)$, around the μ^+ and the Cu^+ ion, via a local pseudo-potential, v_{ps}. $\delta n(r - r_\mu)$ is given by

$$\delta n(r - r_\mu) = n(r - r_\mu) - n_0(r - r_\mu). \quad (4.2)$$

$n(r - r_\mu)$ is the total electronic charge distribution around the position r_μ with the μ^+ present, and $n_0(r - r_\mu)$ the corresponding distribution with the μ^+ absent. $\delta n(r - r_\mu)$ is therefore the excess electronic charge density distribution built up in response to the presence of the μ^+ charge, and screening this charge within a certain radius. $\delta n(r-r_\mu)$ characterises the local electronic structure formed around the μ^+. Taking $\delta n(r-r_\mu)$ from a non-linear screening jellium calculation, and adopting suitable functions for Z_{eff} and v_{ps}, Teichler (1978) directly calculated $g = 2.0$ eV a^{-1} for μ^+ in Cu. This is in fairly good agreement with experimentally derived values quoted in §3.1 ($g = 3.39$ eV a^{-1}, Schilling et al (1982)) and §3.8 ($g = 2.71$ eV a^{-1}, Teichler (1977)).

Furthermore, from this example it is clear that a good understanding of the electronic structure around the μ^+ is a prerequisite to the derivation of other properties of the μ^+ state in the metal lattice. *The problem of the local electronic structure around the μ^+ is not specific to an implanted μ^+, but is the same as for hydrogen in metals.* Proton and muon both carry a single elementary positive charge, and introduce the same strongly disturbing Coulomb potential into the host lattice.

The electronic structure around the μ^+ also determines the magnitude of a possible contact hyperfine field at the μ^+. Most generally we can write (see e.g. Winter 1971)

$$B_{\text{hf}}(r_\mu) = \frac{8\pi}{3}[n^+(r_\mu) - n^-(r_\mu)]. \tag{4.3}$$

$n^+(r_\mu)$ is the density of spin-up electrons at the μ^+ and $n^-(r_\mu)$ the corresponding spin-down density. The difference, $[n^+(r_\mu) - n^-(r_\mu)]$ equals the resulting net spin density at the μ^+. A non-zero spin density may be the result of spontaneous magnetic ordering, or it may be induced by an external magnetic field (e.g. via the Pauli spin paramagnetism of the conduction electrons). In the first case, the ordered local electronic moments induce, via exchange interactions, a spin polarisation of the conduction electrons, which in turn produce a contact hyperfine field at the μ^+. Of course, the conduction electron distribution around the μ^+ is strongly disturbed because of its presence, and a new local electronic structure is established.

In the second case, the polarising effect of ordered local electronic moments is replaced by the applied field. We write

$$B_{\text{hf}}(r_\mu) = KB_{\text{ext}} \tag{4.4}$$

$B_{\text{hf}}(r_\mu)$ is also called 'Knight shift' and K is the 'Knight shift' constant.

4.1.2 Isotropic Knight shift

According to equation 4.4, the Knight shift constant is defined as

$$K_\mu = B_{\text{hf}}(r_\mu)/B_{\text{ext}} \tag{4.5}$$

$$= \frac{8\pi}{3}\mu_B \frac{n^+(r_\mu) - n^-(r_\mu)}{B_{\text{ext}}}. \tag{4.6}$$

At this point it is necessary to consider which types of host metal electrons will enter into the problem. In simple (non-transition) metals, the Pauli spin paramagnetism will rest on the conduction electrons, which are mainly of s symmetry. In transition metals, d and f electrons will contribute to the bulk spin paramagnetism in addition to the s/p conduction electrons. The d and f electrons may be more or less localised at the host atomic positions.

In contrast to the s conduction electron Pauli spin paramagnetism which is temperature independent, the contribution from d and f electrons will to some extent exhibit a temperature dependence. In the so-called pure Van Vleck paramagnets, no temperature dependence is observed, while in systems with local magnetic moments a Curie behaviour emerges.

For simple metals, using a homogeneous electron gas picture, equation (4.6) may be rewritten

$$K_\mu = \frac{8\pi}{3} \mu_B \frac{[n^+(r_\mu) - n^-(r_\mu)]}{(n_0^+ - n_0^-)} \frac{(n_0^+ - n_0^-)}{B_{ext}} \quad (4.7)$$

where

$$n_0^+ - n_0^- = (1/\mu_B)\chi_s B_{ext} \quad (4.8)$$

is the undisturbed average conduction electron spin density and χ_s is the (enhanced) Pauli spin susceptibility. We then obtain

$$K_\mu = \frac{8\pi}{3} \frac{[n^+(r_\mu) - n^-(r_\mu)]}{n_0^+ - n_0^-} \chi_s = \frac{8\pi}{3} \rho_s(r_\mu)\chi_s. \quad (4.9)$$

$\rho_s(r_\mu)$ is called the spin density enhancement factor. If it is assumed that application of an external field leads only to a repopulation of the spin-up and spin-down states (the Pauli spin paramagnetism), equation (4.9) for a spherical Fermi surface becomes (Winter 1971)

$$K_\mu = \frac{8\pi}{3} \langle |\psi_{k_F}(r_\mu)|^2 \rangle_F \frac{\chi_s}{n_0} = \frac{8\pi}{3} \eta_F(r_\mu)\chi_s. \quad (4.10)$$

$\langle |\psi_{k_F}(r_\mu)|^2 \rangle_F$ is an average Fermi surface electron density at the μ^+, $n_0 = \Omega_e^{-1}$ the average undisturbed conduction electron density and $\eta_F(r_\mu)$ is called a charge density enhancement factor (with respect to the Fermi surface electrons).

For future use we shall also quote equation (4.10) in a different, more general form (see Winter 1971), which will allow us to discuss the effect a particular feature of the electronic band structure has on the μ^+ Knight shift

$$K_\mu = \frac{8\pi}{3} \mu_B^2 \frac{\Omega_e}{(2\pi)^3} \iint_{FS} \frac{dS_F}{|\text{grad}_k E_F(k)|} |\psi_{k_F}(r_\mu)|^2. \quad (4.11)$$

Here $|\psi_{k_F}(r_\mu)|^2$ is the density of the electron state with Fermi momentum k_F at the μ^+, dS_F is an element of area on the Fermi surface (FS) and $1/|\text{grad}_k E_F(k)|$ is the density of states with $k = k_F$. Since the μ^+ destroys the periodicity of the lattice, the $\psi_{k_F}(r_\mu)$ are not necessarily well defined band states, and the density of states cannot be identified with the host-metal density of states. Instead, the density of states introduced here is a local density of states with respect to the μ^+ cell.

However, on the basis of the homogeneous electron gas model the presence of the μ^+ will only result in a spatial rearrangement of the electron distribution, and not in a change of the density of state properties. This will lead us back to equation (4.10) noting that

$$\chi_s = \frac{2}{(2\pi)^3} \mu_B^2 \iint_{FS} \frac{dS_F}{|\mathrm{grad}_k\, E_F(\boldsymbol{k})|} = \mu_B^2 N(E_F) \tag{4.12}$$

for a non-interacting electron gas.

So far we have considered the effect of the field induced repopulation of spin-up and spin-down states on the local spin density. In addition, the local spin density is also influenced by exchange interactions between the spin-polarised Fermi surface electrons, and electron states below E_F. This may lead to wavefunction distortions such that the states below E_F also produce a non-zero spin polarisation at the μ^+, although there will be no net effect on the bulk spin polarisation. This mechanism is known as core polarisation. Since the driving source is the spin polarisation of the Fermi surface conduction electrons, this core polarisation contribution to the Knight shift will be proportional to χ_s as well. Hence the effect of core polarisation may be absorbed into $\rho_s(\boldsymbol{r}_\mu)$ and we therefore have to expect that, in general, $\rho_s(\boldsymbol{r}_\mu) \neq \eta_F(\boldsymbol{r}_\mu)$.

It is important to realise that the total contact Knight shift is dependent not only on Fermi surface states around the μ^+, but also on the states below the Fermi energy. The measurement of K_μ therefore contains, in a rather condensed form, a lot of information on the local electronic structure of the hydrogen-like impurity.

In transition metals d and/or f electrons will also contribute to the total spin susceptibility so that

$$\chi_{\mathrm{spin}} = \chi_s + \chi_d + \cdots \tag{4.13}$$

Since the d-electron orbitals (and more so the f-electron orbitals) are rather confined to the host atomic sites, no direct spin density at the interstitial μ^+ position may occur. In that case the d electrons could still contribute indirectly to the Knight shift by producing a core polarisation of those states overlapping with the μ^+. Another possibility is that the hydrogen-like 1s orbitals around the μ^+ and the host d states hybridise to form bonding states, by means of which a direct admixture of d electron states into the local hydrogen states is facilitated. If the tails of the d wavefunctions reach far enough into the interstitial volume, a direct contact hyperfine interaction with the μ^+ spin will also become possible. These are also the mechanisms that have to be considered with respect to the spontaneous hyperfine field in magnetically ordered metals. Whatever the mechanism is by which the d electrons contribute to the Knight shift, the contribution will be proportional to the d-electron related susceptibility χ_d. We write

$$K_\mu = K_s + K_d \tag{4.14}$$

$$= \frac{8\pi}{3} \rho_s(r_\mu)\chi_s + \frac{8\pi}{3} \rho_d(r_\mu)\chi_d. \tag{4.15}$$

$\rho_d(r_\mu)$ has an analogous meaning as $\rho_s(r_\mu)$ and relates the average undisturbed d-electron spin density to the d-electron induced spin density at the μ^+ site. The spontaneous hyperfine field at a μ^+, in for example, a ferromagnetic 3d metal, is related to the domain magnetisation, M_d, in the same way

$$B_{hf}(r_\mu) = \frac{8\pi}{3} \rho_d(r_\mu)M_d. \tag{4.16}$$

Since the local electronic structure should be independant, to the first order, of the magnetic phase of the host-metal, it is expected that $\rho_d(r_\mu)$ should also be the same in the paramagnetic and the ferromagnetic phase. Any deviations could be the source of interesting information, either concerning changes in the electronic structure, or else changes in the strength of the involved exchange mechanisms.

It is sometimes convenient to express K_s and K_d in terms of a hyperfine field per unpaired spin per atom (or atomic volume)

$$K_\mu = \Omega_a[B_{hf,s}^{\Omega_a}(r_\mu)\chi_s + B_{hf,d}^{\Omega_a}(r_\mu)\chi_d]/\mu_B \tag{4.17}$$

with Ω_a = atomic volume. These normalised hyperfine fields must be distinguished from the actual hyperfine fields at the μ^+ site. The s- and d-electron related contributions to the total contact hyperfine field can be separated if the d-electron susceptibility becomes temperature dependent. This will be demonstrated in §4.4.

The conduction electron spin polarisation may not be induced only directly by an applied magnetic field, but also by (dilute) local magnetic moments which polarise the conduction electrons in an oscillatory manner by the famous Rudermann–Kittel–Kasuya–Yosida (RKKY) interaction. This effect can be accounted for by replacing the applied field B_{ext} in equations (4.4) and (4.7), by an effective field (Yosida 1957, Jaccarino et al 1960)

$$B_{eff} = B_{ext}\left(1 + J_{sf} \frac{(g_J - 1)}{g_J} \frac{\chi_f}{2\mu_B^2}\right) \tag{4.18}$$

with

$$J_{sf} = 6\pi z J(0) \sum_i F(2k_F|R_i - r_\mu|). \tag{4.19}$$

g_J equals the Lande g-factor of the local moment, χ_f is the local moment susceptibility, z the number of conduction electrons per atom and $J(0)$ is an electron exchange integral for wave-vector $q=0$. The function $F(2k_F|R_i - r_\mu|)$ accounts for the spin-density oscillations originating from each site, R_i, of a local impurity moment. The Knight shift constant is hence altered to

$$K_s = \frac{8\pi}{3} \rho_s(r_\mu)\chi_s \left(1 + J_{sf} \frac{(g_J-1)}{g_J} \frac{\chi_f}{2\mu_B^2}\right) = K_s^0 \left(1 + J_{sf} \frac{(g_J-1)}{g_J} \frac{\chi_f}{2\mu_B^2}\right). \quad (4.20)$$

4.1.3 Anisotropic Knight shift

The preceding discussion was concerned with the various contributions to the contact Fermi hyperfine field. But what about the more distant electrons, surely they must produce a magnetic dipole field at the μ^+ site? If a net magnetic field results from the superposition of all dipole field components at the μ^+ site, we would have a further contribution to the Knight shift or, in the case of magnetically ordered systems, a contribution to the spontaneous μ^+ hyperfine field. The latter situation will be discussed in §4.6.1. Here we are interested in the dipolar field contribution to the Knight shift.

Suppose the electron distribution around the μ^+ is represented by the density function

$$n(r) = \sum_i |\psi_i(r)|^2 \quad (4.21)$$

where the sum runs over all the electrons. At each position, r, we find a magnetic moment density of

$$\mu(r) = g_e \mu_B n(r) S(r) \quad (4.22)$$

where $S(r)$ is an effective spin per electron. Each point in space is the source of a dipole field. The net dipole field at the μ^+ position ($r=0$) is then given by the integral

$$H_{eff} = g_e \mu_B \int_{\Omega_L} n(r) \left(-\frac{S(r)}{r^3} + 3\frac{r[S(r) \cdot r]}{r^5}\right) dv. \quad (4.23)$$

The integration extends, in principle, over the total volume of the sample, but we will assume the probe to be of a spherical shape so that the demagnetisation field and the Lorentz field cancel, and only the integration over the Lorentz sphere remains. In the presence of an external field, B_{ext}, in the z direction, each electronic spin will on the average acquire a small static component in the z direction which is given by

$$\langle S_z \rangle = (1/g_e \mu_B) \chi_e B_{ext} \quad (4.24)$$

where χ_e is the spin susceptibility per electron (assumed to be independent of position r). As a result we find a non-vanishing H_{eff} in the z direction. The total field acting on the μ^+ is then given by

$$B_{tot} = B_{ext} + H_{eff} = B_{ext}(1 + K_{dip}) \quad (4.25)$$

where $K_{dip} = H_{eff}/B_{ext}$ is the dipolar field contribution to the Knight shift constant. With equation (4.23) K_{dip} can be written as

$$K_{dip} = \left(\int_{\Omega_L} n(r) \frac{(3\cos^2\theta - 1)}{r^3} dv \right) \chi_e \qquad (4.26)$$

where θ is the angle between the z axis and r. K_{dip} vanishes for a charge distribution of spherical and cubic symmetry. Usually the field direction is defined with respect to the natural crystal axes. Denoting one of them as the new z' axis, we want to express K_{dip} in terms of the polar angles ϑ and φ, where ϑ is the angle between B_{ext} and the new z' axis. Making use of the spherical harmonic addition theorem

$$\tfrac{1}{2}(3\cos^2\theta - 1) = P_2^0(\cos\theta)$$

$$= \frac{4\pi}{5} \sum_{m=-2}^{+2} Y_{2m}^*(\vartheta, \varphi) Y_{2m}(\vartheta', \varphi') \qquad (4.27)$$

where ϑ' and φ' are the polar angles of r with respect to the coordinate system x', y', z'. Equation (4.26) can be recast as

$$K_{dip} = 2\frac{4\pi}{5} \sum_{m=-2}^{+2} \left(\int_{\Omega_L} \frac{n(r')}{r^3} Y_{2m}(\vartheta', \varphi') dv' \right) Y_{2m}^*(\vartheta, \varphi). \qquad (4.28)$$

The integrals are the tensor components of the second derivative of the electrostatic potential of a charge distribution, $n(r)$, at the μ^+ position. It is zero, of course, for a spherical charge distribution. For an axially symmetric charge distribution with respect to the z' axis, only the $m=0$ component contributes, leading to the so-called axial Knight shift.

$$K_{ax} = \left(\int_{\Omega_L} \frac{P_2^0(\cos\vartheta')}{r^3} n(r') dv' \right)(3\cos^2\vartheta - 1)$$

$$= \tfrac{1}{2} K_1 (3\cos^2\vartheta - 1). \qquad (4.29)$$

In the case of lower than axial symmetry, choosing the coordinate system x', y', z', such that the field gradient tensor reduces to its principal axis components, equation (4.28) reduces to the expression (see e.g. Carter et al 1977)

$$K_{dip} = \tfrac{1}{2} K_1 (3\cos^2\vartheta - 1) + \tfrac{1}{2} K_2 \sin^2\vartheta \cos 2\varphi \qquad (4.30)$$

where

$$K_2 = \frac{3}{2} \int_{\Omega_L} \frac{x'^2 - y'^2}{r^5} n(r') dv'. \qquad (4.31)$$

The dipole–dipole interaction of the μ^+ with distant electrons therefore leads, in the case of a non-spherical charge distribution, to an anisotropic contribution to the total Knight shift.

An anisotropy may also arise from the contact term if the electron Landé or

g-factor possesses an anisotropy. More complicated angular dependences in the presence of spin–orbit coupling are discussed by Boon (1964) and Rubens et al (1973).

4.1.4 Diamagnetic screening

Quite a different shift of the local magnetic field at a μ^+ stems from diamagnetic screening. This effect is usually neglected in the discussion of nuclear Knight shift data because of its relative smallness. For interstitial μ^+ or protons, on the contrary, diamagnetic screening is a sizable fraction of the total field shift. Two different contributions have to be considered. First there is the contribution from the Landau diamagnetism of the conduction electrons at the Fermi surface (see e.g. Das and Sondheimer 1960)

$$\sigma_{\text{Landau}} = -\frac{1}{3}\frac{8\pi}{3}\left(\frac{m}{m^*}\right)^2 \chi_s. \tag{4.32}$$

χ_s is, as before, the enhanced Pauli spin susceptibility, m the free electron mass and m^* the effective electron mass. The diamagnetic screening constant amounts typically to a few ppm. The second and more important contribution stems from the screening charge piled up around the μ^+ and is analogous to the chemical shift in atoms and molecules. This contribution can be calculated from the Lamb expression (see e.g. Davies 1967)

$$\sigma_{\text{Lamb}} = (-e^2/3mc^2)\langle 0|1/r|0\rangle. \tag{4.33}$$

It involves the calculation of the expectation value of $1/r$ for the ground state configuration $|0\rangle$. In contrast to the contact hyperfine field contribution to the paramagnetic field shift, the entire charge distribution around the μ^+ determines the diamagnetic shielding constant.

4.1.5 Total Knight shift

The total Knight shift constant can now be written as

$$K_\mu = K_s + K_d + \cdots + K_{\text{dip}} + \sigma_{\text{Landau}} + \sigma_{\text{Lamb}}. \tag{4.34}$$

The local electronic structure enters the different terms to varying degrees. Taken altogether, there is probably no aspect of the local electronic structure that is not reflected in the Knight shift. The practical problem is to isolate the different terms in a Knight shift measurement. We mentioned before that K_s and K_d may be distinguished by the different temperature dependence of χ_s and χ_d. σ_{Landau} is usually negligibly small and we need not be concerned with its presence. σ_{Lamb} may be isolated in certain cases by comparing the Knight shift in the solid and the liquid state of the host metal. If χ_s changes upon melting, and if one assumes that the charge distribution around the μ^+ remains essentially the same in both phases, σ_{Lamb} is determined from the experimental K_μ^{liquid} and K_μ^{solid} as follows

$$\sigma_{\text{Lamb}} = \frac{K_\mu^{\text{liquid}} - (\chi_s^{\text{liquid}}/\chi_s^{\text{solid}})K_\mu^{\text{solid}}}{1 - (\chi_s^{\text{liquid}}/\chi_s^{\text{solid}})}. \tag{4.35}$$

$(\chi_s^{\text{liquid}}/\chi_s^{\text{solid}})$ may be estimated from the corresponding ratio of the host nuclear Knight shift values.

4.2 Short Survey on Theoretical Calculations of Spin Densities at the μ^+

The electronic structure of hydrogen in metals has been treated theoretically with various techniques involving various model assumptions (see e.g. Jena 1981, Schenck 1982). The simplest model is the so-called *jellium model* in which the conduction electrons are treated as a free electron gas, and the positive ion cores form a homogeneously charged compensating background. In all of the more recent calculations† (see Zaremba *et al* 1977, Jena 1979, Nieminen 1981), the response of the electron gas to the presence of the positive impurity charge has been treated on the basis of a non-linear screening approach, by using the charge density functional (HKS) formalism (Hohenberg and Kohn 1964, Kohn and Sham 1965) or the spin density functional (SDF) formalism (Barth and Hedin 1972, Rajagopal and Callaway 1973, Gunnarsson and Lundquist 1976).

These treatments were also applied to the calculation of the charge density enhancement factor, $\eta_F(r_\mu)$, (Jena and Singwi 1978b, Munjal and Petzinger 1978), and the spin density enhancement factor $\rho_s(r_\mu)$ (Petzinger and Munjal 1977, Munjal and Petzinger 1978, Jena 1979). The various results for $\eta_F(r_\mu)$ and $\rho_s(r_\mu)$ agree very well with each other. $\eta_F(r_\mu)$ and $\rho_s(r_\mu)$ are plotted in figure 4.1 as a function of the electron density parameter, r_s ($r_s = (3/4\pi n_0)^{1/3}/a_B$, $n_0 =$ conduction electron density, $a_B =$ Bohr radius). It is seen that, except for the very high electron density metals, $\rho_s(r_\mu) < \eta_F(r_\mu)$. Munjal and Petzinger (1978) trace the deviation between $\rho_s(r_\mu)$ and $\eta_F(r_\mu)$ back to the occurrence of bound states. Below $r_s = 1.9$, where no bound states are shown to exist, $\rho_s(r_\mu) = \eta_F(r_\mu)$. Above $r_s = 1.9$, the minority spin-bound state is always more tightly bound than the majority spin-bound state. This results in a negative contribution to the total spin density at the μ^+. Obviously, a 'core polarisation' effect is at work. The total charge and spin density distribution around the μ^+ is shown in figure 4.2 for an electron density of $r_s = 2$ (e.g. Al, $r_s = 2.07$).

The diamagnetic screening of the μ^+ in jellium has been calculated by Zaremba and Zobin (1980). The results for the diamagnetic screening constant, σ, are shown in figure 4.3 as a function of r_s. In the range of the usual

† See also Meier (1975), De Renzi and Dallacasa (1981) for other approaches.

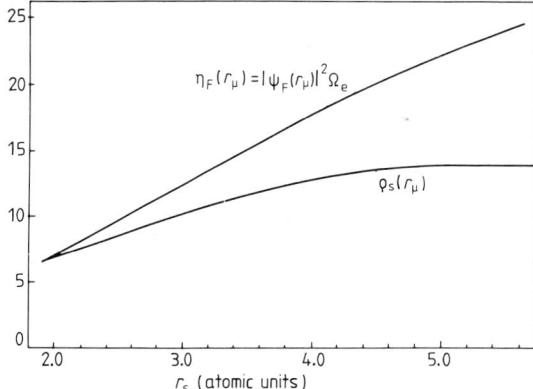

Figure 4.1 Plot of the charge density enhancement factor, $\eta_F(r_\mu) = \Omega_e \langle |\psi(r_\mu)|^2 \rangle_F$, and the spin density enhancement factor, $\rho_s(r_\mu)$ against the electron density parameter r_s ($r_s = (3/4\pi n_0)^{1/3}/a_B$, n_0 = electron density) (Munjal et al 1978).

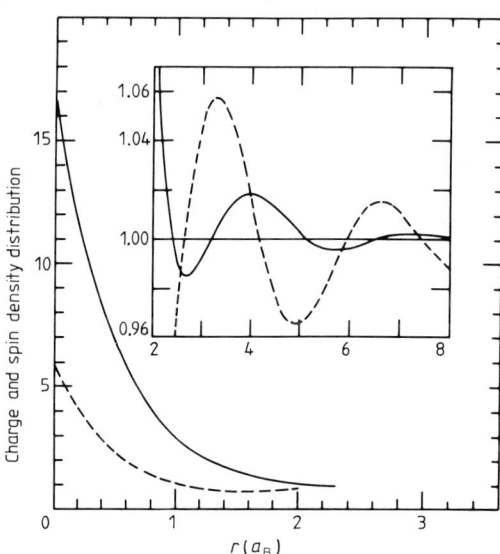

Figure 4.2 Total charge density and spin density distribution around a positive muon in a spin polarised electron gas, with $r_s = 2$ and an average polarisation of 0.17. The full and broken curves correspond to $n(r)/n_0$ and $\rho_s(r_\mu)$, respectively (Jena 1979).

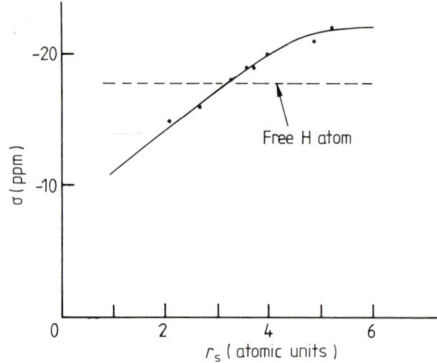

Figure 4.3 The diamagnetic screening constant, σ, as a function of the electron density parameter, r_s (Zaremba and Zobin 1980).

metallic electron densities, σ changes from -15 to -22 ppm and stays close to the free hydrogen atom value of $\sigma_{H^0} = -17.8$ ppm.

The discrete structure of the lattice environment of a proton or a muon has been completely neglected so far. The pure jellium calculations are only, therefore, of a certain limited value, even for nearly free-electron gas metals, since any possible site dependence of the electronic structure will be outside the focus of such calculations. On the other hand it is precisely this correlation of site and electronic structure that will allow us to understand why hydrogen occupies a certain lattice site and enable us to predict answers to such questions as the corresponding heat of solution.

A step forward in this direction has been taken by Manninen and Nieminen (1979) on the basis of non-linear response jellium calculations. The influence of the neighbouring ion potentials was approximated by forming a spherical average (called the spherical solid model (Almbladh and von Barth 1976)) with the proton or muon at the centre, and incorporating it into the SDF equations as an external host potential. These calculations were applied to muons or protons in a variety of metals including all alkali and alkaline earth metals, assuming the muon or proton to occupy the octahedral or tetrahedral interstitial position or a substitutional position (a host vacancy) (Manninen 1983). The results for $\rho_s(r_\mu)$ at interstitial sites are displayed in figure 4.4. As can be seen, the spin density enhancement factor now depends sensitively on the assumed μ^+ site, except in the alkaline metals. Compared with figure 4.2, $\rho_s(r_\mu)$ shows quite a different behaviour.

More sophisticated microscopic model calculations of the electronic structure of hydrogen in metals are based on cluster (Mainwood and Stoneham 1976, Adachi *et al* 1978, Jena *et al* 1979, Keller 1979) and band structure techniques (Switendick 1978, Faulkner 1976, Papaconstantopoulos *et al* 1978, Gelatt *et al* 1978, Gupta and Freeman 1978, Katayama *et al* 1979, Jepsen *et al* 1980, Kanamori *et al* 1981, Klein and Pickett 1984a,b). Some

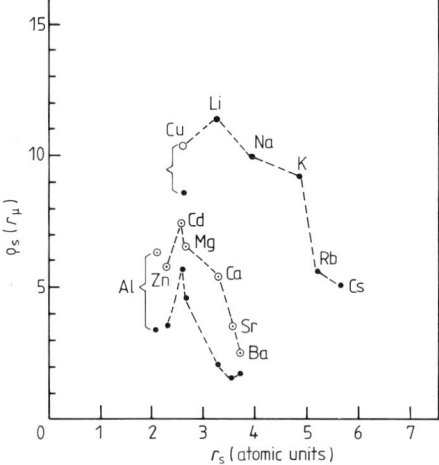

Figure 4.4 Plot of the spin density enhancement factor, $\rho_s(r_\mu)$, resulting from the spherical solid model jellium calculations of Manninen and Nieminen (1979) and Manninen (1983) for, ●, tetrahedral and, ○, octahedral μ^+ sites. $\rho_s(r_\mu)$ is practically independent of the assumed sites in the alkali metals.

relevant results for the μ^+ or proton Knight shift in non transition metals are collected in table 4.1. These calculations are particularly important in transition metals themselves, where more than one type of electron (s, p, d, f) has to be considered, and where a homogeneous free-electron gas picture for the conduction electrons certainly is a bad approximation. In particular, these calculations have served to also predict spontaneous hyperfine fields in

Table 4.1 Theoretical predictions for the μ^+ or proton Knight shift, K_s in non transition metals from cluster and band structure calculations. Note that the chemical shift is not included in K_s. Oct = octahedral, Tet = tetrahedral, Subst = substitutional.

Metal	μ^+ site	K_s (positive ppm)	Reference
Be	Oct	9	Keller and Schenck (1979)
Cu	Oct	71	Castro et al (1979)
		55	Kanamori et al (1981)
PdH	Oct	83	Gupta and Freeman (1978)†
		65.7	Klein and Pickett (1984b)
Al	Tet	102	
Al	Oct	58.4	Klein and Pickett (1984a)
Al	Subst	179.4	

† Calculated with $\chi_s = 16 \times 10^{-6}$ emu mol^{-1} and $\langle |\psi(r_\mu)|^2 \rangle_F = 0.18 |\psi(0)|^2_{H_0}$.

ferromagnetic Ni, Fe, Co, Gd and Dy (see table 4.9). As an example, figure 4.5 shows a plot of the local density of states at a μ^+ in ferromagnetic Ni resulting from the Kohn–Korringa–Rostoker (KKR) band structure calculations of Katayama *et al* (1979). The arrows below the bottom of the conduction band indicate the position of bound local states, whose spectral weights are indicated. The total hyperfine field at the μ^+ calculated from this histogram turns out to be -720 G. The sign is determined by the dominating negative contribution from the bound states, while the continuum states produce a net positive hyperfine field. The validity of cluster calculations with respect to predicting bulk properties has recently been scrutinised by Hintermann and Manninen (1983). Only very large clusters, of at least a 100 atoms seem to be sufficient for a good approximation of the bulk.

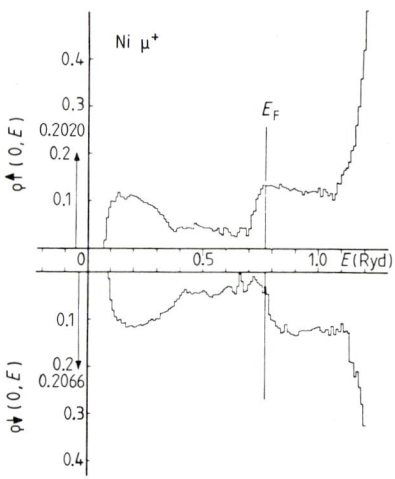

Figure 4.5 The local density of states for spin-up and spin-down electrons, as seen by the μ^+ at an octahedral interstitial site in ferromagnetic Ni. Arrows near the bottom of the conduction band indicate bound states and the attached numbers their spectral weights (Katayama *et al* 1979).

More phenomenological calculations of spontaneous hyperfine fields at the μ^+ in ferro and antiferromagnetic hosts have recently been presented by Manninen and Jena (1980), Manninen and Nieminen (1981) and Estreicher and Meier (1982). These calculations are based in principle on the jellium picture and the spherical solid model. The new feature is that the exchange scattering of the conduction electrons from the host ion magnetic moments (RKKY mechanism) is included by forming a spin dependent spherical average of the host ion potentials. Some results are included in table 4.9.

In all the calculations mentioned so far, the μ^+ or the proton have been assumed fixed in space, for example, at the centre of an interstitial site. The local electronic structure is then identical around both particles and no isotope effect on hyperfine fields and the Knight shift is predicted. In reality both particles are not fixed, but rather perform vibrations in the interstitial relaxed potential. Owing to its smaller mass ($m_\mu \simeq m_p/9$), the μ^+ vibration amplitude will be larger by a factor of $(m_p/m_\mu)^{1/2}$ in the harmonic approximation. Hence the muon and proton will sample the electronic environment to a slightly different extent and as a consequence the Knight shift should display an isotope effect.

No theoretical calculations so far exist which fully take into account the possibility of an isotope effect. Approximate treatments have been done by Jena *et al* (1979), Manninen and Jena (1980), Manninen and Nieminen (1981), Rath *et al* (1979) and Estreicher and Meier (1982, 1984). Jena *et al* (1979) considered the mass dependence of the spin density at interstitial hydrogen isotopes in Pd. The steps in the approach were as follows: first, the ambient charge density, $n(r)$, was calculated by superimposing the atomic charge distributions centred at the host atom sites in the Pd matrix. Secondly, the Fermi charge density enhancement factor, $\eta_F(r)$, at some lattice position r was calculated by applying the HKS density formalism, whereby the local ambient charge density, $n(r)$, was assumed to correspond to the density of a homogeneous free-electron gas (pseudo-jellium model). In the interstitial site centre $\eta_F(r=0) = 10.8$. Finally $\eta_F(r)$ was folded with the probability distribution of the hydrogen isotope, given by its zero-point vibration. Jena and Singwi (1978) found a difference of 1% in the spin density between the deuteron and the proton, that at the deuteron being the larger of the two. This compares favourably with a 2% difference from NMR measurements of the Korringa relaxation rate in PdH (Wiley and Fradin 1978). Of course, this pseudo-jellium approach is rather crude as it neglects the fact that different types of electrons are present, and these may respond differently to the presence of a positive impurity charge. Nevertheless, it might provide some qualitative insight as far as s electrons are concerned into the trends of the corresponding isotope effect.

The effect of the zero-point vibration amplitude on the average spin density at the μ^+ was also considered in the calculations of Manninen and Nieminen (1981) and Estreicher and Meier (1982, 1984) mentioned above. Here, the averaging has to be done over the inhomogeneous spin polarisation induced by the electronic moments. It turns out that, for example in Gd, the hyperfine field experienced by a proton is expected to be as much as 50% smaller in magnitude than the field experienced by a μ^+ (Manninen and Nieminen 1981).

4.3 Experimental Results of μ^+ Knight Shift Investigations

4.3.1 Systematics in simple metals

All available results on the isotropic μ^+ Knight shift (K_μ) in simple non-transition metals are collected in table 4.2. Most of the data were obtained at

Table 4.2 Compilation of μ^+ Knight shift data in non-transition metals. r_s = electron density parameter ($r_s = (3/4\pi n_0)^{1/3} a_B$); χ_s = experimental spin susceptibility (from (2) and Carter et al 1977). The references are numbered as follows: (1) Camani et al (1979), (2) Schenck (1982), (3) Gygax et al (1984a), (4) Studer et al (1984a), (5) Studer (1983), (6) Studer et al (1984b), (7) Gygax et al (1983), (8) Studer et al (1984c), (9) Hartmann et al (1979), (10) Brewer et al (1983), (11) Gygax et al (1984d). Data at room temperature unless indicated otherwise.

Metal	r_s	$10^6 \chi_s$ (emu cm^{-3})	K_{iso} (ppm)	K_{ax} (ppm)	References
Li	3.25	2.10	8.6 ± 4.0		(1), (2)
Na	3.93	1.09	76.5 ± 5.0		(1), (2)
K	4.86	0.90	63.2 ± 4.5		(1), (2)
Rb	5.20	0.86	66.1 ± 4.3		(1), (2)
Cs	5.63	0.81	31.0 ± 4.0		(1), (2)
Cu	2.67	1.36	60.0 ± 2.5		(1), (2)
Ag	3.02	0.88	94.0 ± 3.5		(2)
Au	3.01	0.99	64.0 ± 10		(3)
Be	1.88	$\simeq 0.2$	-9.0 ± 1.6	$-3.0 \pm 1.9(?)$	new value (4), (5), (3)
Mg	2.65	$\simeq 1.58$	43.3 ± 3.5	0.0 ± 2.0	new value (3)
Ca	3.27	$\simeq 1.58$	21.2 ± 6.2		(1), (2)
Sr	3.56	$\gtrsim 2.69$	-18.4 ± 6.8		(1), (2)
Ba	3.69	$\simeq 0.99$	29.6 ± 5.0		(1), (2)
Zn	2.31	$\simeq 0.99$	60.0 ± 5†	3.0 ± 1.3†	(4), (5), (6)
Cd	2.59	$\simeq 0.73$	55.0 ± 5†	-3.3 ± 1.3†	(4), (5), (6), (7), (8)
Hg	2.75	$\simeq 1.73$	117.0 ± 11‡		(2)
Al	2.07	1.77	79.6 ± 4.0		(2)
Ga	2.19	$\simeq 0.7$	-17.0 ± 3.5		(2)
Pb	2.30	$\simeq 2.36$	105.2 ± 3.5		(2)
Sb		$\simeq 0.082$	8700§	2200§	(9), (10)
Bi		$\simeq 0.014$	$-276\|$	$9 \pm 6\|$	new value (11)

† 20 K. ‡ $-50\,°C$. § 15 K. $\|$ 4.2 K (preliminary data).

room temperature (Schenck 1982). It is argued (Schenck 1982) that the measurements at room temperature should reflect mainly the real intrinsic μ^+ Knight shift, unaffected by trapping at impurities or other defects. In fact, K_μ in Cu proved to be independent of temperature between 30 and 800 K. Recent measurements in high purity single crystals of HCP Be and Mg, however, yielded results (table 4.2) which were significantly different, although not drastically so, from previous ones (Camani et al 1979) in less pure samples.

How much the K_μ values in the marginally pure materials investigated (i.e. in Ca, Sr, Ba) are affected by trapping at impurities must be considered open. Nevertheless, it appears that the experimental values in Ca, Sr and Ba also cannot be too far away from the true intrinsic ones, in view of the systematics discussed below.

In a few cases K_μ has also been measured in the liquid phase of the metal (Rb, Cs, Ga, Hg, Cd, Zn) (see table 4.3). Generally the Knight shift in the liquid phase is larger than in the solid phase. The most remarkable features are the change from a negative Knight shift in solid Ga to a positive one in liquid Ga, and the huge jump of the Knight shift in Hg upon melting. In Cs, Rb, Cd and Zn the Knight shift increases only little upon melting. In simple non-transition metals one does not expect to find a significant temperature dependence of K_μ, since the electron spin susceptibility for a degenerate electron gas is, to a first-order approximation, temperature independent. This is reflected in the largely temperature independent nuclear Knight shift (see e.g. Carter et al 1977). Indeed, K_μ in Cu, Al and Zn showed none or only a slight temperature dependence. The K_μ data in Zn are displayed in figure 4.6. A notable exception is Cd, in which the nuclear Knight shift also is strongly temperature dependent. The data on Cd and CdMg (3.4 atomic%) and CdHg (1.2 atomic%) will be discussed separately in §4.3.2.

Table 4.3 Comparison of μ^+ Knight shift, K_s, in the solid and liquid phases of non-transition metals. Kelvin temperatures are given in brackets following K_s, and room temperature denoted RT.

Metal	T_M (K)	K_s liquid (ppm)		K_s solid (ppm)	
Rb	312.1	80 ± 5	(≈322)	66.0 ± 4.3	(RT)
Cs	301.9	32 ± 6	(≈312)	31 ± 4	(RT)
Ga	303.0	98 ± 4	(≈313)	−17.0 ± 3.5	(RT)
Hg	234.8	205 ± 6	(RT)	117 ± 11	(223)
Zn	693	76 ± 2.5	(⩾710)	68 ± 3	(670)
Cd	593	130 ± 2.5	(⩾610)	132 ± 5	(580)

Table 4.2 further shows that a negative Knight shift is also observed in Be and Sr. This fact demonstrates that it is not sufficient to consider the direct Knight shift alone, equation (4.10), but that diamagnetic screening and core polarisation effects do have to be taken into consideration. Measurements in the HCP metals Be, Zn and Cd using single crystal samples revealed an axial Knight shift and this will be discussed in §4.3.3.

The isotropic Knight shift data have been extensively discussed by Schenck (1982, 1983) and Gygax et al (1984a) in relation to the local electronic structure

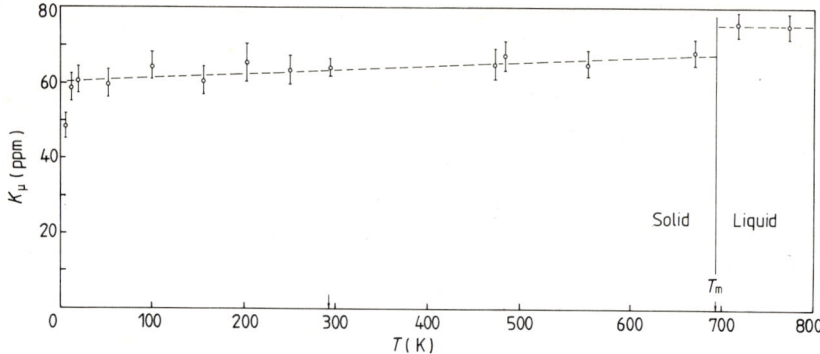

Figure 4.6 Temperature dependence of K_μ in polycrystalline Zn, showing the transition into the liquid phase (Studer et al 1984c).

of a hydrogen-like impurity in a metal matrix. Here we would just like to discuss an interesting systematic behaviour of almost all of the K_μ data in the cubic mono-, di- and trivalent metals, and the divalent HCP metals Be, Mg, Zn and Cd. The same systematic behaviour has recently been reproduced in calculations by Manninen (1983) adopting the spherical solid model (see §4.2).

This systematic behaviour shows up when the logarithm of the induced hyperfine field per unpaired electron per atom, $B_{hf,s}^{\Omega_a} = \mu_B K_s / \chi_s^{at}$, is plotted against the molar electronic specific heat, γ_{cp}^m, the latter being proportional to the density of states at the Fermi energy (figure 4.7). In calculating $B_{hf,s}^{\Omega_a}$, experimental values for χ_s^{at} were used throughout (see table 4.2). K_s was calculated from K_μ by assuming a diamagnetic screening of -25 ± 5 ppm for all metals (see §4.1.4). As can be seen from figure 4.7, $B_{hf,s}^{\Omega_a}$ follows an exponential dependence on γ_{cp}^m. The only clear exceptions are Li and Pb. The deviation of $B_{hf,s}^{\Omega_a}$ in Pb from the empirical curve is also found in the calculations by Manninen (1983), while the Li result is still unaccounted for. Interestingly, the nuclear Li Knight shift is also much smaller than is expected on the basis of the experimental spin susceptibility. It must be noted that the empirical correlation between $B_{hf,s}^{\Omega_a}$ and γ_{cp}^m is apparently independent of crystal structure and of the different μ^+ sites in these crystal structures (e.g. tetrahedral sites in the BCC metals, octahedral sites in the FCC metals).

One can try to calculate a spin density enhancement factor $\rho_s(r_\mu)$ from the empirical correlation by use of the identity (see equation (4.17))

$$K_s = \frac{8\pi}{3} \rho_s(r_\mu) \chi_s = \frac{1}{\mu_B} B_{hf,s}^{\Omega_a} \Omega_a \chi_s \qquad (4.36)$$

from which follows

$$\rho_s(r_\mu) = \frac{3}{8\pi} \Omega_a \frac{1}{\mu_B} B_{hf,s}^{\Omega_a} \qquad (4.37)$$

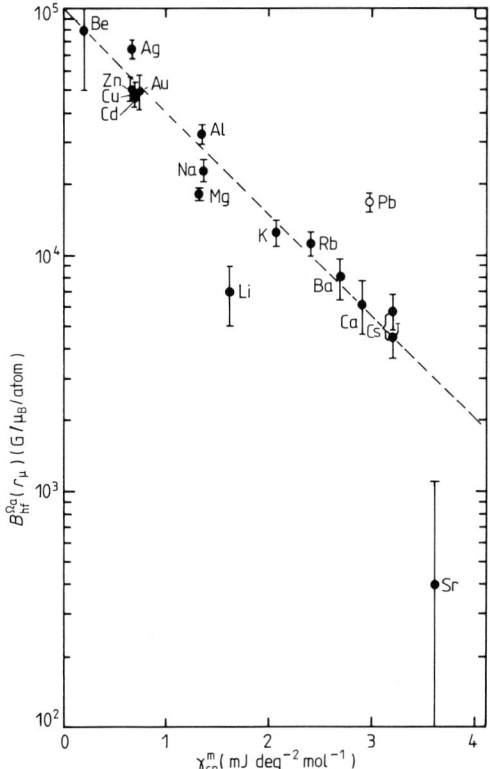

Figure 4.7 Semilogarithmic plot of the induced hyperfine field, $B_{hf}^{\Omega a} = \mu_B K_s / \chi_s^{at}$ against the molar electronic specific heat, γ_{cp}^m (Gygax *et al* 1984a). The broken curve is represented by $B_{hf}^{\Omega a} \propto \exp(-2.34 N^{at}(E_F))$ when $N^{at}(E_F)$ is the electronic density of states at the Fermi energy in units of (eV/atom) (see text).

where the atomic volume, Ω_a, can be expressed as

$$\Omega_a = Z^{4/3} \pi r_s^3 a_B^3. \tag{4.38}$$

Z is the valency of the metal and r_s the electron density parameter $(r_s = (3/4\pi n_0)^{1/3} / a_B)$. From figure 4.7 one obtains

$$\begin{aligned} B_{hf,s}^{\Omega a} &= B_{hf,s}^{\Omega a}(0) \exp(-\alpha \gamma_{cp}^m) \\ &= B_{hf,s}^{\Omega a}(0) \exp(-2.34 N^{at}(E_F)). \end{aligned} \tag{4.39}$$

$N^{at}(E_F)$ is the atomic density of states at the Fermi energy and

$$\gamma_{cp}^m = \tfrac{1}{3} \pi^2 k_B^2 L N^{at}(E_F) \tag{4.40}$$

where L, the Loschmidt number, is used. Since the spin density enhancement factor is usually calculated as a function of the density parameter r_s, it would be useful if the empirical $\rho_s(r_\mu)$ could also be expressed as a function of r_s for purposes of comparison. This can be done on the basis of a free-electron gas model in which

$$N^{\text{at}}(E_F) = \frac{m^*}{m} Z \frac{3}{\hbar^2} \left(\frac{4}{9\pi}\right)^{3/2} a_B^2 r_s^2. \qquad (4.41)$$

m^* is the effective electron mass. $N^{\text{at}}(E_F)$ or γ_{cp}^m itself is not only a function of r_s, but also of the valency Z and the ratio m^*/m. Figure 4.8 shows a plot of experimental γ_{cp}^m versus Zr_s^2. It is seen that all alkalis (except Li) plus Cu, Ba and Al have roughly the same linear dependence on Zr_s^2, implying very similar m^*/m ratios ($\simeq 1.34$) for these metals. It is thus possible to express $\rho_s(r_\mu)$ as a function of r_s with Z as an additional parameter. Figure 4.9 shows the result, together with experimental values for $\rho_s(r_\mu)$, obtained directly from the experimental K_s. Similar curves can, of course, be calculated for other values of m^*/m. Also shown in figure 4.9 is the theoretical $\rho_s(r_\mu)$ stemming from non-linear jellium calculations using the spin density functional formalism (Munjal and Petzinger 1978). The main consequence of this systematic behaviour is that K_s shows a distinctive dependence of the metal valency and only to a lesser degree on the total conduction electron concentration, in contrast to the predictions from the various jellium calculations (Jena 1981, Schenck 1982).

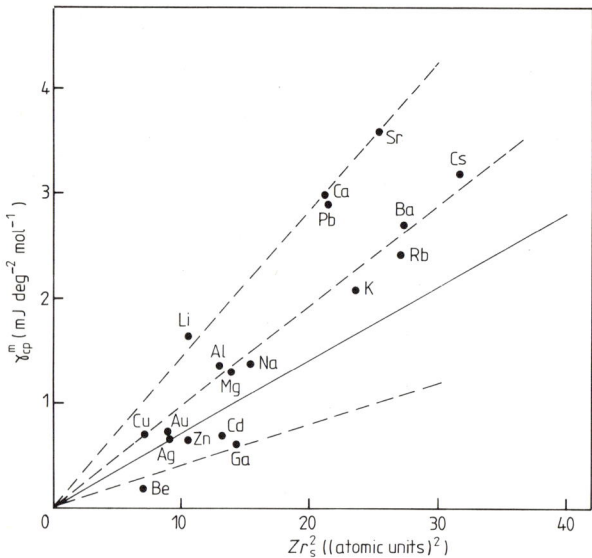

Figure 4.8 Plot of experimental molar electronic heats γ_{cp}^m versus Zr_s^2, where Z is the valency of the metal and r_s the electron density parameter. The full line corresponds to $m_e = m_e^*$ (Schenck 1982).

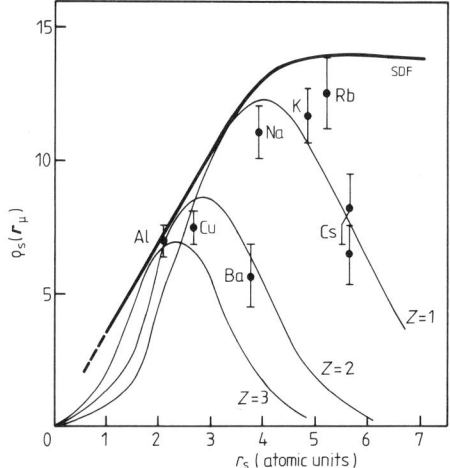

Figure 4.9 Plot of the spin density enhancement factor, $\rho_s(r_\mu)$, following from the empirical correlation expressed by equation (4.49) (light curves). The heavy curve is as in figure 4.1.

In the meantime, considerable progress has been made in understanding the gross features of the observed systematics by Manninen (1983). Essentially it is also a jellium calculation, but the host lattice potential is taken into account by the so-called spherical solid model, which replaces the compensating positive background charge in the conventional jellium approach (see §3.3.1).

This model is used to calculate the spin density enhancement factor, $\rho_s(r_\mu)$ displayed in figure 4.4. Figures 4.10 and 4.11 show the ensuing predictions for K_μ in monovalent and divalent metals respectively. The diamagnetic contribution was calculated for the same model using the formula derived by Zaremba and Zobin (1980). The spin susceptibility, needed to calculate K_s, was taken to be that of a homogeneous interacting electron gas of the appropriate density, rather than the experimental value. This is justified on grounds of internal consistency in using equation (4.9). It is evident from figures 4.10 and 4.11 that the trend of the data with the density parameter, r_s, is reproduced quite well, as is the data grouping with reference to the valency of the host metals. However, if experimental values for the spin susceptibility were used, less agreement would have been found. This is somewhat surprising in view of the systematic behaviour displayed in figure 4.7 using experimental χ_s values (see also Gygax *et al* 1984a). It is also clear from figure 4.10 and 4.11 that the model fails to account for some of the results in detail. Large deviations between the predictions and the experimental results exist, for example, in Cu and Rb, while the data are excellently accommodated by the empirical correlation. Clearly, further improved theoretical calculations are necessary.

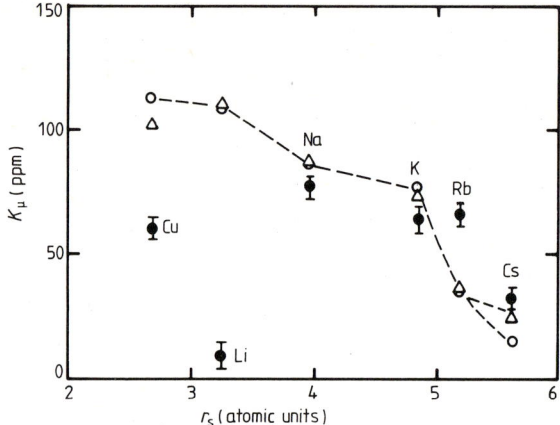

Figure 4.10 Spherical solid model predictions for K_μ in monovalent metals and comparison with: ●, experimental values; ○, μ^+ at octahedral interstices; △, tetrahedral interstices (Manninen 1983, Gygax *et al* 1984a).

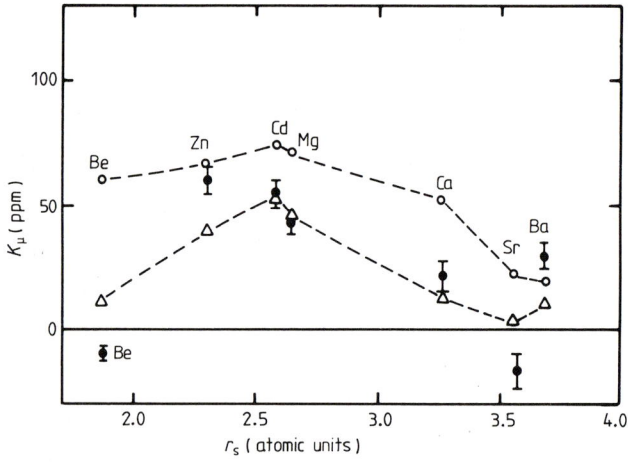

Figure 4.11 Spherical solid model predictions for K_μ in divalent metals and comparison with experimental data for μ^+ at various interstices: ●, experimental values; ○, octahedral; △, tetrahedral (Manninen 1983, Gygax *et al* 1984a).

4.3.2 Temperature dependence in Cd and CdMg and CdHg alloys

The unusual results obtained in polycrystalline samples are shown in figures 4.12, 4.13 and 4.14 (Studer *et al* 1984c). In pure Cd one notes a drastic increase of K_μ with temperature up to the melting point, which exceeds by far the

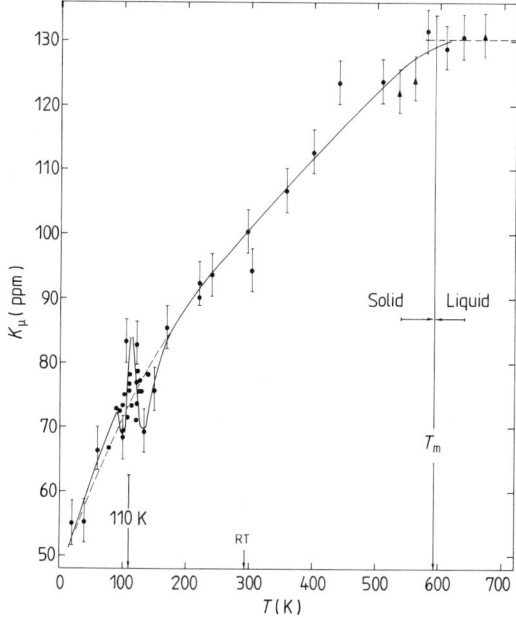

Figure 4.12 Temperature dependence of the μ^+ Knight shift, K_μ in high purity polycrystalline Cd (Studer et al 1984c).

increase of the nuclear Knight shift observed in Cd. There is, however, practically no change in K_μ when passing into the liquid state, again in contrast to the nuclear Cd Knight shift which displays a $+33\%$ jump upon melting. Even more striking is a cusp-like singularity in K_μ at about 110 K, which has been observed also in monocrystalline Cd samples of different origin (see also figure 4.15). Besides the cusp, there are also dips to the left and right of the singularity. In CdHg (1.23 atomic %) the singularity is seen once more, but shifted to lower temperatures. In CdMg (3.38 atomic %) one finds a step-like singularity at $\simeq 32$ K and again a cusp-like singularity at $\simeq 280$ K.

These observations have been interpreted as a band structure effect related to critical points in the vicinity of the K symmetry point of the hexagonal Brillouin zone (Studer 1983, Studer et al 1984b,c). The band structure in the vicinity of the K symmetry point close to the Fermi energy can be described by a three-band model (McClure and Martyniuk 1972) and is shown in figure 4.15. The critical points concern the maxima, minima and the saddle point in the energy momentum dispersion relation, $\mathrm{grad}_k E(k) = 0$. If the Fermi level passes through these points, the corresponding density of states, $N(E)$, displays so-called van Hove singularities (see e.g. Kittel 1971). The concomitant change in the Fermi surface topology is called an electronic, or Lifshitz, phase transition. For a two-dimensional topology of the critical points, that is $E = E(k_x, k_y)$, maxima and minima will lead to step-like changes

150   *Hyperfine fields at the μ^+ in metals*

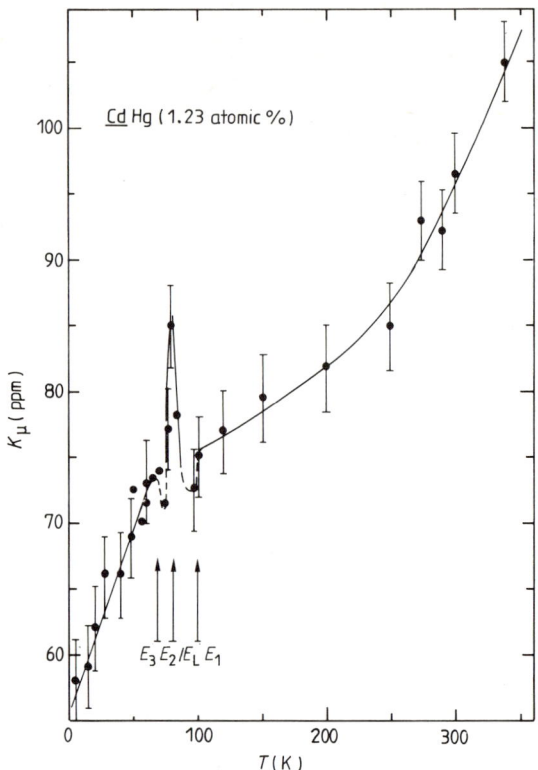

Figure 4.13 Temperature dependence of K_μ in the alloy CdHg (1.23 atomic %) (Studer *et al* 1984c). The various E values are also indicated in figure 4.14 and explained in figure 4.15.

in the density of states $N(E)$, while a saddle point leads to a logarithmic singularity of the form $N(E) \propto \ln(E/E_{cr} - 1)$ (van Hove 1953). $N(E)$ for the band model is shown in the insert in figure 4.14. This seems to be exactly what is observed in the CdMg (3.38 atomic %) sample: a step-like change at $\simeq 32$ K, and the cusp-like singularity at $T_{cr} \simeq 280$ K is well fitted by the expression

$$K_\mu = C \ln[(T/T_{cr}) - 1] \tag{4.42}$$

which is represented in figure 4.14 by curve A for $T < T_{cr}$, and by curve B for $T > T_{cr}$. It appears, therefore, that K_μ precisely maps the density of states curve for electrons near the K symmetry point with two-dimensional characteristics. The temperature here has the effect of shifting the Fermi energy across the critical point region, via the change of the c/a ratio with temperature. The c/a ratio sensitively determines the position of the Fermi energy (McClure and Martyniuk 1972). This mapping of $N(E)$ by K_μ can be understood on the basis of equation (4.11). We split the integral into two terms, one involving only

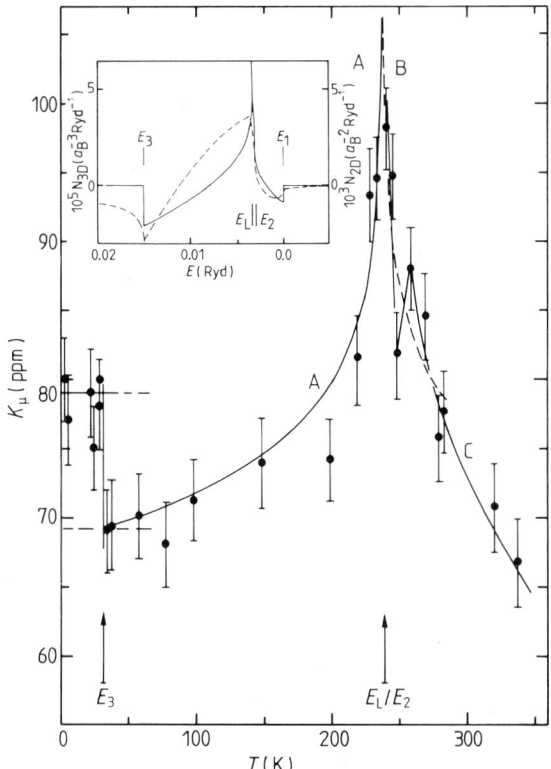

Figure 4.14 Temperature dependence of K_μ in the alloy $\underline{Cd}Mg$ (3.38 atomic %). A and B represent fits to the data of equation (4.42), and C is simply a guide to the eye. The insert shows the two-dimensional (full curve) and three-dimensional (broken curve) electron density of states curve, resulting from the three-band model for electrons near the symmetry point K in the hexagonal Brillouin zone of Cd (McClure and Martyniuk 1972, Studer *et al* 1984c).

integration over a small area of the Fermi surface around the K symmetry point, and the other accounting for the remainder

$$K_s = \frac{8\pi}{3} \mu_B^2 \frac{\Omega_e}{(2\pi)^3} \left(\int_{(K)} \frac{dS_F}{|\text{grad}_k E_F(\boldsymbol{k})|} |\psi_{k_F}(\boldsymbol{r}_\mu)|^2 + \int_{(S-K)} \frac{dS_F}{|\text{grad}_k E_F(\boldsymbol{k})|} |\psi_{k_F}(\boldsymbol{r}_\mu)|^2 \right).$$

(4.43)

Assuming that $\psi_{k_F}(\boldsymbol{r}_\mu)$ does not change appreciably over the small surface area surrounding the K symmetry point, we further write

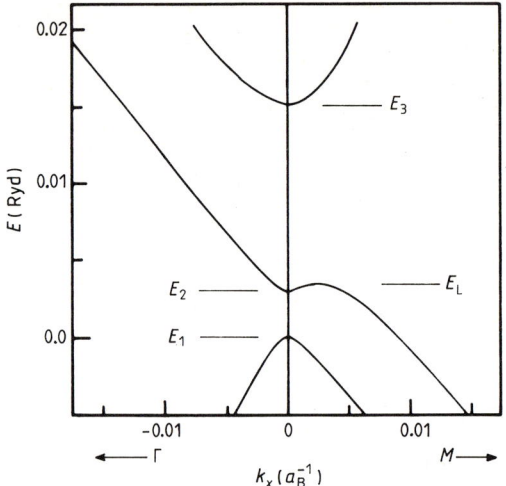

Figure 4.15 Part of the Cd band structure near the symmetry point, K, of the hexagonal Brillouin zone of HCP Cd. The critical points E_1, E_2, E_L and E_3 are marked by the horizontal lines. Γ and M indicate directions in the Brillouin zone towards the symmetry points Γ and M.

$$K_s \simeq \frac{8\pi}{3} \mu_B^2 \Omega_e \langle |\psi_{k_F}(r_\mu)|^2 \rangle_{(K)} \frac{1}{(2\pi)^3} \int_{(K)} \frac{dS_F}{|\text{grad}_k E_F(k)|} + K_{BG}$$

$$\simeq \frac{8\pi}{3} \mu_B^2 \Omega_e \, \delta N(E_F) + K_{BG}. \tag{4.44}$$

$\delta N(E_F)$ is the partial density of states of the electron states near the K symmetry point and K_{BG} is a background contribution from all other electrons.

Further analysis of the data in CdMg (3.38 atomic %) leads to identification of the step-like singularity as the critical point E_1 (see figure 4.15), and of the logarithmic singularity with the saddle point, E_L. The other critical point, E_3 can only be reached at higher temperatures. The observed complex singularities in pure Cd and CdHg (1.23 atomic %) can be interpreted as consisting of the logarithmic singularity in the centre and step-like discontinuities to the left and right, implying that the Fermi energy is shifted through E_1, E_L and E_3 over a very small temperature range. The critical point E_2 is masked by E_L.

The position of E_1, E_L, E_3 with respect to E_F at 0 K in pure Cd and the alloys CdHg (1.23 atomic %) and CdMg (3.38 atomic %) is known from band structure calculations and from the analysis of magnetic susceptibility data (McClure and Martyniuk 1972). The temperature dependence of the Fermi energy position can also be deduced from the susceptibility data, and allows the splitting between E_1, E_L and E_3, obtained from the K_μ measurements, to be expressed in energy units. This analysis shows that the presence of the μ^+ leads to an appreciable lowering of these states at the interstitial site which is different in the various samples. Also, the splitting parameters of the three-band model are altered compared with the undisturbed system. This feature is not understood at present, since the screened, short range potential of the μ^+ is expected to act as an energy conserving elastic scattering centre for the Bloch electrons of the host, simply shifting their phases. Nevertheless, the striking resemblance of the behaviour of K_μ with the electron density of states curve, as shown in figure 4.14, indicates that the three-band model provides an excellent basis for the phenomenological description of the μ^+ Knight shift data. How it relates to the three-band model for electrons near the symmetry point K in the bulk Cd band structure needs to be further investigated.

The fact that critical points are manifest at all in the K_μ properties implies that the corresponding electron states must have a large density at the μ^+ interstitial site. Therefore one immediately obtains some information on the spatial distribution of these electrons. It is interesting to note that the Cd nuclear Knight shift does not display any singularities, implying the absence of such electrons at the nuclear or substitutional site (see Carter *et al* 1977). Indeed, the theoretical analysis by Kasovski (1969) shows that the Cd nuclear Knight shift involves predominantly electron states near the M (monsters) and the Γ (lenses) symmetry points and is unaffected by those electron states near the K symmetry point. The use of the μ^+ as an interstitial probe therefore provides access to electronic properties of the host metal which cannot be obtained by substitutional probes.

4.3.3 Axial Knight shift

Careful measurements of the angular dependence of the μ^+ Knight shift in single crystals of the HCP metals Cd and Zn (Studer 1983, Studer *et al* 1984a,c) showed the clear presence of an axial Knight shift, while no axial Knight shift could be observed in Mg. The anisotropy observed in Be is probably caused by the non-spherical shape of the sample used and the highly anisotropic diamagnetic susceptibility of Be. The axis of axial symmetry generally coincided with the crystalline c axis. In Zn, an anisotropic Knight shift could only be observed at low temperatures (<80 K). Its size and functional dependence on angle varied between different samples and seemed to depend on the sample history, such as the rate of temperature change. In Cd, an axial Knight shift (see figure 4.16) was observable at all temperatures almost up to the melting point, although a change of sign of K_{ax} occurred at 110 K. Figure 4.17 shows the angular dependence of K_μ at 60 K and at room temperature (Gygax *et al* 1983a). The peculiar temperature of 110 K is thus also manifest in the behaviour of the anisotropic Knight shift. If the axial Knight shift in Cd is caused by a non-spherical charge distribution around the μ^+, i.e. by the presence of p-like states, one would have observed a change of the charge

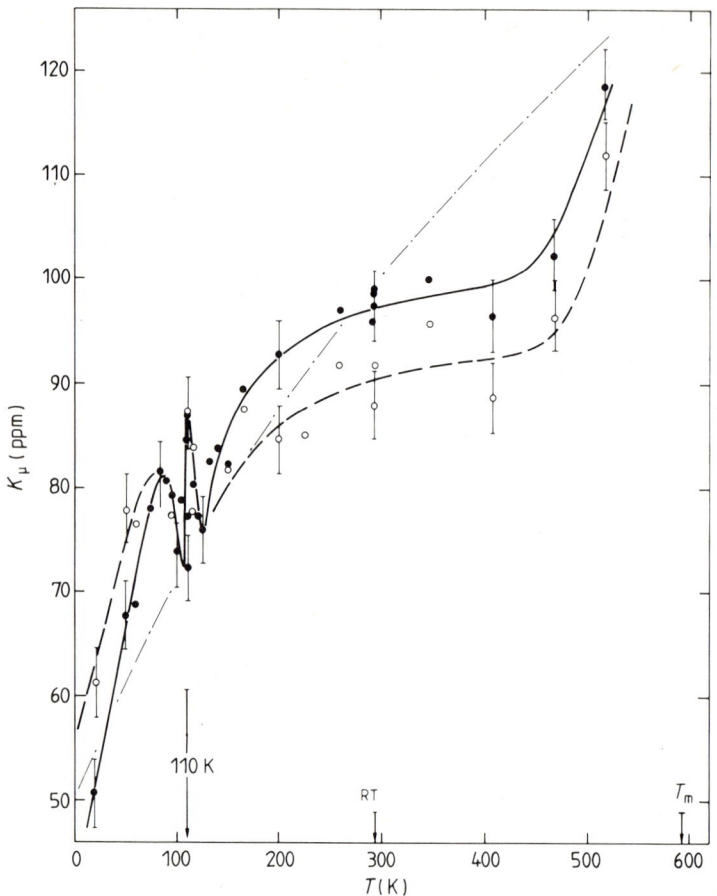

Figure 4.16 Temperature dependence of the μ^+ Knight shift in monocrystalline Cd with the hexagonal c axis: ●, parallel; ○, perpendicular to the applied field (Studer *et al* 1984a).

distribution from oblate-like (p_x, p_y components dominant) at low temperatures, to prolate-like (p_z component dominant) at high temperatures. On the other hand, the observed anisotropy may also be caused by an anisotropic χ_s as a result of some of the electrons contributing to the contact density having an anisotropic Landé (g_e factor). This is not an unreasonable possibility, since it is known that spin–orbit coupling effects play an important role in the Cd band structure.

4.4 μ^+ Knight Shift in Transition Metals and Alloys/Compounds

Table 4.4 contains a compilation of all systems studied so far. The table also indicates the main objective of each study which can be classified into two

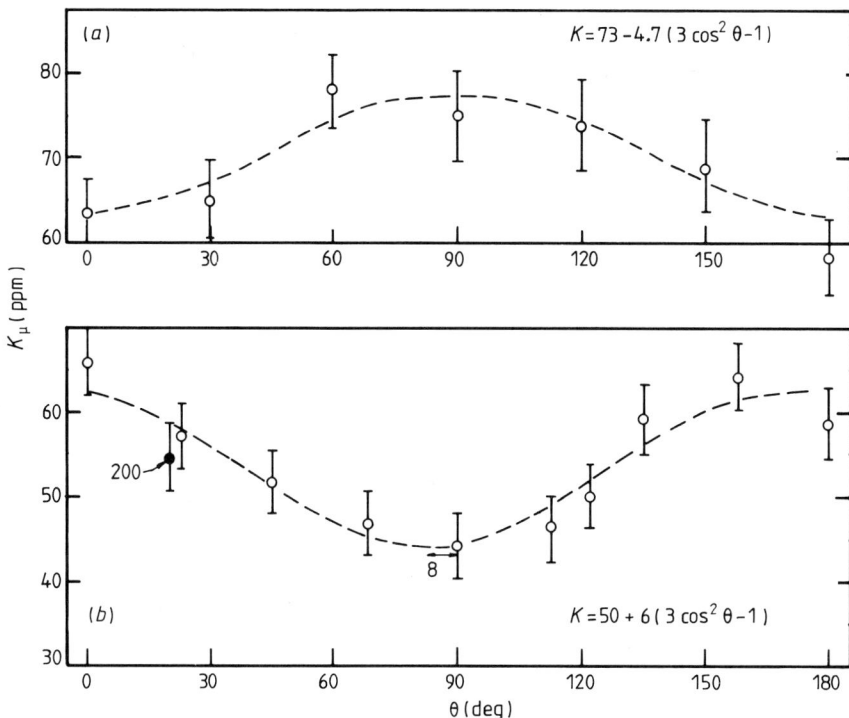

Figure 4.17 Angular dependence of the μ^+ Knight shift in monocrystalline Cd at (a) 60 K and (b) 293 K (Gygax *et al* 1983a).

main categories. Those concerning the local electronic structure problem of hydrogen in metals, and those with the study of the host's properties. The studies in the group VB metals and in Ni, Pd, Pt, βPdH belong to the first category, while the studies in Ce, CeTh, CeSn$_3$ and LaAg$_x$In$_{1-x}$ belong to the second. In the following we will discuss some of the results as representative examples of the observable effects.

4.4.1 Group VB metals: V, Nb, Ta

Muon Knight shift results in these metals are interesting because they can be compared with proton Knight shift data, obtained under very similar conditions (same temperature range, samples with similar purity) and extrapolated to zero hydrogen concentration (Kazama and Fukai 1977). μ^+ and proton Knight shift data are compiled in table 4.5. As far as a possible temperature dependence of K_μ has been investigated no such dependence has been observed. As table 4.5 shows, the μ^+ Knight shift and proton Knight shift are quite close to one another. There are, however, significant differences,

Table 4.4 Compilation of Knight shift measurements in transition metals and alloys/compounds.

Compound	Temperature range investigated (K)	Composition	Objective	References
Ni	637–906		local electronic structure 'H' problem	Gygax et al (1980)
Pd	20–880			Gygax et al (1981b)
Pt	20–785			Gygax et al (1983b, 1984b)
V	293		Isotope effect by comparison with proton K_s data	
Nb	100, 293, 423			Schenck (1982)
Ta	293			
MnSi	28–300		$B_{\text{hf,d}}^{\Omega_a}$	Yasuoka et al (1978)
LaAg$_{1-x}$In$_x$	30–293	$0.0 \leq x \leq 0.4$	correlation with position of E_F	Wehr et al (1983)
LaNi$_5$H$_x$	20–293	$0 \leq x \leq 6$	x dependence	Gygax et al (1981a)
β-PdH$_x$	20–293	$0.59 \leq x \leq 0.86$	isotope effect, x dependence	Gygax et al (1984a,b)
Ce	30–700		study of valence transition	
Ce$_{1-x}$Th$_x$	30–500	$x = 0.26$		Wehr et al (1981)
CeSn$_3$	10–500			Wehr et al (1984)

Table 4.5 Comparison of μ^+ and proton Knight shift data in the group VB metals.

Metal	K_p (ppm)†	Temperature range (°C)	K_μ (ppm)‡	$10^6 \chi_p$ (emu mol^{-1})
V	−66	$100 \leq T \leq 200$	−(88±8)	133
Nb	−25	$34 \leq T \leq 180$	−(15.6±3)	109
Ta	0	$34 \leq T \leq 100$	+(5.5±10)	83
Reference	Kazama and Fukai (1977)		Schenck (1982)	Carter et al (1977)

† Extrapolated to zero hydrogen concentration.
‡ For the temperatures investigated see table 4.4.

particularly in V, pointing to the presence of an isotope effect in the Knight shift. Obviously then, the isotope effect does not display a systematic trend. No theoretical calculations for the proton or μ^+ Knight shift in the group VB metals exist.

4.4.2 Ni, Pd, Pt

In all three metals the μ^+ Knight shift has been measured as a function of temperature over a wide temperature range (see table 4.4). A strong temperature dependence is observed which exactly follows that of the bulk magnetic susceptibility χ. This is clearly visible when the data are plotted against the susceptibility with temperature as an implicit parameter (figure 4.18) (Clogston–Jaccarino plot). Since the temperature dependence of the total susceptibility is associated with the d electrons' susceptibility term, the slope $dK_\mu/d\chi$ can be identified with the induced hyperfine field at the μ^+ per unpaired d electron per atom, $B_{hf,d}^{\Omega_a}$ according to equation (4.17). Extrapolating the susceptibility to its high temperature limit, that is, the sum of the temperature independent s electron susceptibility, χ_s, the orbital or Van Vleck susceptibility, χ_{VV}, and the diamagnetic susceptibility χ_{dia}, one can find the temperature independent Knight shift from the Clogston–Jaccarino plot. In the present case this is just the s-electron induced Knight shift, K_s. A compilation of the thus derived values of K_s and $B_{hf,d}^{\Omega_a}$ is presented in table 4.6. As can be seen, the d electron induced hyperfine field increases from Ni to Pd to Pt by roughly a factor of two each time. Also, K_s shows a tendency to increase, but the involved errors are quite large.

$B_{hf,d}^{\Omega_a}$ in Ni can be compared with the induced hyperfine field per d electron per atom in the ferromagnetic state (see §4.6.2), which is $B_{hf}^{\Omega_a}(r_\mu) = B_{hf}(r_\mu)/M_s = -1.14(2)\,\text{kg}/\mu_B/\text{atom}$ (Schenck 1982). This value is roughly 10% smaller than in the paramagnetic state. It is interesting to note that the same relative deviation is also observed for the Ni nuclear Knight shift in the paramagnetic state and the spontaneous hyperfine field in the ferromagnetic state (Segransan et al 1976, Shaham et al 1977). No explanation has yet been found for these deviations. Likewise, the systematics of the μ^+ data in Ni, Pd, Pt have only been discussed in a very preliminary way (Gygax et al 1983b, 1984b).

4.4.3 Ce and $Ce_{1-x}Th_x$—valence transition study

The measurements performed in these compounds, belonging to the class of so-called intermediate valence systems, were aimed at a study of the valence transition, a bulk phenomenon which has attracted much interest in recent years (Wehr et al 1981, Wehr 1982).

FCC cerium exists in two phases, the γ and the α phase. The cell volume of the α phase is about 16% lower than that of the γ phase. At room temperature the transition to the collapsed phase occurs at a pressure of 8 kbar. Similar to the liquid–gas system, the phase separation line terminates at a critical point, which is at $T_c = 600$ K, $p_c = 20$ kbar (see e.g. Koskimaki and Gschneider 1978). The γ–α transition is also observed in $Ce_{1-x}Th_x$ alloys. Here the Th concentration can be regarded as being equivalent to a pressure. The phase

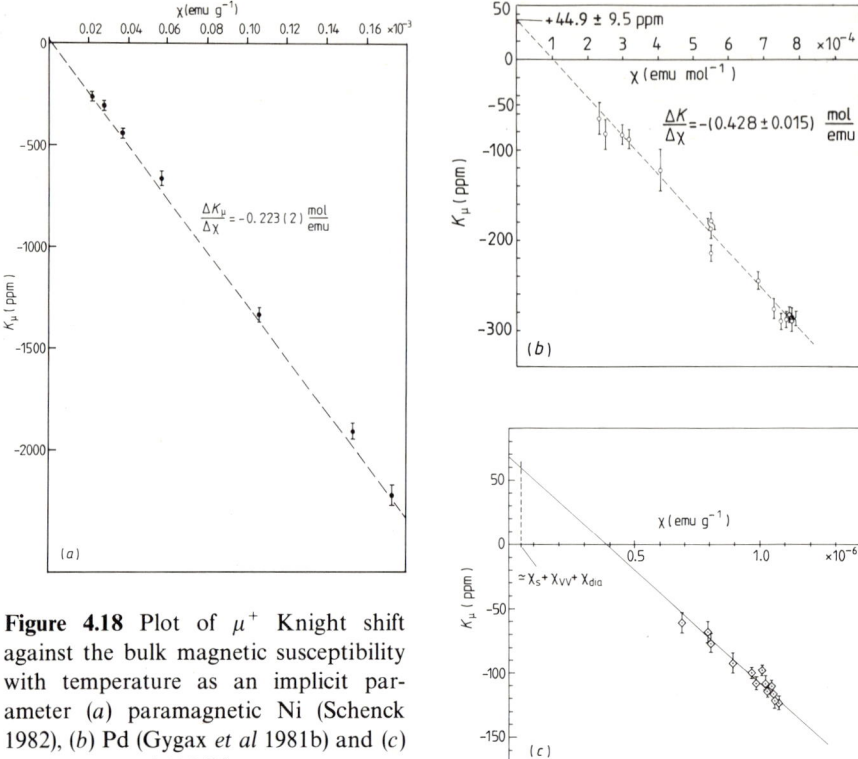

Figure 4.18 Plot of μ^+ Knight shift against the bulk magnetic susceptibility with temperature as an implicit parameter (a) paramagnetic Ni (Schenck 1982), (b) Pd (Gygax et al 1981b) and (c) Pt (Gygax et al 1984b).

separation line ends at $T_c = 150$ K and a critical concentration $x_c = 0.269$ at zero applied pressure (Lawrence et al 1975).

The contraction of the lattice is connected with a change of the magnetic susceptibility from a Curie-type behaviour to a temperature-independent behaviour. The transition was explained by a delocalisation of the 4f' electron. In the γ phase the Ce ions exist in the usual tripositive valence state of the rare earth ions, giving rise to ionic magnetism, whereas in the α phase the Ce ions are formally tetravalent, non-magnetic, and the former 4f' electron is converted into a band electron. The exact nature of this conversion is still a matter for discussion; the electron may be either promoted to the (s, d) bands or may attain dispersion on its own via a Mott transition.

The μ^+ Knight shift measurements were performed in the hope of learning more precisely about the nature of this transition. As long as the 4f' electron is a localised electron, being responsible for the Curie behaviour of the magnetic susceptibility, it will only indirectly contribute to the spin density at the μ^+ site. This is via exchange polarisation of the conduction electrons (RKKY

Table 4.6 Compilation of μ^+ Knight shift and other relevant parameters (lattice constant, s electrons per atom, s-electron spin susceptibility) in Ni, Pd and Pt. The s-electron induced Knight shift is compared with the one in the neighbouring elements with closed d shells. The d-electron induced hyperfine field at the μ^+ is compared with the corresponding hyperfine field at the nuclei (Gygax et al 1983b).

	a (Å)	s electrons per atom	$10^6 \chi_s$ (emu cm^{-3})	K_s (ppm)	$B_{hf,d}^{\Omega_a}$ (μ^+) (kG/μ_B/atom)	$B_{hf,d}^{\Omega_a}$ (nucleus) (kG/μ_B/atom)
Cu	3.61	1	1.36	80(3)	—	—
Ni	3.52	0.6	0.62	31(27)	−1.224(23)	−137
Ag	4.09	1	0.85	114(4)	—	—
Pd	3.89	0.36	0.64	62(13)	−2.39(11)	−345
Au	4.08	1	0.99	64(10)	—	—
Pt	3.92	0.2	0.53	73(15)	−5.03(43)	−1180

mechanism, see equation (4.20)). The corresponding $B_{hf,4f}^{\Omega_a}(r_\mu)$ can be determined from the slope of a plot of K_μ against susceptibility. If, in the α phase the former local 4f' electron became delocalised, one would now expect a different induced $B_{hf}^{\Omega_a}(r_\mu)$. Figure 4.19 shows the Knight shift results obtained in $Ce_{0.74}Th_{0.26}$ as a function of temperature. A sudden drop of the Knight shift occurs at 150 K, the transition temperature from the α to the γ phase. The light full curve represents the temperature dependence of the susceptibility normalised to the highest temperature Knight shift point. We notice that the Knight shift mirrors the susceptibility, except in the very vicinity of the transition temperature. Otherwise there is no difference between the α and the γ phase. This is even more evident in figure 4.20 where the Knight shift is plotted against the magnetic susceptibility, both in Ce and $Ce_{0.74}Th_{0.26}$. This result suggests that the 4f' electron is still localised in the α phase and largely responsible for the α phase magnetic susceptibility, so that the induced hyperfine field at the μ^+ per unpaired electron per atom remains unaltered. One might even argue that the actual spatial distribution of the 4f' electron must be the same in both phases, since otherwise, the exchange coupling constant J_{sf} should be different leading to different $B_{hf}^{\Omega_a}(r)$. (See also Wuilloud et al 1983.)

Figure 4.20 also includes results from measurements on the intermetallic compound, $CeSn_3$, which crystallises in the Cu_3Au structure (Wehr 1982). $CeSn_3$ is regarded as an intermediate valence system displaying valence fluctuations in a broad temperature range (Tsuchida and Wallace 1965). The magnetic susceptibility displays a broad maximum around 140 K. In contrast, the μ^+ Knight shift peaks at about 180 K. Evidently only the data above 180 K

160   *Hyperfine fields at the μ^+ in metals*

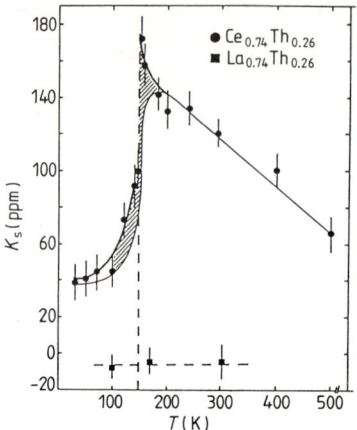

Figure 4.19 Temperature dependence of the μ^+ Knight shift in $Ce_{0.74}Th_{0.26}$ and $La_{0.74}Th_{0.26}$. The vertical broken line separates the α and the γ phases in $Ce_{0.74}Th_{0.26}$. The light full curve represents the magnetic susceptibility normalised to the μ^+ Knight shift at 500 K (Wehr et al 1982).

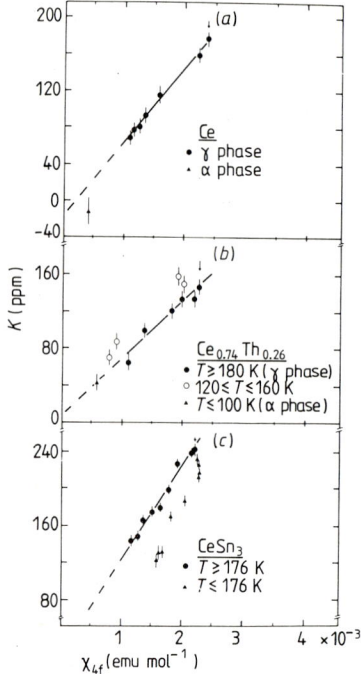

Figure 4.20 Plot of the μ^+ Knight shift in (a) Ce, (b) $Ce_{0.74}Th_{0.26}$ and (c) $CeSn_3$ against the $4f'$ electron magnetic susceptibility with temperature as an implicit parameter (Wehr et al 1981, 1984).

mirror the magnetic susceptibility, which in this temperature region displays a Curie behaviour. It is as yet not clear what causes the anomalous behaviour of the μ^+ Knight shift below 200 K.

4.5 μ^+ Knight Shift in Semi-metals

Measurements have been performed in As, Sb and Bi (Hartmann et al 1979). While in As and Bi no shift could be detected within an accuracy of ± 100 ppm, an abnormally huge shift was observed in monocrystalline Sb, which displayed a strong temperature and orientation dependence, the latter with respect to the rhombohedral symmetry axis (see figures 4.21 and 4.22). At low temperatures and with the applied field parallel to the c axis, the shift amounts to 1.5% which is outrageously large compared with the shifts found in simple and transition metals. It is clear that the shift cannot be an ordinary Knight shift but must be of a different origin. Considering that the conduction electron density in these semi-metals is several orders of magnitude smaller than in ordinary metals (e.g. As: 2×10^{20} e$^-$ cm^{-3}, Sb: 5×10^{19} e$^-$ cm^{-3}; Bi: 3×10^{17} e$^-$ cm^{-3}) one cannot exclude the possibility that a very shallow muonium state is formed (i.e. an electron bound very weakly to the μ^+) involving a local electronic magnetic moment and a corresponding hyperfine field. Such a local moment picture would not be in contradiction with the temperature dependence of the Knight shift which can be described by a Brillouin function with $J = \frac{1}{2}$ (or possibly $J = \frac{3}{2}$) and an exchange field of

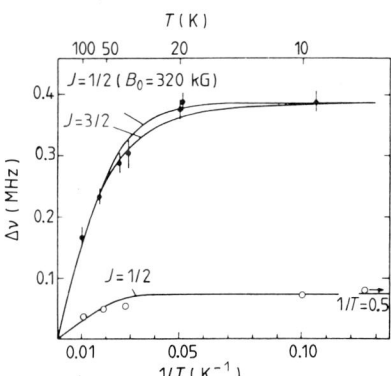

Figure 4.21 Temperature dependence of the frequency shift, Δv, of the Larmor frequency in monocrystalline Sb: ●, applied field of 0.2T; ○, applied field of 0.04T. The full curves are Brillouin functions fitted to the data (Hartmann et al 1979).

≃320 kG for the data obtained at 2 kG. The anisotropy in the frequency shift would imply a large fraction of p character of the local electronic state around the μ^+. On the other hand, the frequency shift rises linearly with the applied field up to 8 kG (see Schenck 1981) showing no tendency of a saturation effect, as would be expected from a local moment picture. Recent measurements by Brewer et al (1983) and Gygax et al (1984c), show that K_μ can be influenced dramatically by the admixture of a few atomic % Bi and Sn. In effect neither the anisotropy, nor the temperature and field dependence has yet found a convincing and consistent explanation. Very recent measurements in high purity monocrystalline Bi, between 10 K and 350 K, revealed a strongly temperature dependent and anisotropic μ^+ Knight shift which, in contrast to Sb is negative everywhere (Gygax et al 1984d).

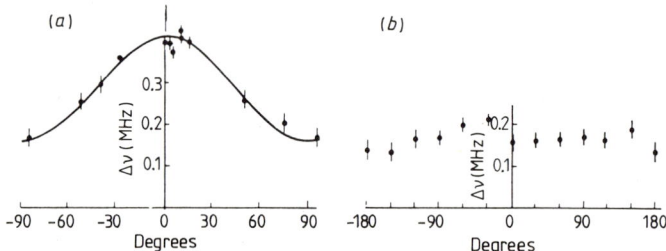

Figure 4.22 Angular dependence of the μ^+ frequency shift in monocrystalline Sb with applied fields as in figure 4.21. (a) B_{ext} in the xz plane, (b) B_{ext} in the xy plane (z axis = crystalline c axis) (Hartmann et al 1979).

4.6 Spontaneous Hyperfine and Dipole Fields in Ferromagnetic and Antiferromagnetic Materials

4.6.1 Contribution to the local field at a μ^+: the dipole field

In a very schematic way we discussed in §2.4 the various contributions to the local field at the μ^+ inside a magnetised sample. Here we would like to discuss the contributions for the specific case of a magnetically ordered system in zero applied field. First, spin polarised conduction electrons will produce a contact hyperfine field, as discussed in §4.1. Secondly, the ordered local moments will give rise to a superposition of dipole field components at the μ^+ site (r_μ). We therefore write

$$B_\mu = B_{hf}(r_\mu) + B_{dip}(r_\mu). \tag{4.45}$$

Following Meier et al (1978) (see also Denison et al 1979, Meier 1980) the ith

($i = x, y, z$) component of the dipole field may be expressed as

$$B_{\text{dip},i}(r_\mu) = \sum_n D_{ij}(n - r_\mu)\mu_j(n). \quad (4.46)$$

The sum extends over all contributions from magnetic moments $\mu(n)$ centred at sites n, and, with $x = n - r_\mu$

$$D_{ij}(x) = (\nabla_i \nabla_j - \tfrac{1}{3}\delta_{ij})x^{-1} - \tfrac{2}{3}\delta_{ij}\Delta x^{-1} \quad (4.47)$$

is the dipole field tensor with respect to each magnetic moment site. The second term in equation (4.47) can be omitted since it is presumed that the μ^+ will be restricted to the interstitial volume. First, we now consider the case of ferromagnetically ordered magnetic moments. Usually the summation in equation (4.46) is separated into two subsums. In the first subsum the summation extends over all moment positions inside a sphere (Lorentz sphere) around the μ^+ position, r_μ and the second subsum contains the rest (see §2.4). The second subsum, which extends up to the surface of the sample, yields the Lorentz field, $B_L = \tfrac{4}{3}\pi M_d$ and the demagnetisation field $B_{\text{DM}} = -(N)M_b$. M_d is the magnetisation of a domain which is supposed to be larger than the Lorentz sphere. N is the demagnetisation factor which depends on the sample shape, and M_b is the bulk magnetisation of the sample. Because of the multi-domain structure of most ferromagnets, the bulk magnetisation will be zero in the absence of an external magnetic field as will the demagnetisation field. The first subsum which depends on the actual microscopic arrangement of the magnetic moments around the μ^+ will be independent of the Lorentz sphere radius provided the Lorentz sphere is chosen large enough.

We can now write

$$B_\mu = B_{\text{hf}}(r_\mu) + B'_{\text{dip}}(r_\mu) + B_L \quad (4.48)$$

with

$$B'_{\text{dip},i}(r_\mu) = \sum_{n \in \Omega_{\text{Lorentz}}} D_{ij}(n - r_\mu)\mu_j(n) = E'_{ij}(r_\mu)\mu_j. \quad (4.49)$$

The dipole tensor E'_{ij} can be evaluated further by using the Ewald method (Meier et al 1978) and the following expression is obtained

$$E'_{ij}(r) = \sum_{n \in \Omega_L} F_{ij}(n - r) + \frac{N}{V} \sum_{g \neq 0} \tilde{D}_{ij}(g) \exp(-i\mathbf{g}\cdot\mathbf{r} - g^2/4G^2) \quad (4.50)$$

with

$$F_{ij}(x) = D_{ij}(x)[\text{erfc}(Gx) + \pi^{-1/2}(2Gx + \tfrac{4}{3}G^3 x^3)\exp(-G^2 x^2)] \quad (4.51)$$

and

$$\tilde{D}_{ij}(g) = -4\pi(g_i g_j - \tfrac{1}{3}\delta_{ij}g^2)/g^2. \quad (4.52)$$

The latter expression is the Fourier transform of $D(x)$, the $\mathbf{g}$ are reciprocal lattice vectors, N/V is the moment density and G a parameter, the appropriate choice of which assures rapid convergence of the sums in equation (4.50). By means of equations (4.50)–(4.52) it is thus possible to calculate the dipole field components at the μ^+ site for any direction of the ferromagnetically ordered magnetic moments, or the domain magnetisation.

It follows from equation (4.47) that $\mathbf{E}'(\mathbf{r}_\mu)$ is a symmetric and traceless tensor. This implies that, if the magnetic moments have an arrangement which is of cubic symmetry with respect to $\mathbf{r}_\mu$, the residual dipolar field will be zero. This is the case for both the tetrahedral and octahedral interstitial sites in a lattice with FCC structure.

In a metal with BCC crystal structure we again find tetrahedral and octahedral interstices (see figure 3.3). These interstitial sites possess tetragonal symmetry with the symmetry axis along either the $\langle 001 \rangle$ or $\langle 110 \rangle$ and $\langle 100 \rangle$ directions. Hence, of each type of interstice there exist three magnetically inequivalent configurations. With respect to the axis of tetragonal symmetry (the z axis), we have, in each case, the following non-vanishing components of $\mathbf{E}'(\mathbf{r}_\mu)$

$$E'_{11} = E'_{22} = -\tfrac{1}{2} E'_{33} \qquad (4.53)$$

with $E'_{33} = 21.414\ a^{-3}$ for the octahedral site, and $-6.073\ a^{-3}$ for the tetrahedral interstitial site, where a is the cubic lattice constant. Table 4.7 lists a number of dipole fields for different interstices and magnetisation directions in ferromagnetic BCC iron.

In an HCP lattice (see figure 3.4) the interstitial sites are also of the tetrahedral and octahedral configuration. The sites of each type are all magnetically equivalent. Again we have (c axis = z direction)

$$E'_{zz} = -2E'_{xx} = -2E'_{yy}. \qquad (4.54)$$

Table 4.7 Compilation of dipole fields at the centre of different interstitial sites, and for different directions of the saturation magnetisation, in BCC Fe (from Meier et al 1978).

Site	Tetragonal axis along	B_{dip}† for $M_s \parallel \langle 001 \rangle$ (kG)	B_{dip}‡ for $M_s \parallel \langle 111 \rangle$ (kG)
Octahedral	$\langle 001 \rangle$	+18.52	13.10
	$\langle 100 \rangle$	−9.26	13.10
	$\langle 010 \rangle$	−9.26	13.10
Tetrahedral	$\langle 001 \rangle$	−5.25	3.71
	$\langle 100 \rangle$	+2.63	3.71
	$\langle 010 \rangle$	+2.63	3.71

† B_{dip} is collinear with M_s with the sign convention + for B_{dip} parallel to M_s and − for B_{dip} antiparallel M_s.
‡ B_{dip} is perpendicular to M_s.

With very good accuracy E'_{zz} can be expressed in the range of $1.46 < c/a < 1.72$ by

$$E'_{zz}(\text{oct}) = [-18.288\,60 - 3.452\,20(c/a) + 22.891\,55(c/a)^2 \\ - 11.251\,32(c/a)^3 + 1.651\,86(c/a)^4]a^{-3} \quad (4.55)$$

and

$$E'_{zz}(\text{tet}) = [461.044\,97 - 1063.786\,37(c/a) + 907.145\,06(c/a)^2 \\ - 340.766\,42(c/a)^3 + 47.990\,15(c/a)^4]a^{-3}. \quad (4.56)$$

It is plotted in figure 4.23 as a function of c/a. Included in this figure is also E'_{zz} for a substitutional site.

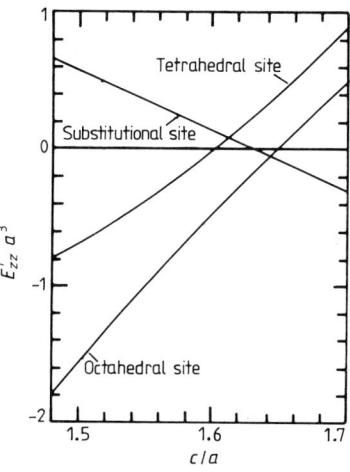

Figure 4.23 Dipolar field tensor component E'_{zz} as a function of the c/a ratio for various sites in the HCP lattice (equations (4.55) and (4.56)) (Meier *et al* 1980).

In a similar way the dipole field components at the μ^+ site can also be calculated for antiferromagnetic systems, or systems with a helical spin arrangement (see Meier *et al* 1978). The procedure will be somewhat different since the separation of the sum in equation (4.46) into a summation over the Lorentz sphere and a summation of the residual volume, will not be very meaningful.

An important question is how the dipole field contribution would change if the μ^+ were not confined to exactly the centre of the interstitial site $r_\mu = 0$, but were allowed to perform vibrations around this site. Indeed, zero-point motion of the μ^+ in the local potential well will always result in an extended wave function $\psi_\mu(r)$ of the μ^+. The average dipole field components seen by the μ^+ are calculated from the expression (Meier 1979, 1980)

$$\langle B_{\text{dip},i}\rangle = \int d^3r |\psi_\mu(r)|^2 B_{\text{dip},i}(n_\mu - r)$$

$$= \sum_n (\nabla_i \nabla_j - \tfrac{1}{3}\delta_{ij}) \int d^3r |\psi_\mu(r)|^2 \frac{1}{|n-r|} \mu_j. \qquad (4.57)$$

If the μ^+ is confined to the interstitial volume and does not overlap with the magnetic moments, that is $\psi_\mu(r) \equiv 0$ for $|r| > |n|$, we can write

$$\frac{1}{|n-r_\mu+r|} = \sum_{l=0}^{\infty} (2l+1) P_l(\cos\theta) \frac{r^l}{|n|^{l+1}}. \qquad (4.58)$$

If $\psi_\mu(r)$ is given by a spherically symmetric wave function, ($\simeq P_0$) the integral in equation (4.57) is non-zero only for the $l=0$ term in the sum of equation (4.58), and we obtain

$$\langle B_{\text{dip},i}\rangle = B_{\text{dip},i}(r_\mu = 0) \qquad (4.59)$$

that is, the same result as in the absence of a finite distribution of the μ^+. The same argument, by the way, also holds for the second moment of the dipolar field spread (see §2.3).

Returning now to equation (4.48), we notice that the three contributions will depend in a different way on the direction of the magnetic moments or the domain magnetisation. While the hyperfine field, $B_{\text{hf}}(r)$ is expected to be parallel or antiparallel to the domain magnetisation (at least for metals with a simple ferromagnetic structure), the dipole field has a more complex dependence on the magnetic moment direction, as follows from equation (4.49). The Lorentz field is always parallel to the domain magnetisation. The different dependences on the domain magnetisation direction can, in principle, be used to distinguish between $B_{\text{hf}}(r_\mu) + B_L$ and $B'_{\text{dip}}(r_\mu)$ by changing the magnetisation direction. This can be achieved by applying strong external magnetic fields, above saturation, in some arbitrary direction, or it becomes possible by a change of the domain magnetisation direction (the easy axis) with temperature as a result of anisotropy effects (see next section). Once the dipole field tensor has been determined, one has, in principle, also determined the interstitial μ^+ site, since the magnitude of the tensor components vary for different sites.

4.6.2 Metals and compounds

The first successful observation of muon spin rotation in a ferromagnet was achieved by the group of Kossler (Foy et al 1973) in ferromagnetic Ni and Fe. The first measurement in Co at room temperature was reported by Gurevich et al (1974) and the first measurement in Gd also by Gurevich et al (1975). Their work is summarised in Gurevich et al (1975). Very detailed measurements were reported by Graf et al (1976) in Co and Graf et al (1977) in Gd. The same group succeeded also in observing muon spin precession both in the ferromagnetic and antiferromagnetic phase of dysprosium (Hofmann et al 1978). Their work in Ni, Fe, Co, Gd and Dy is summarised in Denison et al (1979). Further measurements in Co and Gd were also reported by Nishida et al (1978). The only successful measurements in magnetic, inter-metallic compounds made so far, were performed in the cubic Laves phase systems GdFe$_2$ and YFe$_2$ (Graf et al 1981). In order to study the origin and the nature of the observed spontaneous hyperfine fields in more detail, Butz et al (1980) have measured the pressure dependence of the hyperfine field in ferromagnetic Ni and Fe. Another approach is to study the dependence of the hyperfine field on the concentration of non-magnetic atoms in dilute alloy systems such as FeAl (Stronach et al 1979), NiCu, NiCo (Kossler et al 1979) and NiCr, FeSi (Nishida et al 1979). Some of the main features are now discussed.

(1) Temperature and external field dependence of B_μ

Figure 4.24 displays the temperature dependence of $|B_\mu|$ in Ni, Fe, Co, Gd and Dy. Figure 4.25 shows the same for the cubic Laves phase compound GdFe$_2$ (Graf et al 1981). It is evident from these figures that the temperature dependence of $|B_\mu|$ may be quite complex. We will see in a moment that the complexity is the result of non-vanishing dipole fields at the μ^+ site. In those instances where no net dipole fields are present at the μ^+ site, for example, in Ni, because of its FCC structure, BCC Fe, because of rapid diffusion among interstitial sites (see §3.8.2), $|B_\mu|$ indeed shows a smooth dependence on temperature and closely follows the temperature dependence of the domain magnetisation. In GdFe$_2$ we also find $|B_\mu|$ behaving smoothly, which may be the result of either the μ^+ residing at an interstitial site with $B'_{\text{dip}} = 0$, or of rapid diffusion between different interstitial sites with $\langle B'_{\text{dip}} \rangle = 0$ (Graf et al 1981).

Co has an HCP structure below 690 K, and changes to an FCC structure above 690 K. This structural phase transition is clearly reflected in the behaviour of $|B_\mu|$ (figure 4.24(c)), which displays an abrupt change of almost a factor of two. Since $B'_{\text{dip}} = 0$ above 690 K, this change reflects the disappearance of B'_{dip}.

For all cases mentioned so far, B_μ is simply given by (in zero applied field)

$$B_\mu = B_{\text{hf}}(r_\mu) + \tfrac{4}{3}\pi M_{\text{d}}. \tag{4.60}$$

The situation is different in the ferromagnetic HCP metals Co and Gd. Here the ordered magnetic moments produce a unique dipole field at each type of interstitial site. This is, therefore, not modified by diffusion, provided that diffusion takes place only among interstitial sites of the same type. The total field at the μ^+ is now given by equation (4.48)

$$B_\mu = B_{\text{hf}}(r_\mu) + B'_{\text{dip}}(r_\mu) + \tfrac{4}{3}\pi M_{\text{d}}. \tag{4.61}$$

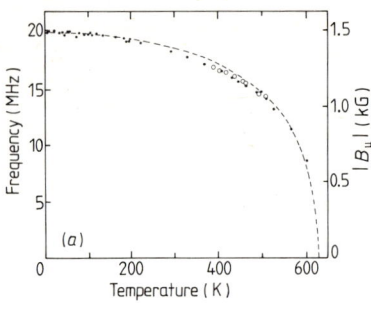
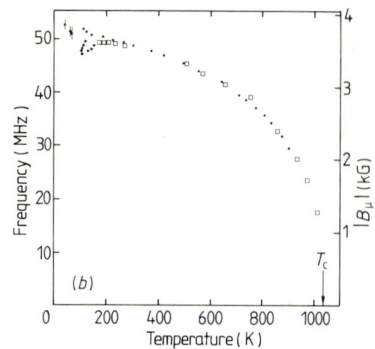
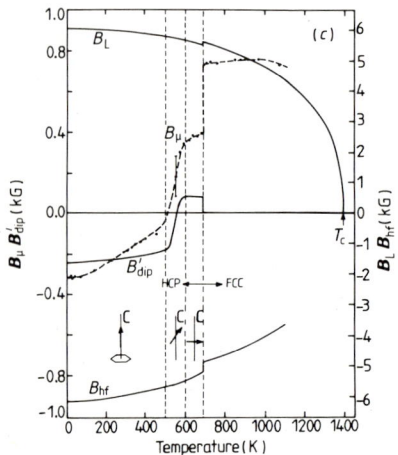
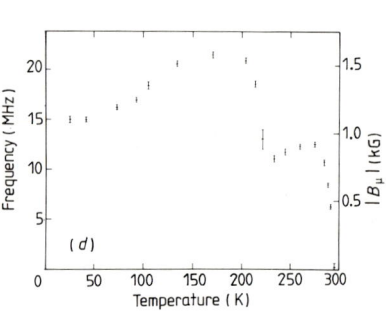
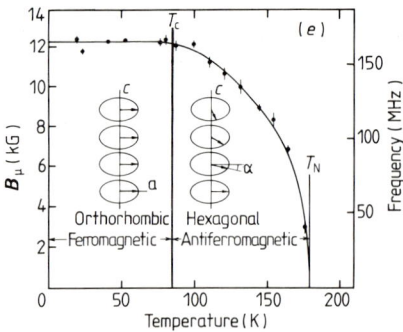

Figure 4.24 (*a*) Temperature dependence of the local field, $|B_\mu|$ ($\omega_\mu = \gamma_\mu B_\mu$) at the μ^+ in zero applied field in (*a*) ferromagnetic Ni, (*b*) ferromagnetic Fe, (*c*) ferromagnetic Co, (*d*) ferromagnetic Gd and (*e*) ferromagnetic ($T \leqslant 85\,\text{K}$) and antiferromagnetic ($85 \leqslant T \leqslant 178\,\text{K}$) Dy (Denison *et al* 1979).

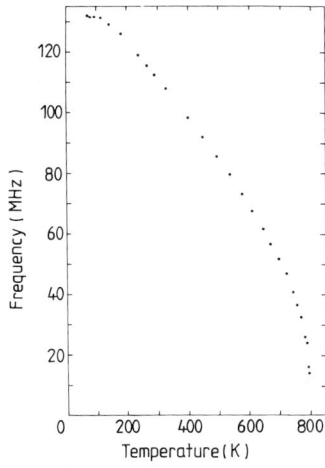

Figure 4.25 Temperature dependence of the local field, $|\boldsymbol{B}_\mu|$ (in frequency units), at the μ^+ in the ferromagnetic phase of the intermetallic compound GdFe$_2$ in zero applied field (Graf *et al* 1981).

$\boldsymbol{B}'_{\text{dip}}$ can be calculated according to the procedure described in §4.6.1. Since

$$|\boldsymbol{B}'_{\text{dip}}(\boldsymbol{r}_\mu)| = |(\mathbf{D})\boldsymbol{\mu}| \simeq M_{\text{d}}(T) \tag{4.62}$$

where $\boldsymbol{\mu}$ is the magnetic moment on one atom, $|\boldsymbol{B}'_{\text{dip}}(\boldsymbol{r}_\mu)|$ should display the same temperature dependence as $M_{\text{d}}(T)$, as long as the direction of $\boldsymbol{\mu}$ relative to the crystal axes is kept fixed. It is, however, well known that in Co and Gd the direction of $\boldsymbol{\mu}$ or $\boldsymbol{M}_{\text{d}}$ undergoes a change with temperature which is manifest macroscopically in a change of direction of the easy magnetisation axis.

In Co the easy axis is parallel to the hexagonal c-axis below 500 K. In the range from 500 to 600 K, however, the easy axis changes slowly from the c axis into the basal plane (Bertaut *et al* 1963). Using the temperature dependence of the easy axis angle with the c axis, from Bertaut *et al* (1963), $\boldsymbol{B}'_{\text{dip}}$ can be calculated. This is provided the type of interstitial site at which the μ^+ resides is known.

In Gd the easy axis of magnetisation is parallel to the c axis above 230 K. Below 230 K the easy axis deviates from the c axis, the angle between the two showing a complicated temperature dependence (see references in Denison *et al* 1979). Taking this change of the direction of the atomic magnetic moments with temperature into account in the calculations of $\boldsymbol{B}'_{\text{dip}}$, a consistent understanding of the complex temperature dependence of $|\boldsymbol{B}_\mu|$ in Co and Gd has been achieved.

Other important information is obtained from the dependence of $\boldsymbol{B}_\mu$ on an external field $\boldsymbol{B}_{\text{ext}}$. It allows us to determine, in particular, the relative sign of $\boldsymbol{B}_\mu$ and $\boldsymbol{M}_{\text{d}}$ for the case when $\boldsymbol{B}_\mu$ and $\boldsymbol{B}_{\text{ext}}$ are collinear. Figure 4.26 shows the results of such a measurement, using a spherical single crystal Ni sample (from Denison *et al* 1979). It is seen that $\boldsymbol{B}_\mu$ is independent of $\boldsymbol{B}_{\text{ext}}$ up to about 2050 G, at which the field corresponds to the saturation field, B_{s}, for the spherical sample at room temperature. As long as $B_{ext} < B_s$ the external field is effectively screened out by the demagnetisation field $NM(B_{ext})$ ($N = \frac{4}{3}\pi$), due to the high permeability of Ni, that is $B_{ext} = \frac{4}{3}\pi M$ below the saturation of the sample. M is the sample magnetisation which must be distinguished from the domain, or saturation, magnetisation M_d. For $B_{ext} > B_s$, $|B_\mu|$ rises linearly with B_{ext}, indicating that $B_\mu(B_{ext} = 0)$ and B_{ext} have the same direction. Since M is parallel to B_{ext} we find that B_μ parallels M_d.

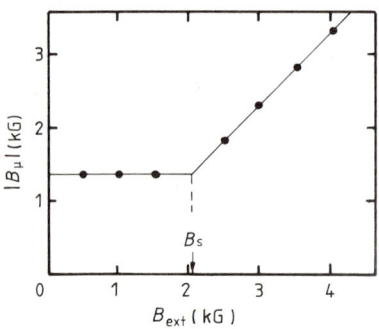

Figure 4.26 Dependence of the local field $|B_\mu|$ on the external field in a single crystal Ni sphere on 293 K, demonstrating that B_μ has the same direction as B_{ext} and hence the saturation magnetisation M_S (Denison et al 1979).

The relative direction of B_μ and M_d in the presence of an external field can also be obtained by measuring the relative phase of the μSR signals in two positron telescopes at different angles. It is not necessary that the applied field be larger than the saturation field. The applied field must be large enough so that the volume with domains with M_d parallel to B_{ext} is sufficiently different (larger) from the volume with domains with M_d antiparallel to B_{ext}. This guarantees that more μ^+ precess, for example left-handed rather than right-handed, so that a precession signal also becomes visible in a positron telescope perpendicular to the initial axis of polarisation, that is the μ^+ beam axis. The amplitude of this signal will be proportional to the sample magnetisation and the sense of precession will be determined with respect to the sample magnetisation or the applied field. This method was applied to determine the sign of B_μ in HCP Co (Denison et al 1979). Because of the non-zero dipole field at the interstitial site, this method requires the use of a single crystal. It was found that $B_\mu \cdot B_{ext} < 0$ at 293 K, $B_\mu \cdot B_{ext} > 0$ at 673 and 875 K. In the first measurement, B_{ext} was applied parallel to the c axis, and in the second it was oriented perpendicular to the c axis in order to leave the magnetic moments unaffected by the applied field and to establish the collinearity condition.

For arbitrary orientation of the applied field the situation is more complicated. One has to consider the interplay between the various contributions to the total magnetic energy in the system as there are the demagnetisation energy, the energy of a domain in an external field and, most crucial, the anisotropy energy. This, for Co, may be written as

$$E_{an} = K_1 \sin^2 \theta + K_2 \sin^4 \theta. \tag{4.63}$$

θ is the angle between the easy axis of magnetisation and the hexagonal c axis.

The theory of the magnetisation process in single crystal Co requires, below saturation, the existence of two domain orientations which are symmetrically disposed with respect to the basal plane (Birss and Martin 1975). Considering this and while determining the angle between the domain magnetisation and the c axis, in the presence of an external field, from minimising the total magnetic energy, Graf et al (1976) (see also Denison et al 1979) derived the following formula for $|B_\mu|$ (valid for a spherical sample)

$$|B_\mu|^2 = [\tfrac{4}{3}\pi M_d + B_{hf}(r_\mu) + \mu E'_{zz}]^2$$
$$+ \{[(2K_1 + 3\mu E'_{xx} M_d)/(2K_1 + \tfrac{4}{3}\pi M_d^2)]B_{ext} \sin \phi\}^2. \tag{4.64}$$

E'_{xx}, E'_{zz} are defined by equations (4.50), (4.54), (4.55) and (4.56). ϕ is the angle of B_{ext} with the c axis. K_1 is the anisotropy constant defined in equation (4.63). Terms coming from K_2 are neglected since $K_2 \ll K_1$ at room temperature, where the field dependence of B_μ has been studied. E'_{xx} and E'_{zz} depend on the μ^+ site, so there is a possibility of learning about the site through careful study of the field dependence of B_μ. As equation (4.64) shows, the sign of B_μ cannot be determined in this way. As in the case of Ni, Fe and GdF$_2$ this requires B_{ext} which exceed the saturation field, since only then is the collinearity condition automatically fulfilled (E' is a diagonal matrix for the systems in question). Figure 4.27 displays the measured field dependence of $|B_\mu|$ at room temperature for two different orientations of B_{ext} with respect to the c axis. The full curves indicated are fits of equation (4.64) to the data. Assuming a tetrahedral site assignment for the μ^+, the fit will yield $K_1 = 5.7(0.4) \times 10^6$ erg cm^{-3}, and $K_1 = 4.8(0.3) \times 10^6$ erg cm^{-3} for an octahedral site assignment. Only the latter value of K_1 is in agreement with other measurements.

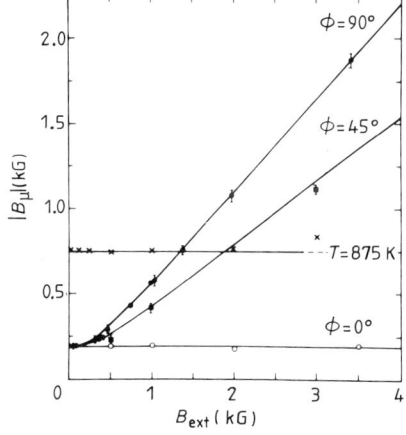

Figure 4.27 Dependence of the local field, $|B_\mu|$, on the external field, B_{ext}, in a single crystal Co sphere at room temperature. As can be seen, the local field depends not only on the magnitude of B_{ext} but also on its direction with respect to the crystalline c axis (ϕ = angle between B_{ext} and c axis) (Denison et al 1979).

The conclusion is, therefore, that the μ^+ in the HCP Co must reside at the octahedral interstitial site.

In the case of Gd, the sign of B_μ was determined from the field dependence of B_μ above saturation, using a polycrystalline sample. The measurements were done at two temperatures (130 K, 250 K) and yielded in each case a positive sign (Gurevich et al 1975). The same conclusion was reached by Graf et al (1977) by a consistent analysis of just the temperature dependence of $|B_\mu|$. In addition, they showed that only an octahedral site assignment of the μ^+ was compatible with the data.

A very special case is presented by dysprosium. Below 85 K, Dy has an orthorhombic crystal structure and orders itself ferromagnetically, with the spins lying along the orthorhombic a axis in the basal plane. Above 85 K, Dy is a helical antiferromagnet and the crystal structure is HCP. The spins lie again in the basal plane of the HCP lattice and the axis of the helix is parallel to the c axis. The helix angle between the ferromagnetically ordered planes is temperature dependent, and varies between 43° just below T_N (178 K) and 26° at 90 K (for references see Denison et al 1979). The dipole fields at the possible μ^+ sites were again calculated by the Ewald method and are shown in figure 4.28 as a function of temperature. Interestingly, the dipole fields change rather smoothly across the orthorhombic hexagonal phase transition. Also there is very little difference between them, and this renders a site determination impossible.

It is also important to realise that, in the antiferromagnetic state, the absolute value of the dipole field is the same everywhere for a given type of site,

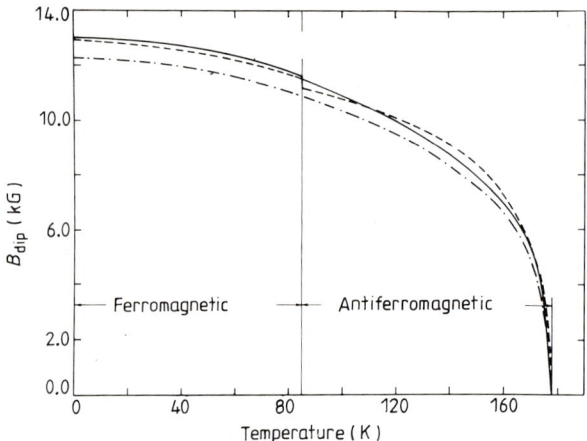

Figure 4.28 Calculated dipole field, B_{dip}, at the μ^+ in Dy for three different μ^+ sites, octahedral (full curve), tetrahedral (broken curve) and substitutional (chain curve). In the ferromagnetic phase the Lorentz field, B_L, is included in B_{dip} (Denison et al 1979).

Spontaneous fields in ferro- and antiferromagnetic materials 173

but the direction changes from one basal plane to the next basal plane, according to the helix angle. The same behaviour is also expected for $B_{hf}(r_\mu)$. If the μ^+ were diffusing in the Dy lattice, it would be exposed to local fields of varying directions which should lead to rapid depolarisation because of dephasing, again making the observation of a μSR signal impossible. If, conversely, diffusion is extremely rapid, the μ^+ would effectively average over all local fields, the average of which would, of course, be zero. So, again, no μSR signal should be visible in the absence of an external field.

The measurements of Hofmann et al (1978) in a polycrystalline sample show, however, that a signal can be observed over the full temperature range 85–178 K (see figure 4.24(e)). It shows a relatively fast, temperature independent, depolarisation of $\lambda \simeq 6.7\ \mu s^{-1}$. One concludes that the μ^+ must be immobile and well localised. The site was assumed to be the octahedral interstitial site, analogous to the findings in Co and Gd. The sign of B_μ has not been determined so far.

Table 4.8 compiles the $|B_\mu|$ results and its sign relative to M_d, as well as $|M_d|$ itself. All numbers are extrapolated to 0 K, except where noted.

(2) Hyperfine field at $T = 0$ K

Knowing the sign of B_μ with respect to M_d, and taking into account dipole fields, if necessary, one can extract the contact hyperfine field $B_{hf}(r_\mu)$ from B_μ (equation (4.60) or (4.61)). Values for $B_{hf}(r_\mu)$ extrapolated to 0 K are collected in table 4.9.

The interpretation of these values leads us back, once more, to the problem of the electronic structure of hydrogen in metals. The fact that the hyperfine field is not just caused by the undisturbed interstitial spin density, as measured

Table 4.8 Compilation of the saturation magnetisation, M_d, and the local fields, B_μ, at the μ^+ in the ferromagnetic systems investigated. All numbers are extrapolated to zero kelvin, except where noted. The sign of B_μ indicates its relative direction with respect to M_d, the domain magnetisation.

Compound	Crystal structure	M_d (kG)	B_μ (kG)	References
Fe	BCC	1.750	-3.67 ± 0.10	Denison et al (1979)
Co	HCP	1.415	-0.317 ± 0.010	
Ni	FCC	0.528	$+1.48 \pm 0.10$	
Gd	HCP	2.010	$+1.10 \pm 0.05$	
Dy	HCP	2.995	$+12.30 \pm 0.20$	
GdFe$_2$	C15	—	$+8.2$†	Graf et al (1981)
YFe$_2$	C15	—	-2.3†	

† At room temperature.

Table 4.9 Compilation of extracted hyperfine fields, B_{hf}, extrapolated to 0 K (except where noted), and comparison with predictions, B_{hf}^0, derived from neutron diffraction results, and with various other theoretical calculations. For the latter the number denotes the references respectively as follows: (1) Jena (1974), (2) Keller (1979), (3) Kanamori et al (1981), (4) Jepsen et al (1980), (5) Manninen and Nieminen (1981), (6) Rath et al (1979), (7) Estreicher et al (1982), (8) Katayama et al (1979). Oct = octahedral, Tet = tetrahedral.

Compound	μ^+ site	B_{hf} (kG)	B_{hf}^0 (kG)	B_{hf}^{theor} (kG) References
Fe	Tet	-11.1 ± 0.2	-1.32	-10.1 (2)
				-13.0 (3)
Co	Oct	-6.1 ± 0.2	-1.62	-5.7 (2)
Ni	Oct	-0.71 ± 0.01	-0.66	-0.6 (1), -0.68 (2)
				-0.72 (8),
				-0.463 (4), -2.2 (6)
Gd	Oct	-6.98 ± 0.10	-2.87	-8.7 (2), -6.0 (5)
				$-(4.0$–$6.0)$ (7)
Dy	Oct	-25.2 or		$-(7.3$–$13.7)$ (2)
		-0.7 (± 1.0)		$-(1.3$–$2.7)$ (5)
				$-(1.5$–$2.0)$ (7)
GdFe$_2$		$+5.9$†		
YFe$_2$		-4.5†		

† At room temperature.

by neutron scattering, can be seen by comparing the experimental $B_{hf}(r_\mu)$ with the hyperfine fields calculated from the undisturbed interstitial spin density or magnetisation $M^0(r_\mu)$ at the μ^+ position

$$B_{hf}^0(r_\mu) = \tfrac{8}{3}\pi M^0(r_\mu). \tag{4.65}$$

These values, using neutron scattering results, are shown in table 4.9. $B_{hf}(r_\mu)$ and $B_{hf}^0(r_\mu)$ are only of the same magnitude for Ni.

It was already mentioned in §4.4.2 that the μ^+ Knight shift in paramagnetic Ni and the μ^+ hyperfine field in ferromagnetic Ni lead to almost the same hyperfine fields per unit magnetisation. This demonstrates that, as far as the spin density at the μ^+ is concerned, it does not matter very much by which mechanism the d spins on the Ni sites become polarised. Rather, it is the strongly disturbed local electronic structure which again determines the sign and magnitude of the spontaneous hyperfine field at the μ^+ site. Nevertheless, the small difference in $B_{hf,d}^{\Omega a}$, as pointed out in §4.4.2 poses an interesting problem which has not yet been seriously investigated.

A compilation of various theoretical calculations of $B_{hf}(r_\mu)$ in various ferromagnetic metals is given in table 4.9. The calculations can be divided into two classes. In the first, the hyperfine field at the μ^+ is assumed to arise solely from the contact interaction with the conduction s electrons, which are treated as a free electron gas. The spatially inhomogeneous spin polarisation of the conduction electrons is the result of exchange interactions, for example of the RKKY-type interaction, with the local electronic moments due to d or f electrons. This approach is thought to be particularly applicable in rare earth metals, where the electronic moments are indeed well localised on the atomic sites. Although corresponding predictions for $B_{hf}(r_\mu)$ in Gd (Manninen et al 1981, Estreicher and Meier 1982) are of the right order of magnitude, the situation in Dy forbids any such conclusion (see table 4.9).

Table 4.10 Pressure dependence of domain magnetisation, M_d, and the compressibility of Fe and Ni at room temperature.

	$\partial(\ln M_d)/\partial p$ $(10^{-3}\,\text{kbar}^{-1})$	$\partial(\ln V)/\partial p$ $(10^{-3}\,\text{kbar}^{-1})$	References
Ni	-0.20	-0.54	Bloch and Pauthenet (1965) Gschneider (1964)
Fe	-0.28	-0.59	Kouvel and Wilson (1961)

The second calculation class involves microscopic cluster and band structure calculations from first principles. These give much more detailed information on the local electronic structure of the μ^+ metal complex and some of the features have already been discussed in §4.2. The spin density at the μ^+ is carried, in part, by local hybridised hydrogen 1s states and various (e.g. 3d) host states and not just conduction s electrons, as in the first class of calculation. As table 4.9 shows, predictions come quite close to the experimental numbers (Keller 1979, Kanamori et al 1981, Jepsen et al 1980). Again, Dy seems to present a problem. It would be extremely beneficial if an unambiguous experimental value of $B_{hf}(r_\mu)$ were available.

No attempts have so far been made to explain the hyperfine fields in GdFe$_2$ and YFe$_2$. It is, however, very interesting to note that GdFe$_2$ is the only example where the hyperfine field shows a positive sign (with respect to M_d). Graf et al (1981) have proposed that this is not in contradiction with the other results if one assumes that the μ^+ hyperfine field is mainly caused by the 3d Fe moments. In YFe$_2$, only the Fe atoms carry magnetic moments and the μ^+ hyperfine field is oriented antiparallel to the Fe moments. In GdFe$_2$ it is known that the Fe moments are oriented antiparallel to the Gd moments, which dominate the total magnetisation (i.e. the magnetisation is parallel to the Gd-moments). Consequently the positive sign of $B_{hf}(r_\mu)$ in GdFe$_2$ implies, again, an antiparallel orientation of $B_{hf}(r_\mu)$ with respect to the Fe moments.

(3) Temperature dependence of $B_{hf}(r_\mu)$

Figure 4.29 shows the temperature dependence of both $B_{hf}(r_\mu)$ and M_d in Ni and Fe. M_d is normalised to $B_{hf}(r_\mu)$ at $T=0$ K. It is evident that $B_{hf}(r_\mu)$ is not strictly proportional to M_d. Similar deviations have also been found for the hyperfine field at other non-magnetic impurities in ferromagnetic hosts. In Ni, the hyperfine field at the μ^+ decreases much slower with rising temperature than the magnetisation (below room temperature) while in Fe, and also in Co and Gd, the opposite is observed (Denison et al 1979).

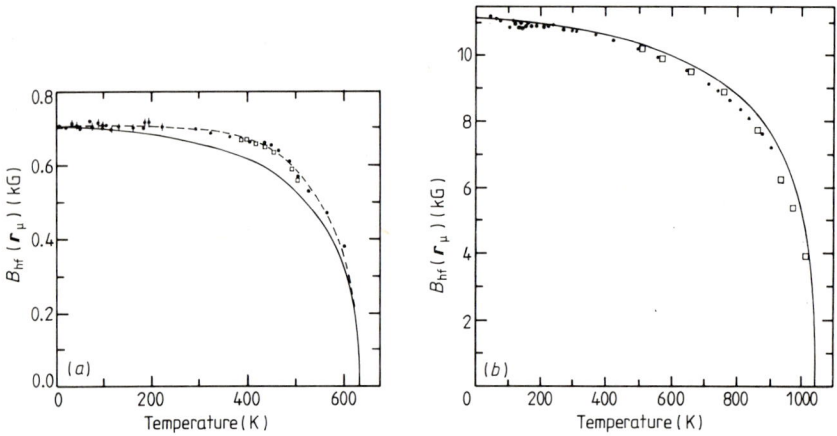

Figure 4.29 Temperature dependence of the extracted hyperfine field $B_{hf}(r_\mu)$ (broken curve, individual points), and the saturation or domain magnetisation, M_d (full curve), in (a) ferromagnetic Ni and (b) ferromagnetic Fe. M_d is normalised to $B_{hf}(r_\mu)$ at zero kelvin (Denison et al 1979).

Among the possible origins of these deviations the following effects have found particular attention:

(i) unperturbed interstitial spin density behaviour is different from the saturation magnetisation;
(ii) non-linear relationship between local spin density and the saturation magnetisation;
(iii) thermal volume expansion;
(iv) lattice vibration and zero-point motion of μ^+.

With respect to (i), Petzinger (1978) has argued that the decrease of exchange splitting of the Ni bands with rising temperature well below T_c will convert the majority s electrons to minority d electrons. Since, in his model (Petzinger and Munjal 1977) the undisturbed interstitial spin density is primarily caused by minority d electrons, this repopulation will keep the negative spin density at the μ^+ site almost unchanged, while the saturation magnetisation decreases. Another explanation of the temperature dependence of $B_{hf}(r_\mu)$ in Ni has been put forward by Kanamori *et al* (1981), where it is dominated by single particle (Stoner) excitations.

With respect to (ii), both Manninen *et al* (1981) and Meier (1978) find in their respective treatments a non-linear dependence between $B_{hf}(r_\mu)$ and M_d.

The effect of thermal volume expansion and the combined effect of lattice vibrations and the zero-point vibration of the μ^+ on $B_{hf}(r_\mu)$ has been considered by Manninen and Nieminen (1981) and Estreicher and Meier (1982, 1984) (see also Rath *et al* 1979, Meier *et al* 1979). The lattice vibrations were taken into account in the Debye–Waller approximation. The results of these calculations show that inclusion of lattice vibrations and the zero-point vibration of the μ^+ changes $B_{hf}(r_\mu)$ considerably from its static value for a point-like probe (see also discussion in §4.2). In general, the calculations provide a qualitatively correct description of the temperature dependence of $B_{hf}(r_\mu)$ (with the exception of Ni), although quantitatively this is still far from being a satisfying situation.

(4) Pressure dependence of $B_{hf}(r_\mu)$

The possibility that the deviation between the temperature dependence of $B_{hf}(r_\mu)$ and M_d originates from the different manner in which thermal volume changes affect the two quantities has been checked further by measurements of the pressure dependence of $B_{hf}(r_\mu)$. Figure 4.30 shows the results of such measurements at room temperature in Ni and Fe (Butz *et al* 1980). From the

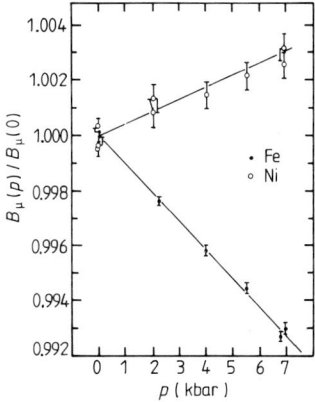

Figure 4.30 Pressure dependence of the local field B_μ at room temperature for ferromagnetic Ni and Fe. B_μ is normalised to unity at zero pressure (Butz *et al* 1980).

slopes one obtains

Fe: $\partial(\ln B_\mu)/\partial p = -1.05(2) \times 10^{-3}$ (kbar^{-1})

Ni: $\partial(\ln B_\mu)\partial p = +0.43(6) \times 10^{-3}$ (kbar^{-1}).

Since in Ni and Fe in zero applied field $B_\mu = B_{hf}(r_\mu) + \frac{4}{3}\pi M_d$, one needs to correct for the pressure dependence of M_d in order to determine $\partial(\ln B_{hf}(r_\mu))/\partial p$. Further, we can write

$$\frac{\partial[\ln B_{hf}(r_\mu)]}{\partial p} = \frac{\partial[\ln B_{hf}(r_\mu)]}{\partial V}\frac{\partial V}{\partial p}. \quad (4.66)$$

The quantities $\partial(\ln M_d)/\partial p$ and $\partial V/\partial p$, taken from the literature, are listed in table 4.10. The resulting volume dependence of $B_{hf}(r_\mu)$ is determined at room temperature as

Fe: $\partial[\ln B_{hf}(r_\mu)]/\partial(\ln V) = +0.92(1)$

Ni: $\partial[\ln B_{hf}(r_\mu)]/\partial(\ln V) = +2.7(2)$.

Comparing these numbers with the temperature dependence of $B_{hf}(r_\mu)$ and $M_d(T)$, one finds that the above number in Ni accounts for about 50% of the deviation between the temperature slope of $B_{hf}(r_\mu)$ and M_d at room temperature. However, in Fe the volume dependence is too small and, in the first place, has the wrong sign to bridge the gap between the behaviour of $B_{hf}(r_\mu)$ and M_d. These results clearly show that the temperature dependence of $B_{hf}(r_\mu)$ is not simply a result of the thermal volume expansion, but other effects, as discussed before, are likely to be very important. Estreicher and Meier (1982) have also calculated $\partial(\ln B_{hf})/\partial(\ln V)$, with the result +0.97 for Fe and +0.47 for Ni. The Fe result is well reproduced, while for Ni again their model seems to fail.

References

Adachi H, Imoto S, Tanabe T and Tsukada M 1978 *J. Phys. Soc. Japan* **44** 1039
Almbladh C O and von Barth U 1976 *Phys. Rev.* B **13** 3307
von Barth U and Hedin L 1972 *J. Phys. C: Solid State Phys.* **5** 1629
Bertaut E F, Delapalme A and Pauthenet R 1963 *Solid State Commun.* **1** 81
Birss R R and Martin J J 1975 *J. Phys. C: Solid State Phys.* **8** 189
Bloch D and Pauthenet R 1965 *J. Appl. Phys.* **36** 1229
Boon H M 1964 *Physica* **30** 1326
Brewer J H, Harshman D R, Koster E, Schilling H, Williams D L and Priestley M G 1983 *Solid State Commun.* **46** 863
Burdick G A 1963 *Phys. Rev.* **129** 138
Butz T, Chappert J, Dufresne J F, Hartmann O, Karlsson E, Lindgren B, Norlin L O, Podini P and Yaouanc A 1980 *Phys. Lett.* **75A** 321

Camani M, Gygax F N, Klempt E, Rüegg W, Schenck A, Schilling H, Schulze R and Wolf H 1979 *Phys. Rev. Lett.* **42** 679
Carter C C, Bennet L H and Kahan D J 1977 *Metallic Shifts in NMR, Progress in Materials Science* vol. 20 (Oxford: Pergamon)
Castro M, Keller J and Schilling H 1979 *Hyperfine Interactions* **6** 43
Das T P and Sondheimer E H 1960 *Phil. Mag.* **5** 529
Davies D W 1967 *The Theory of the Electric and Magnetic Properties of Molecules* (New York: Wiley)
Denison A B, Graf H, Kündig W and Meier P F 1979 *Helv. Phys. Acta* **52** 460
De Renzi R and Dallacasa V 1981 *Nuovo Cimento* B **61** 353
Estreicher S and Meier P F 1982 *Phys. Rev.* B **25** 297
—— 1984 *Hyperfine Interactions* **17–19** 327
Faulkner J S 1976 *Phys. Rev.* B **13** 2391
Foy M L G, Heimann N, Kossler W J and Stronach C E 1973 *Phys. Rev. Lett.* **30** 1064
Gelatt C D, Ehrenreich H and Weiss J A 1978 *Phys. Rev.* B **17** 1940
Graf H, Balzer G, Möslang T, Recknagel E, Weidinger A and Buschow K H J 1981 *Hyperfine Interactions* **8** 605
Graf H, Hofmann W, Kündig W, Meier P F, Patterson B D and Reichart W 1977 *Solid State Commun.* **23** 653
Graf H, Kündig W, Patterson B D, Reichart W, Roggwiller P, Camani M, Gygax F N, Rüegg W, Schenck A, Schilling H and Meier P F 1976 *Phys. Rev. Lett.* **37** 1644
Gschneider K 1964 *Solid State Phys.* **16** 275
Gunnarsson O and Lundquist B I 1976 *Phys. Rev.* B **13** 4274
Gupta M and Freeman A J 1978 *Phys. Rev.* B **17** 3029
Gurevich I I, Klimov A I, Maiorov V N, Meleshko E A, Muratova I A, Nikolskii B A, Roganov V S, Selivanov V I and Suetin V A 1974 *Zh. Eksp. Teor. Fiz.* **66** 374 (Engl. transl. 1975 *Sov. Phys.–JETP* **39** 178)
Gurevich I I, Klimov A I, Maiorov V N, Meleshko E A, Nikolskii B A, Selivanov V I and Suetin V A 1975 *Zh. Eksp Teor. Fiz.* **69** 439 (Engl. transl. 1976 *Sov. Phys.–JETP* **42** 222)
Gygax F N, Hintermann A, Rüegg W, Schenck A and Studer W 1981b *Solid State Commun.* **38** 1245
Gygax F N, Hintermann A, Rüegg W, Schenck A, Studer W, Schlapbach L and Stucki F 1981a *Recent Developments in Condensed Matter Physics* ed. J T DeVreese, L F Lemmerns, V E Van Doren and J Van Royen (New York: Plenum)
Gygax F N, Hintermann A, Rüegg W, Schenck A, Studer W and van der Wal A J 1984b *Hyperfine Interactions* **17–19** 377
Gygax F N, Hintermann A, Rüegg W, Schenck A and van der Wal A J 1984d *SIN Newsletter No* 16 (Swiss Institute for Nuclear Research, Villigen, Switzerland)
——1984a *J. Less-Common Metals* **101** 97
Gygax F N, Hintermann A, Rüegg W, Schenck A, Studer W, van der Wal A J and Schlapbach L 1983b *Hyperfine Interactions* **15–16** 533
Gygax F N, Hintermann A, Rüegg W, Schenck A, Studer W, van der Wal A J and Wehr H 1983a *Electronic Structure and Properties of Hydrogen in Metals* ed. P Jena and C B Satterthwaite (New York: Plenum) p 407
Gygax F N, Hintermann A, Schenck A, Studer W, van der Wal A J, Brewer J H, Harshman D R, Koster E, Schilling H, Williams D Ll and Priestley M G 1984c *Hyperfine Interactions* **17–19** 387

Gygax F N, Rüegg W, Schenck A, Schilling H, Studer W and Schulze R 1980 *J. Magn. Magn. Mater.* **15–18** 1191
Hartmann O, Karlsson E, Norlin L O, Pernestal K, Borghini M and Niinikoski T O 1979 *Hyperfine Interactions* **6** 47
Hintermann A and Manninen M 1983 *Phys. Rev.* B **27** 7262
Hofmann W, Kündig W, Meier P F, Patterson B D, Rüegg K, Echt O, Graf H, Recknagel E, Weidinger A and Wichert T 1978 *Phys. Lett.* **65A** 343
Hohenberg P and Kohn W 1964 *Phys. Rev.* **136** B864
van Hove L 1953 *Phys. Rev.* **89** 1189
Jaccarino V, Matthias B T, Peter M, Suhl H and Wernick J H 1960 *Phys. Rev. Lett.* **5** 251
Jena P 1974 *Solid State Commun.* **15** 1509
—— 1979 *Hyperfine Interactions* **6** 5
—— 1981 *Treatise on Materials Science and Technology* vol. 21 (New York: Academic) p 351
Jena P, Fradin F Y and Ellis D E 1979 *Phys. Rev.* B **20** 3543
Jena P and Singwi K S 1978 *Phys. Rev.* B **17** 3518
Jepsen O, Nieminen R M and Madsen J 1980 *Solid State Commun.* **39** 575
Kanamori J, Yoshida H R and Terakura K 1981 *Hyperfine Interactions* **8** 573
Kasowski R V 1969 *Phys. Rev.* **187** 891
Katayama H, Terakura K and Kanamori J 1979 *Solid State Commun.* **29** 431
Kazama S and Fukai Y 1977 *J. Less-Common Met.* **54** 25
Keller J 1979 *Hyperfine Interactions* **6** 15
Keller J and Schenck A 1979 *Hyperfine Interactions* **6** 39
Kittel C 1971 *Introduction to Solid State Physics* 4th edn (New York: Wiley)
Klein B M and Pickett W E 1984a *Phys. Rev.* **29** 1597
—— 1984b preprint, *Naval Research Laboratory, Washington DC*
Kohn W and Sham L J 1965 *Phys. Rev.* **140** A1133
Koskimaki D C and Gschneider K A 1978 *Handbook on the Physics and Chemistry of Rare Earths* ed. K A Gschneider and L. Eyring (Amsterdam: North Holland)
Kossler W J, Fiory A T, Lankford W F, Lynn K G, Minnich R P and Stronach C E 1979 *Hyperfine Interactions* **6** 93
Kouvel J S and Wilson R H 1961 *J. Appl. Phys.* **32** 435
Lawrence J M, Croft M C and Parks R D 1975 *Phys. Rev. Lett.* **35** 289
Mainwood A and Stoneham A M 1976 *J. Less-Common Met.* **49** 271
Manninen M 1983 *Phys. Rev.* B **27** 53
Manninen M and Jena P 1980 *Phys. Rev.* B **22** 2411
Manninen M and Nieminen R M 1979 *J. Phys. F: Met. Phys.* **9** 1333
—— 1981 *J. Phys. F: Met. Phys.* **11** 1213
McClure J W and Martyniuk J 1972 *Phys. Rev. Lett.* **29** 1095
Meier P F 1975 *Helv. Phys. Acta* **48** 227
—— 1978 *Solid State Commun.* **27** 1163
—— 1980 *Exotic Atoms '79* ed. K M Crowe, J Duclos and G Fiorentini (New York: Plenum)
Meier P F, Kündig W, Patterson B D and Rüegg K 1978 *Hyperfine Interactions* **5** 311
Meier P F, Kündig W and Rüegg K 1979 *Hyperfine Interactions* **6** 77
Munjal R L and Petzinger K G 1978 *Hyperfine Interactions* **4** 301
Nieminen R M 1981 *Hyperfine Interactions* **8** 437

Nishida N, Nagamine K, Hayano R S, Uemura Y J, Imazato J, Yamazaki T, Miyajima H, Chikazumi S, Fleming D G and Brewer J H 1979 *Hyperfine Interactions* **6** 87
Nishida N, Nagamine K, Hayano R S, Yamazaki T, Fleming D G, Duncan R A, Brewer J H, Ahktar A and Yasuoka H 1978 *J. Phys. Soc. Japan* **44** 1131
Papaconstantopoulos D A, Klein B M, Faulkner J S and Boyer L L 1978 *Phys. Rev. B* **18** 2784
Petzinger K G 1978 *Hyperfine Interactions* **4** 307
Petzinger K G and Munjal R L 1977 *Phys. Rev. B* **15** 1560
Rajagopal A K and Callaway J 1973 *Phys. Rev. B* **7** 1912
Rath J, Manninen M, Jena P and Wang C S 1979 *Solid State Commun.* **31** 1003
Rubens A, de Castro B and Schuhmacher R T 1973 *Phys. Rev. B* **7** 105
Schenck A 1981 *Hyperfine Interactions* **8** 445
—— 1982 *Helv. Phys. Acta* **54** 471
—— 1983 *Electronic Structure and Properties of Hydrogen in Metals* ed. P Jena and C B Satterthwaite (New York: Plenum)
Schilling H, Camani M, Gygax F N, Rüegg W and Schenck A 1982 *J. Phys. F: Met. Phys.* **12** 875
Segransan P J, Clark W G, Chabre Y and Carter G C 1976 *J. Phys. F: Met. Phys.* **6** 153
Shaham M, Barak J, El-Hanany E and Warren W W 1977 *Phys. Rev. Lett.* **39** 570
Stronach C E, Kossler W J, Lindemuth J, Petzinger K G, Fiory A T, Minnich R P, Lankford W F, Singh J J and Lynn K G 1979 *Phys. Rev. B* **20** 2315
Studer W 1983 *Doctoral thesis* Eidg. Tech. Hochsch. Zürich
Studer W, Gygax F N, Hintermann A, Rüegg W, Schenck A and van der Wal A J 1984b *Hyperfine Interactions* **17–19** 359
Studer W, Gygax F N, Hintermann A, Rüegg W, Schenck A, van der Wal A J and Wehr H 1984a *Hyperfine Interactions* **17–19** 299
Studer W, Gygax F N, Hintermann A, Schenck A and van der Wal A J 1984c *Phys. Rev. B* **30** 2414
Sugimoto H and Fukai Y 1980 *Phys. Rev. B* **22** 670
Switendick A C 1978 *Hydrogen in Metals I* ed. G. Alefeld and J Völkl *Topics in Applied Physics* vol. 28 (Berlin: Springer)
Teichler H 1977 *Phys. Lett.* **64A** 78
—— 1978 *Phys. Lett.* **67A** 313
Tsuchida T and Wallace W E 1965 *J. Chem. Phys.* **43** 3811
Wehr H 1982 *Doctoral thesis* University of Mainz
Wehr H, Knorr K, Gygax F N, Hintermann A, Schenck A and Studer W 1983 *J. Phys. F: Met. Phys.* **13** 885
Wehr H, Knorr K, Gygax F N, Schenck A and Studer W 1981 *Rev. Rev. B* **24** 4041
—— 1984 *Phys. Rev. B* **29** 6381
Wiley C L and Fradin F Y 1978 *Phys. Rev. B* **17** 3462
Winter J 1971 *Magnetic Resonance in Metals* (Oxford: Clarendon)
Wuilloud E, Moser H R, Schneider W-D and Baer Y 1983 *Phys. Rev. B* **28** 7354
Yasuoka H, Hayano R, Nishida N, Nagamine K, Yamazaki T and Ishikawa Y 1978 *Solid State Commun.* **26** 745
Yosida K 1957 *Phys. Rev.* **106** 893
Zaremba E, Sander L M, Shore H B and Rose J H 1977 *J. Phys. F: Met. Phys.* **7** 1763
Zaremba E and Zobin D 1980 *Phys. Rev. B* **22** 5490

5

Study of Dynamic Effects in Magnetism

5.1 Dynamic Effects and Critical Phenomena

Since μSR offers the possibility of measuring very short relaxation times, down to the order of 10 ns, the study of dynamical effects and critical phenomena close to phase transitions in magnetic systems appears particularly attractive. In a longitudinal field experiment one would be able to measure the spin–lattice relaxation rate, $1/T_1$, which should diverge on approach of the phase transition temperature (see e.g. Shaham *et al* 1977). From such measurements, a determination of dynamical critical exponents should become possible (see e.g. Halperin and Hohenberg 1969).

5.1.1 Spin fluctuation of itinerant electrons in MnSi

Despite such prospects, however, very little has actually been done. The only systematic μSR study of spin fluctuations has been performed in the ordered and paramagnetic phase of the weak itinerant helimagnet, MnSi (Hayano *et al* 1978, 1980, Takigawa *et al* 1980). The results for $1/T_1$ are displayed in figure 5.1. Clearly a divergent behaviour of $1/T_1$ at the transition temperature, $T_c = 29.5$ K, is visible. The data above T_c are shown once more in figure 5.2 and compared there with various theoretical predictions. The data are best reproduced by a temperature depencence of the form

$$1/T_1 \simeq (\gamma_\mu B_{hf}^\mu)^2 T/(T - T_c). \tag{5.1}$$

This also follows from a self consistent renormalisation (SCR) theory of electron spin fluctuations for a weak intinerant ferromagnet (Moriya 1977, Usami and Moriya 1980). B_{hf}^μ is the effective hyperfine field at the μ^+. Also shown are $1/T_1$ results from NMR measurements using the Mn nucleus, which have been rescaled by the factor, $F^2 = (\gamma_\mu B_{hf}^\mu / \gamma_{Mn} B_{hf}^{Mn})^2$ (Yasuoka *et al* 1978). The hyperfine fields, B_{hf}^μ and B_{hf}^{Mn} are taken from Knight shift measurements (Hayano *et al* 1980, Yasuoka *et al* 1978). These rescaled data, are nicely described by the ferromagnetic SCR prediction too. Other theoretical

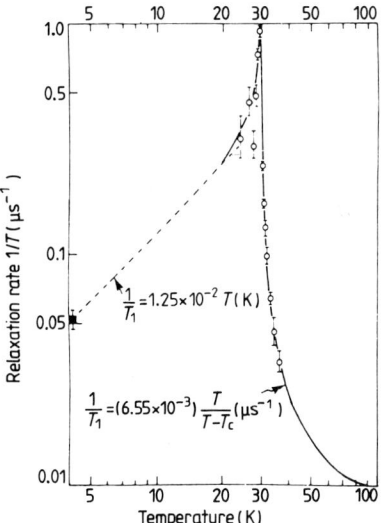

Figure 5.1 Temperature dependence of the μ^+ spin–lattice relaxation rate, $1/T_1$ in MnSi from longitudinal field measurements at 70 mT. The full curve is a fit of equation (5.1) to the data above $T_c = 29.5$ K (Hayano et al 1980). The broken curve is equation (5.2).

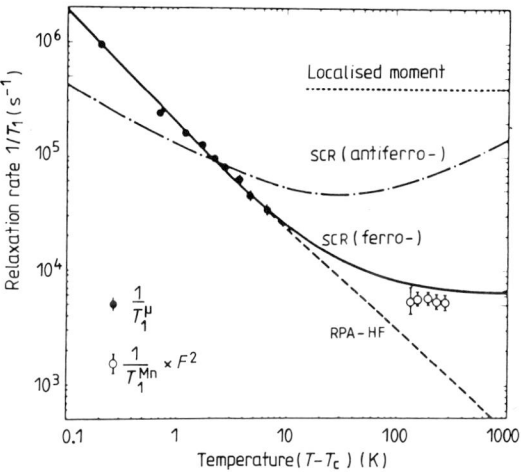

Figure 5.2 Plot of spin–lattice relaxation rates, $1/T_1$ from: ●, μSR and ○, NMR measurements in MnSi, against $(T - T_c)$. The NMR results have been rescaled with $F^2 = (\gamma_\mu B_{hf}^\mu / \gamma_{Mn} B_{hf}^{Mn})^2$ as described in the text. The various curves represent theoretical predictions (RPA–HF is the random phase approximation–Hartree Fock one) of which only the ferromagnetic SCR curve seems compatible with the data (Hayano et al 1978).

predictions (localised moment picture, RPA–Hartree Fock calculations) are clearly ruled out by the combined μSR and NMR data sets.

Below T_c the ferromagnetic SCR theory predicts a spin–lattice relaxation rate of

$$1/T_1 \propto T/M_0(T) \qquad (5.2)$$

where $M_0(T)$ is the uniform magnetisation. In the present case, $M_0(T)$ was taken to be $M_Q(T)$, the helical component of the magnetisation. Equation (5.2) provides a very good description of the relaxation rate up to about 20 K. Above this temperature the data indicate a change to an antiferromagnetic behaviour, possibly associated with the helical spin structure. The SCR prediction for weakly antiferromagnetic metals is

$$1/T_1 \propto T/M_Q(T).$$

The helical spin structure of MnSi displays a very long period of $\simeq 180$ Å in the $\langle 111 \rangle$ direction, associated with a small wave vector of $Q \simeq 0.035$ Å. The divergent behaviour of $1/T_1$ on approaching T_c from below is then explained as caused by spin fluctuations along Q in a small region around Q.

Critical phenomena have also recently been studied by μSR in Ni below and above T_c. Preliminary results on various critical exponents have been published by Nishiyama et al (1983).

5.2 Spin Glasses

5.2.1 μ^+ relaxation functions in zero, longitudinal and transverse fields

Spin glasses are more or less dilute spin systems, realised in metals and metal alloys containing randomly distributed magnetic ions, and in certain insulating compounds. A characteristic feature of such a system is an apparent freezing of the spin dynamics below a critical temperature, the so-called freezing or glass temperature, T_g, leading to a random but static order of the spin orientations. It was detected in 1972 by Cannella and Mydosh that the low field AC susceptibility showed a cusp like behaviour at T_g suggesting the possible occurrence of a phase transition. Ever since the nature of this phase transition has been the subject of controversial theories and also the interpretation of various experiments has met with considerable difficulties (for recent reviews see Blandin 1978, Murani 1978, 1981, Mydosh 1981, McLaughlin 1981). In this section we will discuss the effect of a random, dilute system of magnetic ions (spins), as present in a spin glass, on the μ^+ relaxation function. In the next section, some of the actual results obtained in the metallic spin glass systems CuMn, AuFe, AgMn by μSR will be described.

Let us consider a system of static, dilute and randomly orientated magnetic ions in some host lattice with the μ^+ occupying random interstitial lattice sites. The magnetic ions are sources of dipole fields and, in metals, sources of the so-called RKKY fields (see §4.1). According to Walstedt and Walker (1974), the local field components, $\tilde{H}_i$ ($i=x, y, z$) acting on the μ^+ show a Lorentzian distribution owing to the r^{-3} dependence of the dipolar and RKKY fields

$$P_L(\tilde{H}_i) = \frac{1}{\pi} \frac{\tilde{\sigma}}{\tilde{\sigma}^2 + \tilde{H}_i^2}. \tag{5.3}$$

This can be contrasted with the local field distribution stemming from host nuclear moments, as discussed in §2.3.1, which is Gaussian.

If a sufficiently strong external field ($B_{\text{ext}} \gg \tilde{H}$) is applied transverse to the initial μ^+ polarisation (e.g. $|B_{\text{ext}}| = B_z \gg \tilde{H}_z$), the resulting μ^+ precession will display an exponential relaxation function

$$G_x(t) = \exp(-\gamma_\mu \tilde{\sigma} t) = \exp(-\sigma t). \tag{5.4}$$

In zero external field the Kubo–Toyabe relaxation function follows from the expression (see equation (2.131))

$$G_z(t) = \tfrac{1}{3} + \tfrac{2}{3} \int \cos(\gamma_\mu \tilde{H} t) P_L(\tilde{H}) \, d\tilde{H} \tag{5.5}$$

$$= \tfrac{1}{3} + \tfrac{2}{3}(1 - \sigma t) e^{-\sigma t}. \tag{5.6}$$

Here

$$P_L(\tilde{H}) = \frac{4}{\pi} \frac{\tilde{\sigma} \tilde{H}^2}{\tilde{\sigma}^2 + \tilde{H}^2} \tag{5.7}$$

is the distribution function for the magnitude of the internal local fields, $\tilde{H}$, following from the Lorentzian distribution for the components $\tilde{H}_i$.

Finally, we quote the static Kubo–Toyabe relaxation function in a longitudinal applied field, B_{ext} (Uemura *et al* 1981b)

$$G_z(t, B_{\text{ext}}) = 1 - (\sigma/\omega_\mu) j_1(\omega_\mu t) \exp(-\sigma t)$$
$$- (\sigma/\omega_\mu)^2 [j_0(\omega_\mu t) \exp(-\sigma t) - 1]$$
$$- [1 + (\sigma/\omega_\mu)^2] \sigma \int_0^t j_0(\omega_\mu t') \exp(-\sigma t') \, dt' \tag{5.8}$$

where $\omega_\mu = \gamma_\mu B_{\text{ext}}$ and j_0 and j_1 denote spherical Bessel functions.

A static arrangement of spins has been assumed so far. We now allow the spins to fluctuate randomly. For a strict Lorentzian distribution of the local fields, no second moment can be defined, in which case no motional narrowing is produced by the spin fluctuations (Uemura 1980, Fiory 1981, Leon 1981, Emmerich 1982a). The transverse field relaxation function, equation (5.4), remains unchanged, while the steep decline at short times ($G_z(t) \simeq \exp(\frac{4}{3}\sigma t)$) of the zero field relaxation function is unchanged. The zero field relaxation function is damped like $\frac{1}{3}\exp(-\frac{2}{3}vt)$ for $v/\sigma \ll 1$ and like $\frac{1}{3}\exp(-\frac{4}{3}vt)$ for $v/\sigma \gg 1$ in the $\frac{1}{3}$ recovery range (i.e. $\sigma t \gg 1$). Here, v is the spin fluctuation rate (Uemura 1981a). Computer simulations by Fiory (1981) are in agreement with these findings.

The reason for the absence of a motional narrowing effect can be traced back to the high field tails of the Lorentzian distribution of local fields. These lead to an infinite second moment. Any distortion of the Lorentzian distribution which makes the second moment finite, will also produce motional narrowing (Emmerich 1982a). In reality there is always a break in the high field tail of the distribution, $P_L(\tilde{H}_i)$, and it will therefore always possess a finite second moment.

To facilitate motional narrowing, Uemura et al (1980) (see also Uemura 1980) postulate locally varying field distributions of a Gaussian form, $P_G(\tilde{H}_i, \tilde{\delta})$ with different widths, $\tilde{\delta}$, and with a weight distribution for $\tilde{\delta}$ of $\rho(\tilde{\delta})$ such that the bulk field distribution $P_L(\tilde{H}_i, \tilde{\sigma})$ is again Lorentzian with width $\tilde{\sigma}$

$$P_L(\tilde{H}_i, \tilde{\sigma}) = \int_0^\infty P_G(\tilde{H}_i, \tilde{\delta})\rho(\tilde{\delta}) \, d\tilde{\delta}. \tag{5.9}$$

$\rho(\tilde{\delta})$ is then determined to be

$$\rho(\tilde{\delta}) = \left(\frac{2}{\pi}\right)^{1/2} \frac{\tilde{\sigma}}{\tilde{\delta}^2} \exp(-\tilde{\sigma}^2/2\tilde{\delta}^2). \tag{5.10}$$

It is now easy to write down the transverse field relaxation function in the presence of spin fluctuations. For simplicity, we assume for the moment that the spin fluctuations are uncorrelated and can be described by a single rate, v, or a single correlation time $\tau_c = 1/v$. For μ^+, subject to the Gaussian field distribution $P_G(\tilde{H}_i, \tilde{\delta})$ the spin fluctuations can be treated as a Gaussian–Markovian process, analogous to the treatment of motional narrowing in §3.3.1. The resulting relaxation function is then given by expression equation (3.45)

$$g_x(t, \tau_c, \delta) = \exp\{-2\delta^2\tau_c^2[\exp(-t/\tau_c) + (t/\tau_c) - 1]\} \tag{5.11}$$

where $\delta = \gamma_\mu \tilde{\delta}$. The total relaxation function for all μ^+ is given by (Uemura 1980)

$$G_x(t) = \int_0^\infty g_x(t, \tau_c, \delta)\rho(\delta) \, d\delta. \tag{5.12}$$

For very fast fluctuations, $g_x(t, \tau_c, \delta)$ assumes the form of equation (3.46) in which case the integral equation (5.12) can be evaluated explicitly

$$G_x(t) = \left(\frac{2}{\pi}\right)^{1/2} \frac{\sigma}{\delta^2} \int_0^\infty \exp\left(-2\delta^2\tau_c t - \frac{\sigma^2}{2\delta^2}\right) d\delta$$

$$= \exp[-2\sigma(\tau_c t)^{1/2}]. \tag{5.13}$$

That is, rapid fluctuations produce an exponential decay, following the square root of time. This has been noted before by McHenry et al (1972) and was also found in recent computer simulations by Fiory (1981).

The zero field relaxation function can be calculated in the same way (Uemura 1980, Uemura et al 1980). The corresponding function to $g_x(t, \tau_c, \delta)$, $g_z(t, \tau_c, \delta)$ is calculated in the strong collision approximation, previously discussed in §§3.3.2 and 3.3.3 (Kehr et al 1978, Hayano et al 1979). In analogy to equation (5.12) we write

$$G_z(t, \tau_c, \sigma) = \int_0^\infty g_z(t, \tau_c, \delta)\rho(\delta)\,\mathrm{d}\delta \tag{5.14}$$

with

$$g_z(t, \tau_c, \delta) = \mathscr{L}^{-1}\{f_z(s)/[1 - \tau_c^{-1} f_z(s)]\} \tag{5.15}$$

where $\mathscr{L}^{-1}$ represents the inverse Laplace transform (see equation (3.57)) and

$$f_z(s) = \int_0^\infty g_z^{\mathrm{KT}}(t)\exp(-st)\,\mathrm{d}t \tag{5.16}$$

(see equation 3.55). $g_z^{\mathrm{KT}}(t)$ is the static Kubo–Toyabe function for a Gaussian field distribution, derived in §2.3.3. Figure 5.3 shows a plot of $G_z(t, \tau_c, \sigma)$ against time for various ratios of v/σ. For slow fluctuations, only the $\frac{1}{3}$ component is affected, and decays exponentially like (Uemura 1980)

$$G_z(t, \tau_c, \sigma) \simeq \tfrac{1}{3}\exp(-\tfrac{2}{3}vt). \tag{5.17}$$

For fast fluctuations ($v \gg \sigma$), equation (5.14) can be evaluated explicitly with the result

$$G_z(t, \tau_c, \sigma) = \exp[-2\sigma(\tau_c t)^{1/2}] \tag{5.18}$$

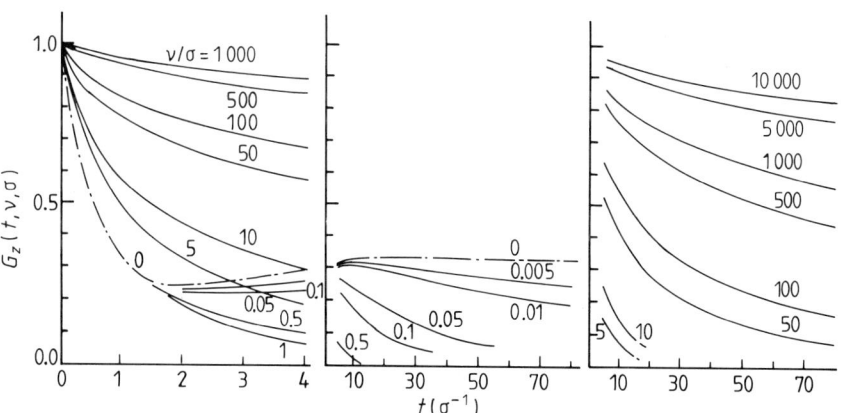

Figure 5.3 The zero field relaxation function, $G_z(t, v, \sigma)$ for various normalised fluctuation rates v/σ. Note that the time is measured in units of σ^{-1} (Uemura et al 1980).

as in the transverse field case. Figure 5.3 clearly shows the motional narrowing effect on the relaxation function.

Essentially, the same results were obtained by Leon (1981) applying a weak collision model (WCM) and Fiory (1981) by means of a computer simulation. The results of Fiory (1981) are interesting because they show that v and τ_c^{-1}, from equations (5.17) and (5.19), and the decay rate, λ, of a properly defined, field correlation function are not simply given by the spin fluctuation rate, v_s, entering into the simulation calculations. For uncorrelated fluctuations Fiory finds

$$\lambda/v_s = 1.13, \qquad \tau_c^{-1}/v_s = 1.32, \qquad v/v_s = 1.54.$$

Application of a sufficiently strong longitudinal external field B_{ext} ($B_{ext} > \tilde{H}$) will decouple the μ^+ spin from the random static local fields $\tilde{H}$ (see equation (5.8)), and in that limit the static $G_z(t, B_{ext})$ becomes time independent. In contrast, for rapidly fluctuating random fields, μ^+ depolarisation will proceed by a spin–lattice relaxation mechanism involving the non-secular terms of the dipolar interaction Hamiltonian, and resulting in an exponential relaxation function

$$G_z(t, B_{ext} > \tilde{H}, \tau_c) = \exp - (\Gamma_1 t)^{1/2}. \tag{5.19}$$

Γ_1 is the spin–lattice, or longitudinal relaxation rate, whose τ_c and B_{ext} dependence can be taken straight over from the NMR literature (see e.g. Abragam 1970):

$$\Gamma_1 = 1/T_1 = c\tau_c/(1 + \omega_\mu^2 \tau_c^2) \qquad \omega_\mu = \gamma_\mu B_{ext}. \tag{5.20}$$

The constant, c, is proportional to the second moment of the internal field distribution. For extremely rapid fluctuations (i.e. $\omega_\mu \tau_c \ll 1$), $1/T_1$ is independent of the applied field and simply proportional to τ_c. For applied fields with $B_{ext} \lesssim \tilde{H}$, and moderately fluctuating random spins, the general derivation of $G_z(t, B_{ext}, \sigma, \tau_c)$ follows the same procedure as the derivation of $G_z(t, B_{ext}=0, \sigma, \tau_c)$, (equation (5.14)), whereby $g_z(t, \tau_c, \delta)$ is replaced by equation (5.8).

So far, it has been assumed that the spin dynamics were characterised by a single fluctuation rate v, or correlation time τ_c, consistent with the assumption of an exponentially damped auto-correlation function for the spin, S, of the magnetic ions

$$\langle S(t)S(0)\rangle / \langle S(0)^2 \rangle = \exp(-vt). \tag{5.21}$$

This auto-correlation function has been measured directly by a neutron spin echo technique in CuMn (0.5 atomic %) (Mezei and Murani 1979). The results indicate that the form of equation (5.21) is not a correct one. In conjunction with other observations (Murani 1978), it must be concluded that a single correlation time model is inadequate. For the spin glass phase below T_g, another form for the auto-correlation function has been proposed by Edwards and Anderson (1975, 1976).

$$\langle S(t)S(0)\rangle/\langle S(0)^2\rangle = (1-Q_{EA})\exp(-vt) + Q_{EA} \quad (5.22)$$

where Q_{EA} is an 'order' parameter, which indicates the presence of some preferred direction for each individual spin. Uemura et al (1981b) has calculated the zero field relaxation function on the basis of equation (5.22).

For rapid fluctuations around the preferred direction, the zero field relaxation function assumes the analytical form (Uemura and Yamazaki 1983a)

$$G_z(t,v) = \tfrac{1}{3}\exp-(4a_d^2tv^{-1})^{1/2}$$
$$+\tfrac{2}{3}\left(1-\frac{a_s^2t^2}{(4a_d^2tv^{-1}+a_s^2t^2)^{1/2}}\right)\exp-(4a_d^2tv^{-1}+a_s^2t^2)^{1/2}. \quad (5.23)$$

This function describes the depolarisation of μ^+ spins in static random fields with spread a_s, and in rapidly fluctuating fields with a static spread a_d, where $a_s^2 + a_d^2 = \sigma^2$. σ is the width of the Lorentzian distribution previously introduced. The Edwards–Anderson parameter Q_{EA} is simply given by

$$Q_{EA} = a_s^2/(a_s^2 + a_d^2). \quad (5.24)$$

Two model approaches have been discussed with reference to the possibility of a wide distribution of correlation times. Emmerich (1982a) considers a model where, at each interstitial lattice site, the internal field correlation relaxes through different channels, characterised by different correlation times, τ_c. The resulting correlation function for the local field $\tilde{H}$ is then given by

$$\psi(t) = \frac{\langle \tilde{H}(t)\tilde{H}(0)\rangle}{\langle \tilde{H}(0)^2\rangle} = \int_0^\infty p(\tau_c)\exp(-t/\tau_c)\,d\tau_c \quad (5.25)$$

where $p(\tau_c)$ is a distribution function for τ_c. Assuming a Gaussian–Markovian description, the transverse field relaxation function follows immediately from equation (3.30)

$$G_x(t) = \exp\left(-2\sigma^2\int_0^t(t-\tau)\psi(\tau)\,d\tau\right). \quad (5.26)$$

Emmerich assumed $p(\tau_c)$ to have the form

$$p(\tau_c)\,d\tau_c = [(1-b)\delta(\tau_c - \tau_{min}) + bp'(\tau_c)]\,d\tau_c \quad (5.27)$$

with

$$p'(\tau_c) = \begin{cases} 1/[\tau_c\ln(\tau_{max}/\tau_{min})] & \tau_{min}\leq\tau_c\leq\tau_{max} \\ 0 & \tau_c<\tau_{min},\ \tau_c>\tau_{max} \end{cases}. \quad (5.28)$$

The form of this model reflects the fluctuation spectrum very schematically as evidenced by the neutron studies. τ_{min} is determined by the position of the Korringa peak in the spectrum, and this is accounted for by the δ function in equation (5.27) and τ_{max} is used as a fit parameter. Evaluating equation (5.26) on this basis, it turns out that the relaxation function shape is practically unchanged from the single correlation time model. The experiment will therefore not permit any distinction between these different models. Hence additional information from other experiments is required. It is, for example, possible to relate the AC susceptibility above T_g to the correlation function $\psi(t)$ when temperatures are not too high (Emmerich, 1982). Following Kubo and Tomita (1954) one can write

$$\frac{T\chi(\omega)}{\theta} = -\mathrm{Re}\left(\frac{1}{V}\int_0^\infty \frac{d\psi(\tau)}{d\tau}\exp(-i\omega\tau)\,d\tau\right). \tag{5.29}$$

For slowly varying $p(\tau_c)$ this can be approximated by

$$T\chi(\omega)/\theta = 1 - \psi(1/\omega). \tag{5.30}$$

In contrast to Emmerich, Uemura (1981a) discusses the possibility that the distribution of τ_c values is of spatial origin which is caused by inhomogeneous samples. Assuming a Gaussian distribution of $\ln v$, $P(\ln v)$ the zero field relaxation function is calculated from

$$G_z^{\mathrm{inh}}(t,\sigma) = \int G_z(t,v,\sigma)P(\ln v) \tag{5.31}$$

where $G_z(t, v, \sigma)$ is given by equation (5.14).

We finish this section by comparing the accessible fluctuation or time range from μSR with the other available experimental methods (see figure 5.4, Uemura

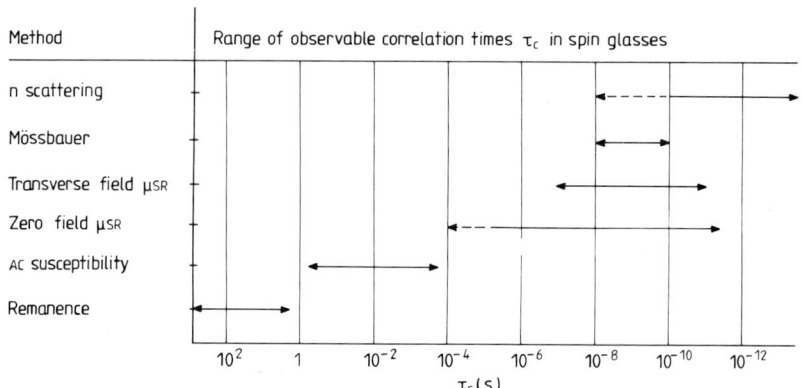

Figure 5.4 Comparison of the range of experimentally accessible spin correlation times, τ_c, by various methods (Uemura 1981a).

(1981a)). It can be seen that μSR, and in particular zero field μSR, is bridging the gap between Mössbauer and neutron spectroscopy on the fast fluctuation side, and AC susceptibility measurements on the slow fluctuation side. This strongly underlines the fact that μSR is well suited to playing an important role in the study of spin glass dynamics.

5.2.2 Short review of μSR results in CuMn, Fe and AgMn

The first μSR study of spin glasses was performed by Murnick et al (1976) using the systems CuMn and AuFe. These transverse field measurements revealed a drastic increase of the relaxation rate whilst approaching and crossing the spin freezing temperature, T_g. In addition, it was found that the observed relaxation rates depended significantly on the applied field strength, B_{ext}, increasing proportionately with it. The limited accuracy of the data did not allow further details of the transition into the spin glass phase to be studied. Since this first experiment, many more studies have been undertaken, applying mainly zero and longitudinal field techniques, and extending the measurements to other spin glass systems. These include the insulators $MnO \cdot Al_2O_3 \cdot SiO_2$ (Bieman et al 1978), $(CoO)_{40} \cdot (Al_2O_3) \cdot (SiO_2)_{50}$ (Uemura et al 1981a) and $(FeTiO_3)_{88} \cdot (Fe_2O_3)_{12}$ (Uemura et al 1983b). In this review however, we will restrict ourselves to a presentation of the results in the 'archetypal' dilute spin glass alloys CuMn, FeAu and AgMn.

Transverse field measurements were performed in CuMn by Emmerich and Schwink (1981) and in AgMn by Brown et al (1981). In agreement with Murnick et al (1976), the μ^+ relaxation rate increased rapidly, but smoothly, as the temperature was lowered across T_g. In CuMn the relaxation rate is also influenced by the Cu nuclear magnetic moments. In the analysis of the μSR spectra obtained in CuMn (Emmerich and Schwinck 1981), an exponential relaxation was assumed with a damping constant

$$\lambda(T) = \frac{1}{T_2(T)} + \sigma_{Cu}. \quad (5.32)$$

$T_2(T)$ is the contribution from the Mn electronic moments and σ_{Cu} is identified with the static inhomogeneous linewidth, originating from the spread in nuclear dipole fields at the μ^+. The latter quantity is kept fixed in the analysis, since from μ^+ diffusion studies in moderately pure Cu (see §3.8), σ_{Cu} is known to be temperature independent below 80 K. Figure 5.5 displays the $T_2(T)^{-1}$ results from various CuMn alloys (Emmerich 1982b). Although the analysis with the ansatz equation (5.32) appears quite crude, (the exact analysis would have implied a relaxation function of the form $P(t) = \exp(-\sigma_{Cu}^2 t^2) G_x(t)$, equation (5.4) or (5.13)) the behaviour of $T_2(T)$ in figure 5.5 provides a good illustration of the overall trends in the relaxation rate induced by the Mn local electronic moments. One first notes that T_2^{-1} keeps increasing slightly below

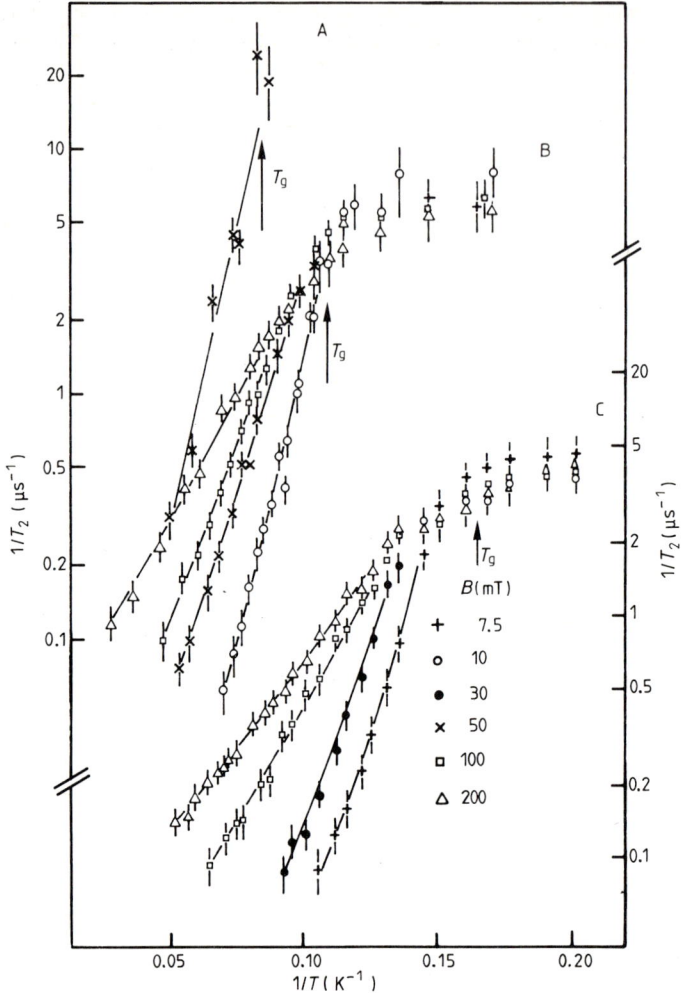

Figure 5.5 Log–log plot of the relaxation rate, $1/T_2$, induced by the Mn moments, against reciprocal temperature in CuMn for various fields, B, and several Mn atomic % concentrations: A, 1.48; B, 0.87; C, 0.54. The fields are all applied in the [100] direction. The arrows indicate the positions of the respective spin freezing temperature T_g (Emmerich 1982b).

T_g until it seems to saturate at temperatures well below T_g. Figure 5.5 further reveals that T_2^{-1} is strongly field dependent above T_g, but appears to be field independent below T_g. The same behaviour is inferred from the measurements in AgMn (Brown *et al* 1981). The saturation value of T_2^{-1} well below T_g is consistent with the field spread from the randomly, frozen spins of the Mn ions, if it was only the dipole–dipole interaction being considered. Obviously the RKKY fields do not contribute significantly to the field spread at the μ^+. Brown et al (1981) have estimated the field spread induced by the applied field above T_g, via the susceptibility, using the expression

$$\lambda = (1/T_2) = (8\pi^2/9\sqrt{3})\rho c \gamma_\mu g \mu_B \langle S_z \rangle \tag{5.33}$$

where

$$\langle S_z \rangle = g\mu_B \tfrac{1}{3} S(S+1) B_{ext}/k_B T \tag{5.34}$$

is the induced static spin component in the direction of $\boldsymbol{B}_{ext}$, ρ is the host site density and c is the magnetic ion concentration. For AgMn one obtains

$$\lambda T/B_{ext} c = 2.9 \times 10^{-3} (\text{K } \mu\text{s}^{-1} \text{ Oe}^{-1}).$$

This value is compared with the corresponding experimental numbers for various applied fields for a AgMn (1.6 atomic %) sample in figure 5.6. As can be seen, the calculated value is below the experimental numbers everywhere, even at temperatures far above T_g ($\simeq 5 T_g$). The experimental scaled relaxation rate, $\lambda T/B_{ext}$, also displays a significant field dependence everywhere. These results have been taken as evidence for a field induced static line broadening above T_g (Uemura and Yamazaki 1983a).

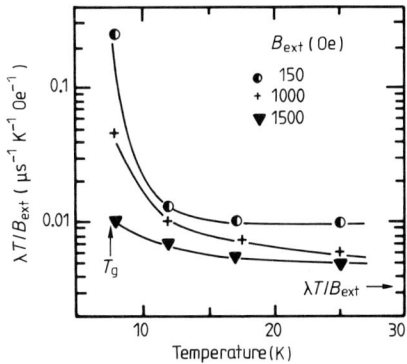

Figure 5.6 Temperature dependence of the scaled relaxation rate, $\lambda T/B_{ext}$ in AgMn (1.6 atomic %) for various B_{ext}. The arrow in the right hand lower corner indicates the expected value induced by the applied field via the Mn spin susceptibility (Brown et al 1981).

The question of static versus dynamic broadening cannot be answered unambiguously using transverse field experiments alone, since T_2 may originate from inhomogeneous line broadening as well as dynamical effects (i.e. homogeneous line broadening). Here zero and longitudinal field experiments are a great advantage, since they allow us to distinguish between static and dynamical relaxation mechanisms (see §3.3.2). Zero field measurements have the additional advantage that the spin glass dynamics can be studied in the absence of any disturbing field. It is well known that even very small fields tend to destroy the sharpness of the AC susceptibility cusp which marks the position of the transition into the spin glass phase.

The first zero field study of spin glasses was performed by Uemura et al (1980) in CuMn and AuFe (see also Uemura and Yamazaki 1982). Later, zero field measurements were also carried out by Heffner et al (1982) in AgMn (see figure 5.7 as an example).

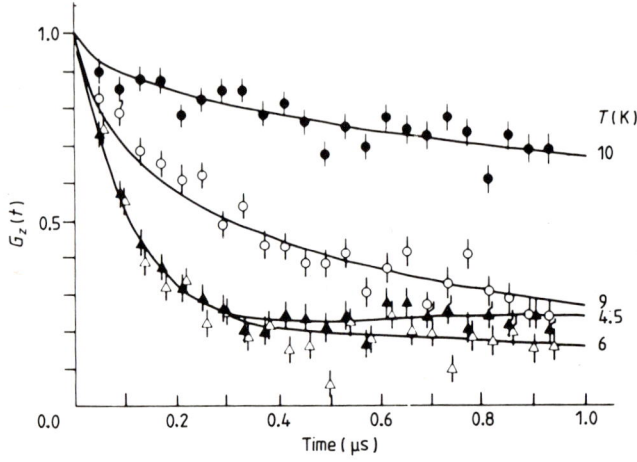

Figure 5.7 Examples of the zero field relaxation behaviour of μ^+ in AuFe (1.1 atomic %) for temperatures below and above T_g ($T_g = 9.1$ K). The full curves represent fits of equation (5.14) to the data (Uemura et al 1980).

The zero field μSR spectra were analysed by fitting equation (5.14), numerically evaluated in tabular form, to the data and hence yielding the parameters σ and τ_c. In this analysis, σ was kept constant at the value, obtained from the quasi-static Kubo–Toyabe signal, for temperatures well below T_g. The temperature dependence of τ_c, multiplied with the corresponding transition temperature T_g, displays a universal behaviour when plotted against T/T_g, as can be seen in figure 5.8. This is to be expected for a spin glass system in which the electronic moments mainly couple via the RKKY exchange mechanism (Souletie and Tournier 1971). Figure 5.8 shows a rapid change of τ_c, by about 5 orders of magnitude, between $T/T_g = 0.6$ and 1.4. Although quite rapid, the change of τ_c around T is nevertheless a smooth one and no discontinuous behaviour is indicated. Uemura et al (1980) find that, on the high temperature side, τ_c is well represented by the power law expression

$$\tau_c \propto [T/(T - T_g)]^2 \tag{5.35}$$

indicating a critical slowing down of the spin dynamics near T_g.

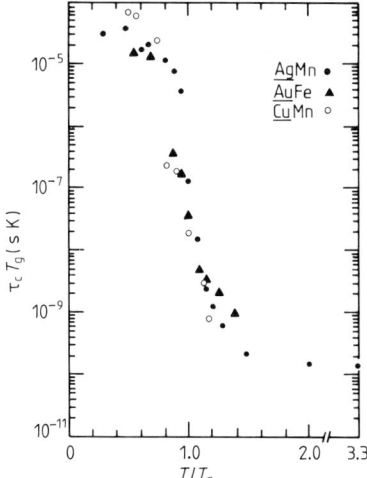

Figure 5.8 Scaled correlation time, $\tau_c T_g$ against scaled temperature, T/T_g, as obtained in AuFe (1.4 atomic %), CuMn (1.1 atomic %) and AgMn (1.6 atomic %) from zero field measurements. τ_c is extracted from the μSR spectra on the basis of the single correlation time model (equation (5.14)) (Uemura et al 1980, Heffner et al 1982).

To investigate the dynamic and static aspects of the spin glass phase transition further, Uemura et al (1981b) performed longitudinal field measurements between 0 and 64 mT. As mentioned in §5.2.1, sufficiently strong longitudinal fields allow the μ^+ spin to be decoupled from the static random field components. The results obtained in a CuMn (1.1 atomic %) ($T_g = 10.8$ K) sample are displayed in figure 5.9. For temperatures $T > 11$ K ($> T_g$) one finds that the relaxation function $G_z(t)$ is practically independent of the field. This implies that the observed relaxation is caused by rapidly fluctuating random fields, where $G_z(t)$ is given by equation (5.19) and $\Gamma_1 \propto \tau_c$ (equation (5.20)), that is, the relaxation is of purely dynamical origin. However, at 10 K, just below T_g, we notice an appreciable field dependence which is surprising, as the zero field results at this temperature yield a $\tau_c^{-1} \gg \sigma$, leading us back to equation (5.19). Taking the field dependence of Γ_1 (equation (5.20)) into account, and using the τ_c value originating from the zero field results at 10 K, one calculates the longitudinal relaxation function for 64 mT indicated by the upper full curve in figure 5.9. This, as can be seen, is not very different from the zero field curve, but quite in contrast to the data. At lower temperatures a static behaviour is clearly in evidence as shown by the pronounced field dependence, and is consistent with the zero field data which imply $\tau_c^{-1} \ll \sigma$. In fact, the full curves in figure 5.9 at 5 and 8 K were calculated on the basis of the zero field result. The strange behaviour at 10 K seems to indicate the presence of both static random field components and rapidly fluctuating components.

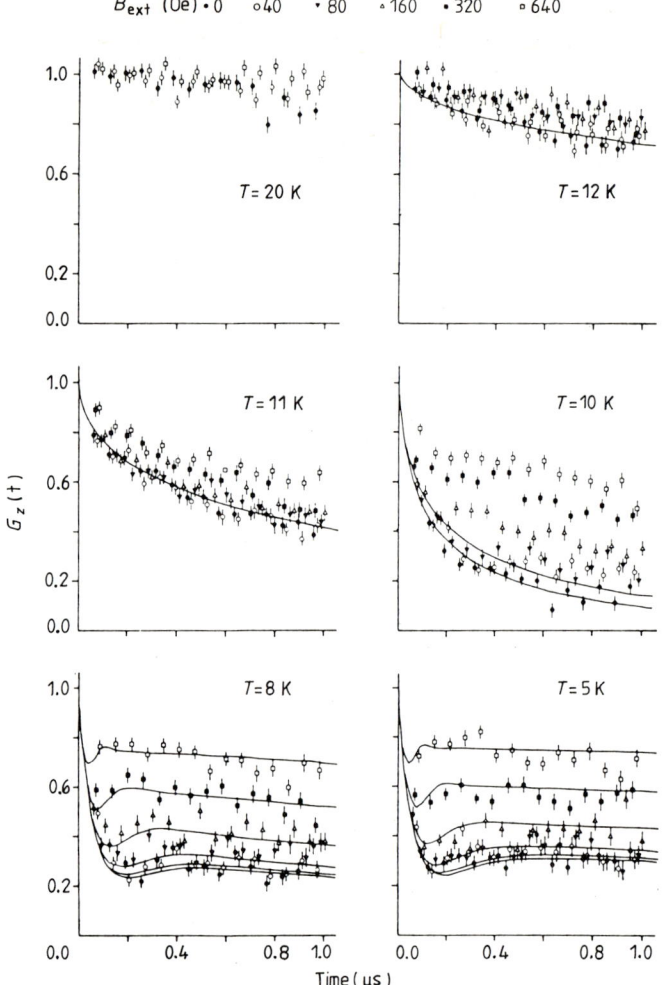

Figure 5.9 Muon spin relaxation behaviour in CuMn (1.1 atomic %) in longitudinally applied fields, $B_{ext} = 0$–64 mT ($\boldsymbol{B}_{ext} \parallel \boldsymbol{P}_\mu(0)$). The full curves represent zero and longitudinal field relaxation functions for the single τ_c model. The zero field relaxation functions equation (5.14) shown are fits to the data, from which τ_c was obtained. The longitudinal field relaxation functions are calculated by using this fitted zero field τ_c (Uemura et al 1981b).

This possibility complies with the Edwards–Anderson model, in which it is postulated that below T_g each impurity spin has its own preferred direction. Restricted fluctuations may occur around this direction such that the time average of the spin component along this direction assumes a non-zero value.

Indeed, Heffner et al (1982) and Uemura and Yamazaki (1983a) have found from a more detailed analysis that $(\sigma = a_s)$ decreases with temperature from a maximum value σ_0 towards zero at T_g. This maximum corresponds to a completely static and random orientation of the impurity spins. The data analysis of Uemura and Yamazaki (1984) proceeded on the basis of equation (5.23), involving the parameters a_s and a_d with $a_s^2 + a_d^2 = \sigma_0^2$. Figure 5.10 shows some recent results for $a_s(T)$ in CuMn for various Mn concentrations (Uemura et al 1984). As can be seen, a_s, the static width of the field distribution, drops to zero above T_g. The analysis also allows us to extract the relaxation rate, $\lambda_d = 4a_d^2 \tau_c$, associated with the dynamic amplitude, a_d. The results of Uemura et al (1984) are shown in figure 5.11. We find that λ_d increases rapidly as T_g is approached both from higher and lower temperatures. Above T_g where $a_d = \sigma_0$ the temperature dependence of λ_d originates solely from τ_c. Below T_g the increase of λ_d is mainly a result of the increase of a_d, fixing τ_c at roughly 10^{-10} s.

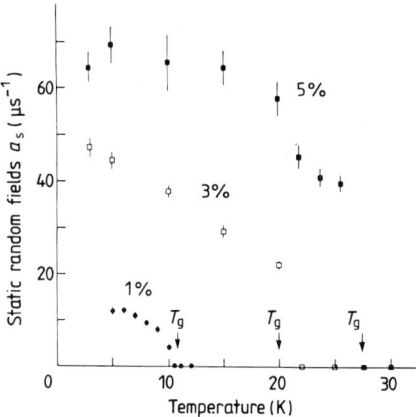
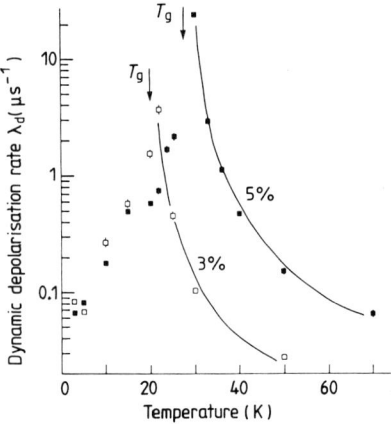

Figure 5.10 Temperature dependence of the static component, a_s, of the Mn-moment induced field spread at the μ^+ in CuMn for various Mn concentrations. The amplitude a_s was obtained by fitting equation (5.23) to zero field data (Uemura et al 1984).

Figure 5.11 Temperature dependence of the dynamic μ^+ relaxation rate, λ_d, associated with the fluctuating component of the Mn spins, from zero field measurements in CuMn. The full curves represent power laws of the form $\lambda_d \propto [T/(T-T_g)]^z$, with $z = 2.9$ for CuMn (5 atomic %), 2.6 for CuMn (3 atomic %). (Uemura et al 1984).

McLaughlin et al (1983) used longitudinal field studies to investigate the field dependence of λ_d in AgMn. It was found that $\lambda_d \propto B_{\text{ext}}^{\nu-1}$ with $\nu = 0.5$ for a wide range of fields, various Mn concentrations and with $0.3 < T/T_g < 0.7$. At $T = 0.9 T_g$ the exponent, ν, dropped to $\nu = 0.24$. The latter value is in disagreement with certain dynamic theories which predict $v(T_g)=\frac{1}{2}$ (see McLaughlin et al (1983) for a discussion and references). The field dependence of λ_d displays a scaling law when $\lambda_d T_g/x^2$ is plotted versus B_{ext}/T_g. x denotes the Mn concentration. The corresponding isotherms are displayed in figure 5.12. Such a scaling behaviour is expected for dilute spin glasses, where the dipolar coupling of the μ^+ spin (which enters the relaxation rate quadratically: $\lambda \propto \langle H^2_{dip} \rangle$) scales with x and spin glass energies scale with T_g. The power law dependence of λ_d on B_{ext} provides evidence for a power law governing the decay of the spin correlation function below T_g as t^{-v}.

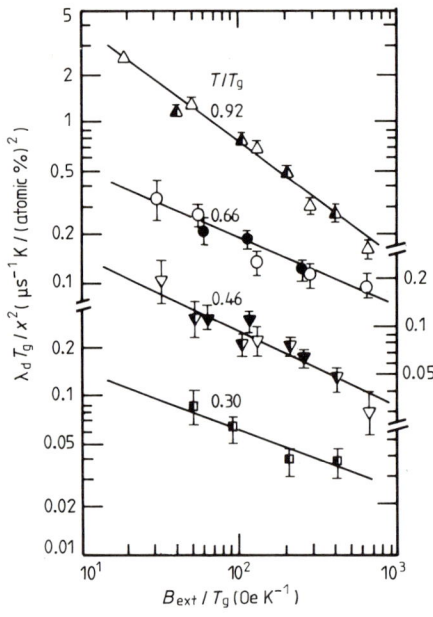

Figure 5.12 Log–log plot of the scaled μ^+ relaxation rate, $\lambda_d T_g/x^2$ against the scaled applied field B_{ext}/T_g from longitudinal field studies in $\underline{AgMn_x}$. Filled symbols represent 6 atomic %; half filled, 3 atomic %; open symbol, 1.6 atomic %. McLaughlin et al 1983.

With reference to the spin dynamics below T_g, there seems to be agreement that a homogeneous cooperative behaviour is indicated, and that each impurity spin has both a static and a dynamic component. There is less unanimity concerning the spin dynamics above T_g. In the analysis referred to so far, it was always assumed that the spin dynamics could be characterised by a single correlation time, τ_c. This is a very questionable assumption in view of other data, notably from neutron spin echo measurements (Mezei and Murani

1979). Nevertheless, most of the data could be fitted reasonably well by adopting the single τ_c model (equations (5.14), (5.18), (5.22), (5.23)), and the application of more extended models (equations (5.26), (5.31)) did not lead to any significantly better fits.

However, Emmerich et al (1983) noted that for both zero and transverse field measurements in CuMn, the fits failed to reproduce the data in, typically, the first 50 ns of the μSR histogram. The data deviated from the fitted theoretical curve, being of consistently higher values. This was reflected in the fit results by a decreasing asymmetry when approaching T_g. These observations led to a new model fit function for $T \gtrsim T_g$ which was free of the deficiency just mentioned. The new fit function has the form

$$G_i(t) = G_i^{Cu}(t)[a + (1-a)G_i^{SG}(t)] \tag{5.36}$$

where $i = x, z$, $G_i^{Cu}(t)$ is the relaxation function observed in pure Cu for a static μ^+, and $G_i^{SG}(t)$ is the relaxation function induced by the dilute electronic moments. The latter function may be given by equations (5.13), (5.14) and

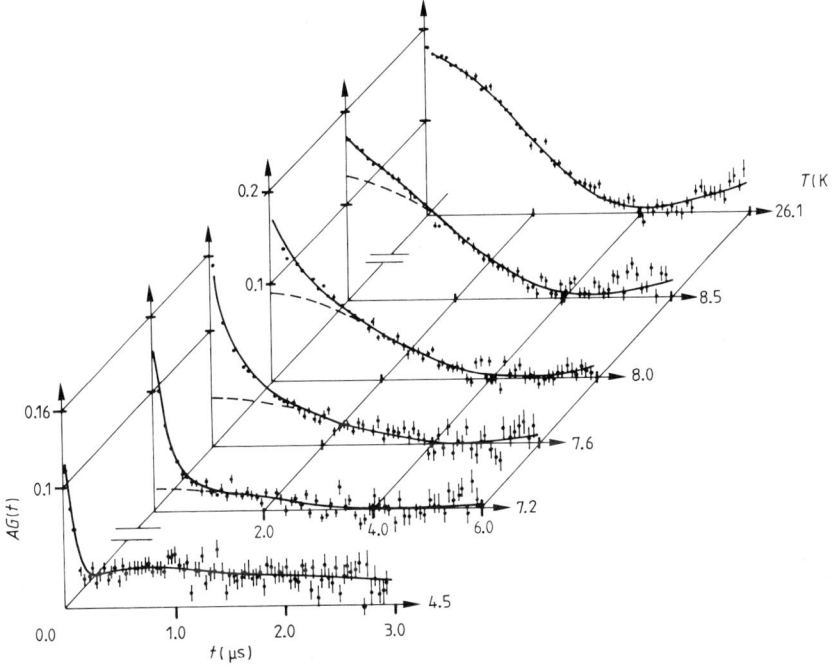

Figure 5.13 Zero field depolarisation behaviour in quenched CuMn (0.54 atomic%). The full curves are fits of equation (5.36) to the data except for $T = 4.5$ K where $G_z(t) = G_z^{SG}(t)$ is used. The spin glass relaxation function, $G_z^{SG}(t)$ is taken from Emmerich (1982a) and Emmerich et al (1983).

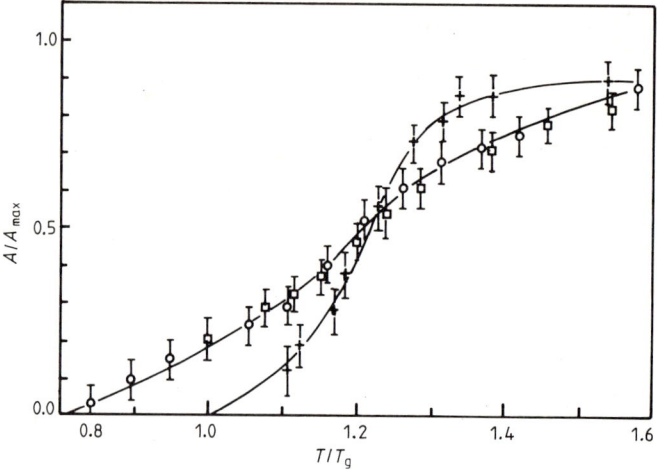

Figure 5.14 Temperature dependence of $a = A/A_{max}$ following from the analysis of zero field data in CuMn according to the inhomogeneous spin freezing model: +, 0.54 atomic % Mn quenched; ○, 0.54 atomic % Mn slowly cooled; □, 0.87 atomic % Mn slowly cooled. The parameter a indicates the fractions of μ^+ outside the islands of spin glass order. This fraction becomes zero at T_g in the quenched sample +, but persists to somewhat lower temperatures in the slowly cooled samples ○, □ (Emmerich et al 1983).

(5.18), depending on the circumstances. In fact, the exact form of $G_i^{SG}(t)$ does not seem to matter much. Figure 5.13 shows zero field data in CuMn (0.54 atomic %) together with fits of equation (5.36) (full curves) for various temperatures, mostly above $T = 6.5$ K. Figure 5.14 displays the fit results for the parameter a which approaches one at high temperatures and declines to zero at or somewhat below T_g (Emmerich et al 1983).

The model behind equation (5.36) and the interpretation of the obtained results presupposes an inhomogeneous nature of the spin glass dynamics above T_g. According to this model, the spin glass above T_g is characterised by the coexistence of islands, with more or less frozen impurity spins, and surrounding areas with rapidly fluctuating ($\gg 10^9$ s^{-1}) impurity spins, leaving the μ^+ spin essentially unaffected. With decreasing temperature, the spin glass islands grow in size and number at the expense of the paramagnetic environment. At T_g all spin glass islands make contact with each other and the paramagnetic volume eventually shrinks to zero. This explains the two component structure of equation (5.36) and the temperature behaviour of the parameter, a. It also leads to a homogeneous picture of the spin glass below T_g. The model described pictures the transition into the spin glass phase as essentially percolative in nature, supporting a suggestion by Mydosh (1981).

References

Abragam A 1970 *The Principles of Nuclear Magnetism* (Oxford: Clarendon)
Bieman L H, de Graaf A M, Kossler W J, Fiory A T, Stronach C E, Lankford W F and Lynn K G 1978 *Hyperfine Interactions* **4** 861
Blandin A 1978 *J. Physique* **39** C6-1499
Brown J A, Heffner R H, Kitchens T A, Leon M, Olsen C E, Schillaci M E, Dodds S A and MacLaughlin D E 1981 *J. Appl. Phys.* **52** 1766
Cannella V and Mydosh J A 1972 *Phys. Rev.* B **6** 4220
Edwards S F and Anderson P W 1975 *J. Phys. F: Met. Phys.* **5** 965
—— 1976 *J. Phys. F: Met. Phys.* **6** 1927
Emmerich K 1982a *Hyperfine Interactions* **12** 59
—— 1982b *Doctoral thesis* Tech. Univ. Braunschweig, FRG
Emmerich K, Gygax F N, Hintermann A, Pinkvos B H, Schenck A, Schwink Ch and Studer W 1983 *J. Magn. Magn. Mater.* **31-34** 1361
Emmerich K and Schwink Ch 1981 *Hyperfine Interactions* **8** 767
Fiory A T 1981 *Hyperfine Interactions* **8** 777
Halperin B I and Hohenberg P C 1969 *Phys. Rev.* **177** 952
Hayano R S, Uemura Y J, Imazato J, Nishida N, Nagamine K, Yamazaki T, Ishikawa Y and Yasuoka H 1980 *J. Phys. Soc. Japan* **49** 1773
Hayano R S, Uemura Y J, Imazato J, Nishida N, Yamazaki T and Kubo R 1979 *Phys. Rev.* B **20** 850
Hayano R S, Uemura Y J, Imazato J, Nishida N, Yamazaki T, Yasuoka H and Ishikawa Y 1978 *Phys. Rev. Lett* **41** 1743
Heffner R H, Leon M, Schillaci M E, McLaughlin D E and Dodds S A 1982 *J. Appl. Phys.* **53** 2174
Kehr K W, Honig G and Richter D 1978 *Z. Phys.* **B32** 49
Kubo R and Tomita K 1954 *J. Phys. Soc. Japan* **9** 888
Leon M 1981 *Hyperfine Interactions* **8** 781
McHenry M R, Silbernagel B G and Wernick J H 1972 *Phys. Rev.* B **5** 2958
McLaughlin D E 1981 *Hyperfine Interactions* **8** 749
McLaughlin D E, Gupta L C, Cooke D W, Heffner R H, Leon M and Schillaci M E 1983 *Phys. Rev. Lett.* **51** 927
Mezei F and Murani A P 1979 *J. Magn. Magn. Mater.* **14** 211
Moriya T 1977 *Physica* **86-88B** 356
Murani A P 1978 *J. Physique* **39** C6-1517
—— 1981 *J. Magn. Magn. Mater.* **22** 271
Murnick D E, Fiory A T and Kossler W J 1976 *Phys. Rev. Lett.* **36** 100
Mydosh J A 1981 in *Disordered Systems and Localisation* ed. J Ehlers *et al* (Berlin: Springer) p 87
Nishiyama K, Nagamine K, Natsui T, Nakajima S, Ishida K, Kuno Y, Imazato J, Nakayama H, Yamazaki T and Yagi E 1983 *J. Magn. Magn. Mater.* **31-34** 695
Shaham M, Barak J, El-Hanany U and Warren W W 1977 *Phys. Rev. Lett.* **39** 570
Souletie J and Tournier R 1971 *J. Physique* **32** C1-172
Takigawa M, Yasuoka H, Uemura Y J, Hayano R S, Yamazaki T and Ishikawa Y 1980 *J. Phys. Soc. Japan* **49** 1760

Uemura Y J 1980 *Solid State Commun.* **36** 369
—— 1981a *Hyperfine Interactions* **8** 739
—— 1981b *PhD thesis* University of Tokyo
Uemura Y J, Harshman D R, Senba M, Ansaldo E J and Murani A P 1984 *Phys. Rev.* B **30** 1606
Uemura Y J, Huang C Y, Clawson C W, Brewer J H, Kiefl R F, Spencer D P and de Graaf A M 1981a *Hyperfine Interactions* **8** 757
Uemura Y J, Nishiyama K, Kadano R, Imazato J, Kuno Y, Nagamine K, Yamazaki T and Ishikawa Y 1983b *J. Magn. Magn. Mater.* **31–34** 1379
Uemura Y J, Nishiyama K, Yamazaki T and Nakai R 1981b *Solid State Commun.* **39** 461
Uemura Y J and Yamazaki T 1982 *Physica* **109–110B** 1915
Uemura Y J and Yamazaki T 1983a *J. Magn. Magn. Mater.* **31–34** 1359
Uemura Y J, Yamazaki T, Hayano R S, Nakai R and Huang C Y 1980 *Phys. Rev. Lett.* **45** 583
Usami K and Moriya T 1980 *J. Magn. Magn. Mater.* **20** 171
Walstedt R E and Walker L R 1974 *Phys. Rev.* **89** 4857
Yasuoka H, Hayano R S, Nishida N, Nagamine K, Yamazaki T and Ishikawa Y 1978b *Solid State Commun.* **26** 745
Yasuoka H, Jaccarino V, Sherwood R C and Wernick J H 1978a *J. Phys. Soc. Japan* **44** 842

6

Applications in Superconductors

6.1 Magnetic Effects

The most dramatic magnetic phenomenon in a superconductor is the well known Meissner effect, by which the magnetic flux is excluded from the interior of a superconductor. In a type I superconductor the Meissner effect always accompanies superconductivity below a critical field, H_c, above which both the superconductivity and the flux exclusion break down. H_c is temperature dependent and vanishes on approaching the transition temperature, T_c. In a type II superconductor the Meissner effect exists below a lower critical field H_{c1}. Agove H_{c1} the superconductor is in the so-called mixed or Shubnikov state in which the sample is penetrated by magnetic flux lines confined to tube-like regions, the so-called vortices. Each vortex carries a flux quantum $\phi = ch/2e = 2 \times 10^{-7}$ G cm^2. If the applied field is B_0 ($> H_{c1}$) the number of vortices is $n = B_0/\phi$. The vortices in general form a two-dimensional array of hexagonal symmetry. The magnetic field distribution inside a type II superconductor in the mixed state will thus turn out to be highly inhomogeneous. We will come back to this point in the next section. The mixed phase extends to an upper critical field, H_{c2}, above which, like in a type I superconductor, superconductivity is destroyed and the sample will be fully penetrated by the external field. H_{c1} and H_{c2} are again temperature dependent and vanish at the transition temperature T_c. Figure 6.1 shows schematically the induced magnetisation, M of a type I and type II superconductor below the transition temperature T_c. The resulting internal field, sensed by the μ^+ and given by $B_0 - 4\pi M$ is also indicated.

So far it has been assumed that the probe is an infinitely long cylinder, with the applied field, B_0, along the cylindrical axis. For this case, the demagnetisation field is zero. If, however, the probe is of ellipsoidal shape a non-zero demagnetisation factor (N) will result leading to a demagnetisation field of $-(N)M$, where M is the magnetisation of the sample. The field in the interior is then given by

$$H_i = B_0 - (N)M \qquad (6.1)$$

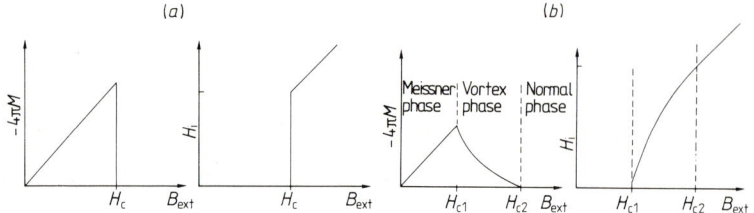

Figure 6.1 Magnetisation and internal field of an infinitely long, cylindrical superconductor of type I (a) and type II (b).

which, because of the Meissner effect, is related to the magnetisation (ideal diamagnetism)

$$M = (1/4\pi) H_i. \tag{6.2}$$

Combining equations (6.1) and (6.2) we obtain

$$H_i = \frac{B_0}{1 - (N)/4\pi}. \tag{6.3}$$

We see from this equation that H_i exceeds the critical field H_c (or H_{c1}) as soon as $B_0 > H_c[1 - (N)/4\pi]$. As a consequence, the Meissner phase becomes unstable. On the other hand, the phase cannot be destroyed altogether, since at that moment $B_0 < H_c$. Instead the superconductor enters the so-called intermediate phase, where superconducting and normal conducting layers parallel to B_0 appear. In the normal conducting layers a magnetic field H_i corresponding to H_c is established independent of B_0.

The ratio of the normal conducting volume to the total volume of the sample (V_n/V) is now of interest. Viewing the sample along B_0 we define cross sections σ_n and σ_s corresponding to the normal and superconducting regions. We have

$$\eta = V_n/V = \sigma_n/(\sigma_n + \sigma_s). \tag{6.4}$$

The flux density in the normal conducting layers is H_c, and zero in the superconducting layers. The average flux density is therefore

$$\bar{B} = \frac{\sigma_n}{\sigma_n + \sigma_s} H_c = \eta H_c. \tag{6.5}$$

On the other hand

$$\bar{B} = H_c + 4\pi \bar{M} \tag{6.6}$$

$$H_i = H_c = B_0 - (N)\bar{M} \tag{6.7}$$

where $\bar{M}$ is the average magnetisation of the sample. Combining equations

(6.5)–(6.7) we find

$$\eta = \frac{B_0 - H_c[1 - (N)/4\pi]}{[(N)/4\pi]H_c} \quad (6.8)$$

for $H_c[1 - (N)/4\pi] \leqslant B_0 \leqslant H_c$. Hence the normal conducting volume rises linearly with the field, starting at $B_0 = H_c[1 - (N)/4\pi]$ and filling the total volume at $B_0 = H_c$.

This behaviour can be nicely monitored by μSR measurements. The fraction of μ^+ stopped in the normal conducting volume will precess with a frequency determined by H_c, and no precession will be observed in the superconducting volume because of the flux exclusion. The precession amplitude, being proportional to the number of precessing μ^+, should then depend on B_0 according to equation (6.8). Such measurements have been performed by Grebinnik et al (1980) in the type I superconductor Pb ($T_c = 7.19$ K). The results are shown in figure 6.2. The temperature dependence of the precession frequency in the intermediate state determines directly the H_c temperature dependence. Corresponding measurements in the type I superconductor Ta, by Gladish et al (1979) are represented in figure 6.3. Here H_c, determined from ω_μ, is plotted against the square of the temperature. We find a linear dependence on T^2 in accordance with the theoretical expectation. These data allow further determination of the transition temperature, T_c, by extrapolating the data to $H_c = 0$.

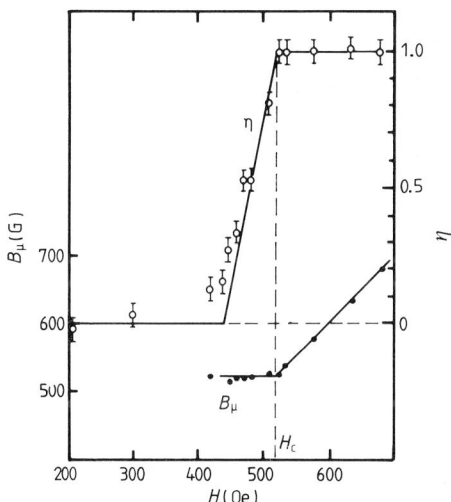

Figure 6.2 Field dependence of the filling factor η, proportional to the μSR amplitude, and the field B_μ at the μ^+, from measurements in the type I superconductor Pb. Note that the field B_μ below the critical field H_c is equal to H_c (Grebinnik et al 1980).

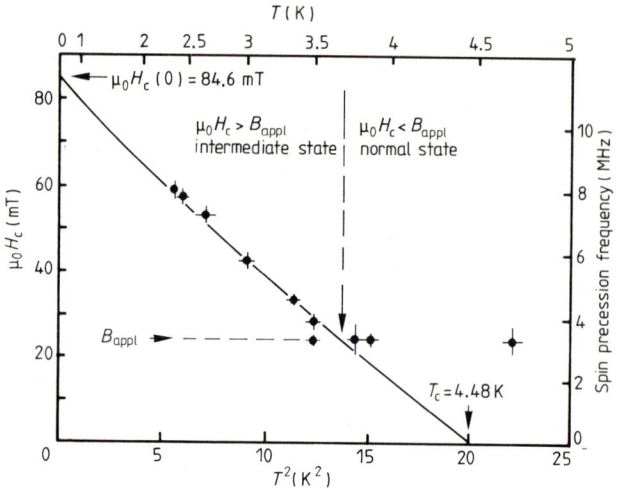

Figure 6.3 Plot of the field, B_μ at the μ^+, against the square of temperature, in superconducting Ta. The observed temperature dependence of B_μ and its absolute values are in excellent agreement with direct measurements of the critical field, H_c (Budnick 1960). Note that B_μ becomes equal to the applied field for $B_{ext} > H_c(T)$ (Gladisch et al 1979).

Grebinnik et al (1980) have also studied the intermediate phase in the type II superconductors V_3Ga ($T_c = 16.5$ K, $H_{c2}(0\,K) = 21.0$ T) and Nb ($T_c = 9.2$ K, $H_{c2}(0\,K) = 0.19$ T). In these samples the transverse field μSR signal had to be described by the following expression

$$N(t) = N_0 \exp(-t/\tau_\mu)[1 + a_0 \exp(-\sigma_0^2 t^2) + a_\perp \exp(-\sigma_\perp^2 t^2)\cos\omega t].$$

The non-precessing component, proportional in amplitude to the superconducting volume in the Meissner state, showed a temperature independent damping which allowed a_0 to be measured unambiguously. (This damping is due to the nuclear dipole fields, and is exactly expressed by the static Kubo–Toyabe relaxation function.) The field dependence of a_0 is thereby given by

$$a_0 \simeq V_s = V - V_n = V(1-\eta). \tag{6.10}$$

The results for a_0 for $B_0 < H_{c1}$, obtained in V_3Ga, are shown in figure 6.4. As can be seen, the data are not in agreement with equation (6.10). Rather, they are characterised by the occurrence of two or more sections with different slopes. This surprising behaviour has not so far been explained. The amplitude of the precessing component was very small over the investigated field range, and did not show the kind of increase observed in Pb. This is probably because the intermediate and the mixed state appear together in a type II super-

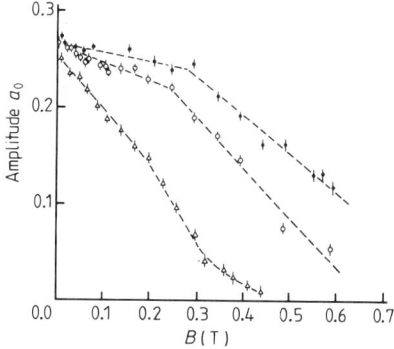

Figure 6.4 Amplitude, a_0, of the non-precessing μ^+ signal in the type II superconductor V_3Ga against applied field for different temperatures. The broken curves are guides to the eye only (Grebinnik *et al* 1980).

conductor. For the same reasons, the results on a_0 and $a_\perp$ in Nb, where H_{c1} and H_{c2} are relatively close together, cannot be interpreted in a straightforward way.

6.2 μ^+ Precession in Mixed Phase of Type II Superconductors

As mentioned in the previous section, the mixed state of a type II superconductor, extending from a lower critical field H_{c1} to an upper critical field H_{c2}, is characterised by a partial flux penetration in the form of flux tubes or vortices arranged in a regular (triangular) fashion. The internal field distribution in the vortex lattice is described by the Ginzburg–Landau equation for an extreme type II superconductor ($\lambda \gg \xi$) (see e.g. Saint-James *et al* 1969)

$$\boldsymbol{h} = \lambda^2 \operatorname{curl} \operatorname{curl} \boldsymbol{h} = \phi \sum_i \delta(\boldsymbol{r} - \boldsymbol{r}_i). \tag{6.11}$$

Here, $\lambda(T)$ is the penetration depth, $\xi(T)$ the coherence length, ϕ the flux quantum and $\boldsymbol{r}_i$ the centre of the ith vortex. The resulting field distribution is shown schematically in figure 6.5(a). The field is at a maximum in the centre of a vortex ($H_{\max}$). In figure 6.5(a) the position of the so-called saddle point field, H_s, and the minimum field $H_{\min}$ is also indicated. Figure 6.5(b) shows the spectral density of the field strength distribution (a). The spectral density diverges at the saddle point field. For an early treatment of the effects of the flux distribution on the μ^+ transverse field relaxation function see Ivanter and Smilga (1968).

The solution of equation (6.11) can be expressed as a sum over Fourier

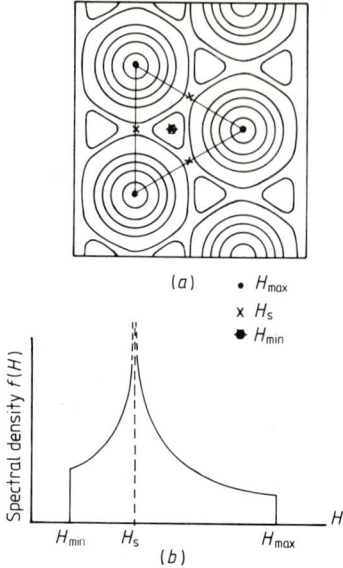

Figure 6.5 Magnetic field distribution in the mixed phase of a type II superconductor. (b) shows the spectral density of (a) the field strength distribution. Indicated are the positions of the saddle point field, H_s, the maximum field H_{max} and the minimum field H_{min}.

components h_k of h

$$h = \sum_k h_k \exp(i\mathbf{k} \cdot \mathbf{r}) = \sum_k \frac{n\phi}{1+\lambda^2 k^2} \exp(i\mathbf{k} \cdot \mathbf{r}) \hat{z}_0. \quad (6.12)$$

The sum extends over the reciprocal lattice vectors of the vortex lattice. n is the number of vortices/cm^2.

With respect to transverse field μSR measurements, we are particularly interested in the average field spread as this determines the transverse field depolarisation rate of the μ^+. The RMS deviation of the field distribution from the average field, $\langle h \rangle$, is given by

$$\overline{\Delta H^2} = \langle h^2 \rangle - \langle h \rangle^2 = \frac{1}{S}\int h^2 \, d\mathbf{r}_\perp - \left(\frac{1}{S}\int h \, d\mathbf{r}_\perp\right)^2 \quad (6.13)$$

where S is the sample cross section perpendicular to the flux tubes. Clearly

$$\langle h \rangle = n\phi = B_0. \quad (6.14)$$

Inserting equation (6.12) into (6.13), one arrives at (Pincus et al 1964)

$$\overline{\Delta H^2} = \sum_k h_k h_{-k} = n^2 \phi^2 \sum_k \frac{1}{(1+\lambda^2 k^2)^2}. \quad (6.15)$$

The sum can be replaced by an integral

$$\overline{\Delta H^2} = \frac{n^2\phi^2}{4\pi n} \int_{k_{min}}^{\infty} \frac{dk^2}{(1+\lambda^2 k^2)^2} = \frac{n\phi^2}{2\pi} \left(\frac{-1}{2\lambda^2(1+\lambda^2 k^2)} \right)_{k_{min}}^{\infty}. \quad (6.16)$$

If the distance between two nearest neighbour vortices is L, the density of vortices for a triangular lattice is

$$n = (2/3^{1/2})L^{-2}. \quad (6.17)$$

The smallest reciprocal lattice vector for a triangular (hexagonal) lattice is

$$k_{min} = 4\pi/3^{1/2}L. \quad (6.18)$$

Finally one obtains

$$\overline{\Delta H^2} = \frac{B_0^2}{2\pi\sqrt{3}} \left\{ \frac{\lambda}{(3^{1/2}/2)L} \left[1 + 4\pi^2 \left(\frac{\lambda}{(3^{1/2}/2)L} \right)^2 \right] \right\}^{-1}. \quad (6.19)$$

Since the actual field distribution, as shown in figure 6.5, is far from being Gaussian, we conjecture that an exponential relaxation function would describe the actual transverse field depolarisation behaviour of μ^+ in the mixed state more adequately, that is

$$P_x(t) = \exp(-t/T_2) \quad (6.20)$$

with

$$1/T_2 = \pi \gamma_\mu (\overline{\Delta H^2})^{1/2}. \quad (6.21)$$

(Note that $(\overline{\Delta H^2})^{1/2}$ is only half of the full RMS width of the field distribution.)

The vortex lattice constant, L, is related to the so-called coherence length $\xi(T)$ by

$$L = 2.7\xi(T). \quad (6.22)$$

Both the penetration depth, λ, and the coherence length, ξ, are temperature dependent, and this is described by the Gorkov expressions (see e.g. Saint-James et al 1969). For a pure superconductor one has

$$\lambda(T) = \frac{1}{2^{1/2}} \lambda_L(0) \left(\frac{T_c}{T_c - T} \right)^{1/2} \quad (6.23)$$

$$\xi(T) = 0.74\xi_0 \left(\frac{T_c}{T_c - T} \right)^{1/2}. \quad (6.24)$$

$\lambda_L(0)$ and ξ_0 are characteristic constants of the material in question.

Early attempts to study the field distribution in the mixed state of the type II superconductors Nb and PbIn by μSR were only partially successful (Fiory et al 1974). While no effect was found in Nb, the Fourier spectrum of the μ^+ precession in PbIn revealed a component corresponding to the saddle point field. The specimen was prepared in the mixed state by cooling the sample in the presence of an applied field. The actual relaxation rate of the μ^+ precession signal was not studied directly.

μ^+ precession in the mixed state of Nb was also investigated by Gladisch et al (1979). Under their experimental conditions the sample was partially in the Meissner and partially in the mixed phase. Fourier analysis of the μ^+ precession signal revealed an average field about 25% lower than the applied field, in qualitative agreement with the expectations (see figure 6.6). Also, no detailed studies of relaxation rates as a function of temperature or applied field were undertaken here. A more detailed study of relaxation rates of the μ^+ precession in superconducting Nb was performed by Grebinnik et al (1980). These data, however, did not find an unambiguous interpretation, since it was impossible to distinguish the formation of the mixed state from the formation of sections already in the normal state.

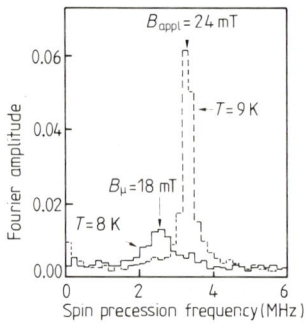

Figure 6.6 Fourier spectra of the μSR signal in Nb at 8 K (full curve) and 9 K (broken curve). At 9 K the sample is in the normal state and the applied field is 24 mT. The μ^+ precession frequency in the superconducting phase corresponds to an average field which is about 25% lower than the applied field (Gladisch et al 1979).

A very detailed study of the field distribution in the mixed phase of a type II superconductor became possible in Vanadium, but only because of a very peculiar and abnormally high flux pinning (Gygax et al 1981). In this experiment it was observed that, by lowering the applied field from a value well above $H_{c2}(T)$ to a value below $H_{c2}(T)$, magnetic flux corresponding to $H_{c2}(T)$ was frozen in, as revealed by the corresponding μ^+ precession frequency. The trapped flux remained constant upon lowering the applied field to zero, and even upon reversing its sign. It was destroyed when the applied field reached about $-H_{c2}(T)$. This procedure was repeated for a series of temperatures below T_c, allowing the temperature dependence of H_{c2}, or the phase boundary in the B_0–T plane to be monitored (see figure 6.7). Another procedure was to

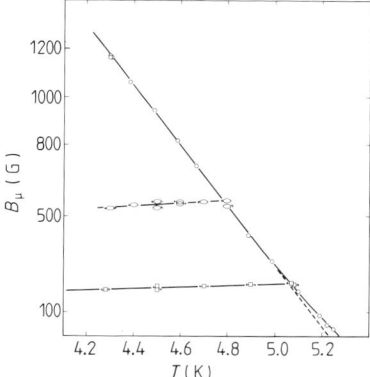

Figure 6.7 Temperature dependence of the field $B_\mu \simeq H_{c2}$, frozen or trapped, in superconducting V as described in the text. The 'horizontal' branches are obtained by lowering the temperature below the temperature at which the field was frozen in (Gygax et al 1981).

freeze the field at a certain temperature as above, and then to lower the temperature. In that case the internal field remained almost constant, showing perhaps a slight tendency to decrease with decreasing temperature but in a reversible fashion (see figure 6.7). This implies that the number of vortices remained constant (see equation (6.14)). The μ^+ precession signal displayed pronounced relaxation, the temperature dependence of which is shown in figure 6.8. Clearly visible are three branches which correspond to the three branches in figure 6.7. In all instances, a linear temperature dependence is indicated (except when close to T_c). The relaxation rates taken on the phase boundary should approach zero on approaching T_c, since, in that limit the internal field inhomogeneity should vanish. This allows the critical temperature, T_c to be determined. A value of 5.17 K was extracted from figure 6.8, which is in clear disagreement with a temperature of 5.24 K at which H_{c2} approaches zero (see figure 6.7). The reason for this inconsistency is not clear, but it might point to non-homogeneous properties of the sample, which might also explain the anomalous amount of flux trapping.

The relaxation rates were analysed on the basis of equation (6.19). For the relaxation rates on the phase boundary ($B = H_{c2}(T)$), equation (6.21) becomes

$$\frac{1}{T_2} = \pi \gamma_\mu \frac{H_{c2}(T)}{(2\pi 3^{1/2})^{1/2}} \left[2\pi \left(\frac{\lambda(T)}{(3^{1/2}/2)L} \right)^2 \right]^{-1}. \quad (6.25)$$

Inserting the Gorkov expressions for $\lambda(T)$ and $\xi(T)$, and using appropriate numbers for $\lambda_L(0)$ and ξ_0 (Gygax et al 1981, Usui et al 1969) one predicts

$$1/T_2 = 1.74 \times 10^3 H_{c2}(T) \, \text{s}^{-1}. \quad (6.26)$$

Figure 6.9 shows a plot of $1/T_2$ against $H_{c2}(T)$. A linear relation is clearly

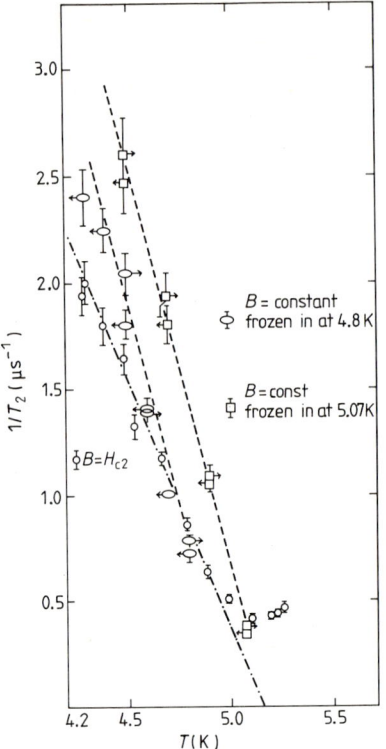

Figure 6.8 Temperature dependence of the transverse μ^+ relaxation rate corresponding to the three branches in figure 6.7 (Gygax et al 1981).

indicated with a slope of 1.7×10^3 (s^{-1} G^{-1}), in excellent agreement with equation (6.26).

The relaxation rates corresponding to the horizontal branches in figure 6.7 must reflect the temperature dependence of $\lambda(T)$ only, since the number of vortices, and therefore the vortex lattice constant, L, remains constant. The change of the field modulation over the sample is therefore only caused by the change of the size of the individual vortices. On this assumption, $\lambda(T)$ was extracted from the relaxation data. The results for two different frozen in fields are shown in figure 6.10, where $\lambda(T)/(3^{1/2}L/2)$ is plotted against $[1-(T/T_c)^4]^{-1/2}$. In agreement with other data (Usui et al 1969), a linear relationship is obtained, but only when $T_c = 5.17$ K is used.

This example shows that the μSR method is indeed capable of revealing much of the magnetic structure of a superconductor in the mixed state. In this example, the unusual flux pinning was a great help and allowed the relaxation measurement to be performed in zero applied field, and therefore avoiding the complications caused by demagnetisation and intermediate state effects. Further, the extraction of conflicting critical temperatures, although not

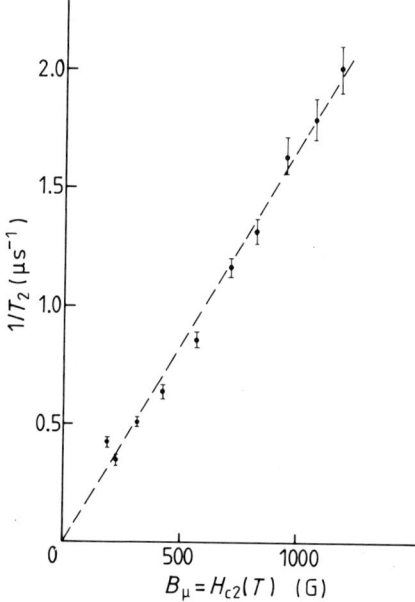

Figure 6.9 Plot of the μ^+ relaxation rate, $1/T_2$ versus $B_\mu \simeq H_{c2}(T)$ from measurements in superconducting V (Gygax et al 1981).

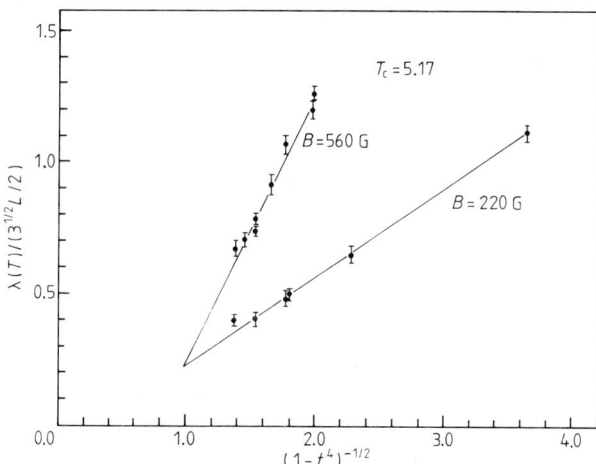

Figure 6.10 Temperature dependence of the penetration depth, $\lambda(T)$ extracted from the μ^+ relaxation rates corresponding to the horizontal branches in figure 6.7 (Gygax et al 1981).

understood at present, indicates that other interesting properties of a superconductor are revealed by the μSR data.

More recently, attempts have been made to investigate using μSR the so-called magnetic superconductors, like the re-entrant ferromagnets $Ho_{0.7}Lu_{0.3}Rh_4B_4$ (Boekema *et al* 1982) and $ErRh_4B_4$ (Huang *et al* 1984) or the antiferromagnetic superconductor $SmRh_4B_4$ (Huang *et al* 1984). Because of the preliminary nature of the results, no discussion is attempted here.

References

Boekema C, Heffner R H, Hutson R L, Leon M, Schillaci M E, Smith J L, Dodds S A and McLaughlin D E 1982 *J. Appl. Phys.* **53** 2625
Budnick J I 1960 *Phys. Rev.* **119** 1578
Fiory A T, Murnick D E, Leventhal M and Kossler W J 1974 *Phys. Rev. Lett.* **33** 969
Gladisch M, Herlach D, Metz H, Orth H, zu Putlitz G, Seeger A, Teichler H, Wahl W and Wigand M 1979 *Hyperfine Interactions* **6** 109
Grebinnik V G, Gurevich I I, Zhukov V A, Klimov A I, Levina L A, Maiorov V N, Manych A P, Melnikov E V, Nikolskii B A, Pirogov A V, Ponomarev A N, Roganov V S, Selivanov V I and Suetin V A 1980 *Zh. Eksp. Teor. Fiz.* **79** 518 (Engl. transl. 1980 *Sov. Phys.–JETP* **52** 261)
Gygax F N, Hintermann A, Ott H R, Rüdigier H, Rüegg W, Schenck A and Studer W 1981 *Hyperfine Interactions* **8** 623
Huang C Y, Ansaldo E J, Brewer J H, Harshman D R, Crowe K M, Rosenblum S S, Clawson C W, Fisk Z, Lambert S, Torikachili M S and Maple M B 1984, *Hyperfine Interactions* **17–19** 509
Ivanter I G and Smilga V P 1968 *Zh. Eksp. Teor. Fiz.* **55** 548 (Engl. transl. *Sov. Phys.–JETP* **28** 286)
Pincus P, Gossard A C, Jaccarino V and Wernick J H 1964 *Phys. Rev. Lett.* **13** 21
Saint-James D, Sarma G and Thomas E J 1969 *Type II Superconductivity* (Oxford: Pergamon)
Usui N, Ogasawara T, Yasukochi K and Tomoda S 1969 *J. Phys. Soc. Japan* **27** 574

7

Muonium in Matter

7.1 Magnetic Properties in the 1S Ground State

7.1.1 Hyperfine structure and Zeeman splitting

The bound system (μ^+e^-), known as muonium (Mu), resembles the hydrogen atom (pe$^-$) in great detail. The groundstate electron wavefunction of Mu in a vacuum is given by

$$\psi_{1s} = \frac{1}{\pi^{1/2}} \frac{1}{a_0^{3/2}} \exp\left(-\frac{r}{a_0}\right) \tag{7.1}$$

with the Bohr radius

$$a_0 = \frac{\hbar^2}{e^2} \frac{1}{m_e^*} \tag{7.2}$$

where m_e^* is the electron reduced mass

$$1/m_e^* = (1/m_e) + (1/m_\mu). \tag{7.3}$$

The electron reduced mass in Mu is only about 0.5% smaller than the reduced mass in H. The binding energy of the electron in Mu is given by

$$E = -m_e^* e^4/2\hbar^2 \simeq -13.55 \text{ eV} \tag{7.4}$$

which is again only 0.5% smaller than in H.

Electron spin (S) and muon spin (I) are coupled through the Fermi contact hyperfine interaction (see e.g. Hughes and Kinoshita 1977)

$$\mathcal{H}_{hf} = \tfrac{8}{3}\pi g_\mu \mu_B^\mu g_e \mu_B^e \, S \cdot I \, \delta(r) = A S \cdot I. \tag{7.5}$$

The hyperfine coupling constant, A, is given by

$$A = \hbar\omega_0 = \tfrac{8}{3} g_\mu \mu_B^\mu g_e \mu_B^e / a_0^3. \tag{7.6}$$

μ_B^μ and μ_B^e are the muon and electron Bohr magnetons, and g_μ and g_e the muon and electron g factors, respectively.

216 Muonium in matter

The hyperfine coupling leads to a splitting of the ground state energy level into two levels with total spin $F=1$ (the triplet state) and $F=0$ (the singlet state). The splitting, ΔE, amounts to

$$\Delta E = A = \hbar \omega_0 \qquad \text{with } \omega_0/2\pi = 4.46 \text{ GHz.}$$

The hyperfine operator, equation (7.5), is spatially isotropic and applies, in particular, to Mu when in a vacuum. It will become necessary to also consider the more general, anisotropic form

$$\mathcal{H}_{hf} = \mathbf{S} \cdot \mathbf{A} \cdot \mathbf{I} \tag{7.7}$$

where **A** is now a tensor. This can be explicitly written as

$$\mathcal{H}_{hf} = \alpha S^+ I^+ + \alpha^* S^- I^- + \beta S^+ I^- + \beta^* S^- I^+$$
$$+ 2\gamma S^+ I_z + 2\gamma^* S^- I_z + 2\delta S_z I^+ + 2\delta^* S_z I^- + 4\kappa S_z I_z \tag{7.8}$$

where

$$\alpha = \tfrac{1}{4}[A_{xx} - A_{yy} - i(A_{xy} + A_{yx})] \qquad \beta = \tfrac{1}{4}[A_{xx} + A_{yy} + i(A_{xy} - A_{yx})]$$
$$\gamma = \tfrac{1}{4}(A_{xz} - iA_{yz}) \qquad \delta = \tfrac{1}{4}(A_{zx} - iA_{zy}) \qquad \kappa = \tfrac{1}{4}A_{zz}. \tag{7.9}$$

In order to obtain the energy eigenvalues and the spin eigenfunctions of the hyperfine Hamiltonian, we introduce as a basis the four possible spin combinations of the muon and electron spin

$$\chi_1 = |\tfrac{1}{2}, \tfrac{1}{2}\rangle \qquad \chi_2 = |\tfrac{1}{2}, -\tfrac{1}{2}\rangle$$
$$\chi_3 = |-\tfrac{1}{2}, \tfrac{1}{2}\rangle \qquad \chi_4 = |-\tfrac{1}{2}, -\tfrac{1}{2}\rangle. \tag{7.10}$$

The first number refers to the magnetic quantum number of the electron, m_e, the second to the muon magnetic quantum number m_μ, respectively. The eigenfunctions are then expressed in terms of these basis functions

$$\varphi = \sum_{i=1}^{4} a_i \chi_i. \tag{7.11}$$

Eigenfunctions and energy eigenvalues follow from the solution of the set of equations

$$\langle \chi_i | \mathcal{H}_{hf} | \varphi \rangle = E \langle \chi_i | \varphi \rangle \qquad (i = 1, 2, 3, 4).$$

In particular the energy eigenfunctions are determined from the secular equation

$$\det |\mathcal{H}_{ij} - \delta_{ij} E| = 0 \tag{7.13}$$

with $\mathcal{H}_{ij} = \langle \chi_i | \mathcal{H}_{hf} | \chi_j \rangle$.

Explicitly, using the general Hamiltonian equation (7.7), the secular equation reads

$$\begin{vmatrix} \kappa - E & \delta & \gamma & \alpha \\ \delta^* & -\kappa - E & \beta & -\gamma \\ \gamma^* & \beta^* & -\kappa - E & -\delta \\ \alpha^* & -\gamma^* & -\delta^* & \kappa - E \end{vmatrix} = 0. \qquad (7.14)$$

In general, we now have four different eigenvalues and the ground state splits into four levels.

Since the Hamiltonian $\mathcal{H}_{hf}$ is invariant under a rotation of the coordinate system, we can choose a coordinate system (x', y', z') in which **A** is diagonal. It follows then, from equation (7.9) that $\gamma = \delta = 0$ and the secular equation (7.14) reduces to

$$[(\kappa - E)^2 - \alpha^2)(-\kappa - E)^2 - \beta^2] = 0 \qquad (7.15)$$

with the solutions

$$E_{1,3} = \kappa \pm \alpha = \tfrac{1}{4}[A_{z'z'} \pm (A_{x'x'} - A_{y'y'})]$$
$$E_{2,4} = -\kappa \pm \beta = -\tfrac{1}{4}[A_{z'z'} \mp (A_{x'x'} + A_{y'y'})]. \qquad (7.16)$$

It can be seen that, although the secular equation is of the fourth order in E, we can always obtain an analytic solution for E. Hence it is easy now to write down the energy eigenvalues for several special cases. The isotropic Hamiltonian, equation (7.5), with $A_{xx} = A_{yy} = A_{zz} = \hbar\omega_0$, possesses the eigenvalues

$$E_1 = E_2 = E_3 = +\tfrac{1}{4}\hbar\omega, \qquad E_4 = -\tfrac{3}{4}\hbar\omega_0. \qquad (7.17)$$

An axially isotropic Hamiltonian with $A_{x'x'} = A_{y'y'} \neq A_{z'z'}$, such as the dipole–dipole Hamiltonian

$$\mathcal{H}_{dip} = \frac{\hbar^2 \gamma_1 \gamma_2}{r^3} [\mathbf{I}_1 \cdot \mathbf{I}_2 - 3(\mathbf{I}_1 \cdot \mathbf{r})(\mathbf{I}_2 \cdot \mathbf{r})/r^2]$$

$$= \frac{\hbar^2 \gamma_1 \gamma_2}{(z')^3} [\mathbf{I}_1 \cdot \mathbf{I}_2 - 3 I_{z',1} I_{z',2}] \qquad (7.18)$$

leads to eigenvalues of

$$E_1 = E_3 = \tfrac{1}{4} A_{z'z'}$$
$$E_2 = -\tfrac{1}{4} A_{z'z'} + \tfrac{1}{2} A_{x'x'} \qquad E_4 = -\tfrac{1}{4} A_{z'z'} - \tfrac{1}{2} A_{x'x'}. \qquad (7.19)$$

Although, of course, the energy eigenvalues must be independent of the particular choice of the coordinate system, the eigenfunctions will depend on the coordinate system with the exception of the isotropic Hamiltonian, equation (7.5), as we shall see later.

As the next step, we now include the Zeeman interaction of the μ^+ and the electron in an external magnetic field, B. The total spin Hamiltonian then reads

218   Muonium in matter

$$\mathcal{H} = \mathcal{H}_{hf} + g_e \mu_B^e \mathbf{S} \cdot \mathbf{B} - g_\mu \mu_B^\mu \mathbf{I} \cdot \mathbf{B}$$
$$= \mathcal{H}_{hf} + \gamma_e h \mathbf{S} \cdot \mathbf{B} - \gamma_\mu h \mathbf{I} \cdot \mathbf{B}$$
$$= \mathcal{H}_{hf} + h\boldsymbol{\omega}_e \cdot \mathbf{S} - h\boldsymbol{\omega}_\mu \cdot \mathbf{I} \quad (7.20)$$

and the secular equation assumes the form

$$\begin{vmatrix} \kappa + \tfrac{1}{2}h(\omega_e^z - \omega_\mu^z) - E & \delta - \tfrac{1}{2}h\omega_\mu^- & \gamma + \tfrac{1}{2}h\omega_e^- & \alpha \\ \delta^* - \tfrac{1}{2}h\omega_\mu^+ & -\kappa + \tfrac{1}{2}h(\omega_e^z + \omega_\mu^z) - E & \beta & -\gamma + \tfrac{1}{2}h\omega_e^- \\ \gamma^* + \tfrac{1}{2}h\omega_e^+ & \beta^* & -\kappa - \tfrac{1}{2}h(+\omega_e^z + \omega_\mu^z) - E & -(\delta + \tfrac{1}{2}h\omega_\mu^-) \\ \alpha^* & -\gamma^* + \tfrac{1}{2}h\omega_e^+ & -(\delta^* + \tfrac{1}{2}h\omega_\mu^+) & \kappa - \tfrac{1}{2}h(\omega_e^z - \omega_\mu^z) - E \end{vmatrix}$$

(7.21)

Note that g_e, g_μ, γ_μ, γ_e and correspondingly ω_μ and ω_e are all positive numbers.

If $\mathcal{H}_{hf}$ is given by the isotropic form equation (7.5), and if the field $\mathbf{B}$ is applied along a chosen z axis, equation (7.21) can be solved analytically with the solutions

$$E_1 = \tfrac{1}{4}A + \tfrac{1}{2}h(\omega_e^z - \omega_\mu^z) \qquad\qquad = \tfrac{1}{4}A(1 + 2dx)$$
$$E_2 = -\tfrac{1}{4}A + [\tfrac{1}{4}A^2 + \tfrac{1}{4}h^2(\omega_e^z + \omega_\mu^z)^2]^{1/2} \qquad = \tfrac{1}{4}A[-1 + 2(1 + x^2)^{1/2}]$$
$$E_3 = \tfrac{1}{4}A - \tfrac{1}{2}h(\omega_e^z - \omega_\mu^z) \qquad\qquad = \tfrac{1}{4}A(1 - 2dx) \quad (7.22)$$
$$E_4 = \tfrac{1}{4}A - [\tfrac{1}{4}A^2 + \tfrac{1}{4}h^2(\omega_e^z - \omega_\mu^z)^2]^{1/2} \qquad = \tfrac{1}{4}A[-1 - 2(1 + x^2)^{1/2}]$$

where

$$x = B(g_e \mu_B^0 + g_\mu \mu_B^\mu)/A. \quad (7.23)$$

(for muonium in vacuum, $x \simeq B/1585$) and

$$d = (1 - g_\mu \mu_B^\mu/g_e \mu_B^e)/(1 + g_\mu \mu_B^\mu/g_e \mu_B^e) = 0.990. \quad (7.24)$$

These are the famous Breit–Rabi equations which describe the energy splitting of atomic hydrogen or muonium in an external magnetic field. A so-called Breit–Rabi diagram is shown in figure 7.1.

Equation (7.21) also has analytic solutions if the external field, $\mathbf{B}$, is applied along one of the principal axes (x', y', z') of the hyperfine Hamiltonian equation (7.7).

The resulting expressions for the energy eigenvalues are listed in table 7.1(a) and (b) for two special orientations of the applied field. Figure 7.2 displays the Breit–Rabi diagram for the dipole–dipole interaction, equation (7.18), for the two orientations $\mathbf{B} \| z'$ and $\mathbf{B} \| x'$.

The eigenfunctions are also easily obtained if $\mathbf{B}$ is directed along one of the principal axes of the hyperfine Hamiltonian equation (7.7). For that purpose we have to go back to equations (7.11) and (7.12), also including the Zeeman terms, and find the solutions for the coefficients a_{ij}, where i refers to the ith eigenvalue, E_i. In explicit form equation (7.12) reads as follows

Magnetic properties in the 1S ground state

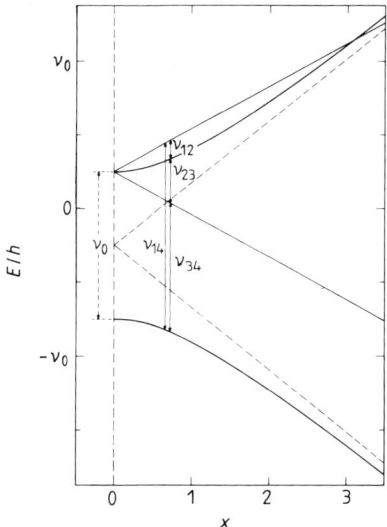

Figure 7.1 Breit–Rabi diagram of the ground state muonium energy eigenvalues, in an external magnetic field, expressed by the dimensionless parameter x (equation (7.23)). For graphical clarity, a non-physical value of m_μ/m_e is used to generate the plot. v_{ij} represent the allowed transitions between the ith and jth states.

$$[\kappa + \tfrac{1}{2}\hbar(\omega_e^z - \omega_\mu^z) - E_i]a_{i1} + (\delta - \tfrac{1}{2}\hbar\omega_\mu^-)a_{i2} + (\gamma + \tfrac{1}{2}\hbar\omega_e^-)a_{i3} + \alpha a_{i4} = 0$$

$$(\delta^* - \tfrac{1}{2}\hbar\omega_\mu^+)a_{i1} + [-\kappa + \tfrac{1}{2}\hbar(\omega_e^z + \omega_\mu^z) - E_i]a_{i2} + \beta a_{i3} + (-\gamma + \tfrac{1}{2}\hbar\omega_e^-)a_{i4} = 0$$

$$(\gamma^* + \tfrac{1}{2}\hbar\omega_e^+)a_{i1} + \beta^* a_{i2} + [-\kappa - \tfrac{1}{2}\hbar(+\omega_e^z + \omega_\mu^z) - E_i]a_{i3} - (\delta + \tfrac{1}{2}\hbar\omega_\mu^-)a_{i4} = 0$$

$$\alpha^* a_{i1} + (-\gamma^* + \tfrac{1}{2}\hbar\omega_e^+)a_{i2} - (\delta^* + \tfrac{1}{2}\hbar\omega_\mu^+)a_{i3} + [\kappa - \tfrac{1}{2}\hbar(\omega_e^z - \omega_\mu^z)E_i]a_{i4} = 0.$$

(7.25)

This simplifies considerably if we work in the coordinate system (x', y', z'), where the tensor **A** is transformed to its principal axes, and also if the external field is applied along one of the principal axes. The coefficients, a_{ij}, are determined up to a common factor which is eliminated by the normalisation condition

$$\sum_{j=1}^{4} a_{ij}^2 = 1.$$

(7.26)

Table 7.1(a) and (b) lists the eigenvalues and eigenstates for the two cases, **B** parallel to (a) the z' axis and (b) the x' axis. Note that the basis states, $|m_e, m_\mu\rangle$ are in both cases defined with respect to the z' axis.

For zero field and $\alpha = 0$, one finds explicitly ($B \parallel z'$)

Muonium in matter

Table 7.1 Eigenvalues and eigenstates of the spin Hamiltonian equation (7.20) with the external field parallel to (a) the z' axis, and (b) the x' axis. These are both principal axes of the tensor **A**. The quantisation axis is the z' axis.
(a) $x_\alpha = \hbar(\omega_e + \omega_\mu)/2\alpha$, $x_\beta = \hbar(\omega_e + \omega_\mu)/2\beta$, $\alpha = \frac{1}{4}(A_{x'x'} - A_{y'y'})$, $\beta = \frac{1}{4}(A_{x'x'} + A_{y'y'})$, $\kappa = \frac{1}{4}A_{z'z'}$.

$$E_1 = \kappa + [\alpha^2 + \tfrac{1}{4}\hbar^2(\omega_e - \omega_\mu)^2]^{1/2} = \kappa + \alpha(1 + (dx_\alpha)^2)^{1/2}$$

$$a_1 = 2^{-1/2}\left(1 + \frac{dx_\alpha}{[1+(dx_\alpha)^2]^{1/2}}\right)^{1/2} = \cos\varepsilon$$

$$a_2 = a_3 = 0$$

$$a_4 = 2^{-1/2}\left(1 - \frac{dx_\alpha}{[1+(dx_\alpha)^2]^{1/2}}\right)^{1/2} = \sin\varepsilon$$

$$E_2 = -\kappa + [\beta^2 + \tfrac{1}{4}\hbar^2(\omega_e + \omega_\mu)^2]^{1/2} = -\kappa + \beta(1 + x_\beta^2)^{1/2}$$

$$a_1 = a_4 = 0.$$

$$a_2 = 2^{-1/2}\left(1 + \frac{x_\beta}{(1+x_\beta^2)^{1/2}}\right)^{1/2} = \cos\delta$$

$$a_3 = 2^{-1/2}\left(1 - \frac{x_\beta}{(1+x_\beta^2)^{1/2}}\right)^{1/2} = \sin\delta$$

$$E_3 = \kappa - [\alpha^2 + \tfrac{1}{4}\hbar^2(\omega_e - \omega_\mu)^2]^{1/2} = \kappa - \alpha[(1+(dx_\alpha)^2]^{1/2}$$

$$a_1 = 2^{-1/2}\left(1 - \frac{dx_\alpha}{[1+(dx_\alpha)^2]^{1/2}}\right)^{1/2} = \sin\varepsilon$$

$$a_2 = a_3 = 0$$

$$a_4 = -2^{-1/2}\left(1 + \frac{dx_\alpha}{[1+(dx_\alpha)^2]^{1/2}}\right)^{1/2} = -\cos\varepsilon$$

$$E_4 = -\kappa - [\beta^2 + \tfrac{1}{4}\hbar^2(\omega_e + \omega_\mu)^2]^{1/2} = -\kappa - \beta(1 + x_\beta^2)^{1/2}$$

$$a_1 = a_4 = 0$$

$$a_2 = 2^{-1/2}\left(1 - \frac{x_\beta}{(1+x_\beta^2)^{1/2}}\right)^{1/2} = \sin\delta$$

$$a_3 = -2^{-1/2}\left(1 + \frac{x_\beta}{(1+x_\beta^2)^{1/2}}\right)^{1/2} = -\cos\delta$$

Table 7.1 Continued
(b) $x_{\alpha'} = \hbar(\omega_e + \omega_\mu)/2\alpha'$, $x_{\beta'} = \hbar(\omega_e + \omega_\mu)/2\beta'$, $\alpha' = \frac{1}{4}(A_{y'y'} - A_{z'z'})$, $\beta' = \frac{1}{4}(A_{y'y'} + A_{z'z'})$, $\kappa = \frac{1}{4}A_{x'x'}$.

$E_1 = \kappa' - [\alpha'^2 + \frac{1}{4}\hbar^2(\omega_e - \omega_\mu)^2]^{1/2} = \kappa' - \alpha'[1 + (dx_{\alpha'})^2]^{1/2}$

$a_1 = a_4 = \frac{1}{2}\left(1 + \frac{1}{[1+(dx_{\alpha'})^2]^{1/2}}\right)^{1/2} = 2^{-1/2}\cos\rho'$

$a_2 = a_3 = -\frac{1}{2}\left(1 - \frac{1}{[1+(dx_{\alpha'})^2]^{1/2}}\right)^{1/2} = -2^{-1/2}\sin\rho'$

$E_2 = \kappa' + [\alpha'^2 + \frac{1}{4}\hbar^2(\omega_e - \omega_\mu)^2]^{1/2} = \kappa' + \alpha'[1 + (dx_{\alpha'})^2]^{1/2}$

$a_1 = a_4 = \frac{1}{2}\left(1 - \frac{1}{[1+(dx_{\alpha'})^2]^{1/2}}\right)^{1/2} = 2^{-1/2}\sin\rho'$

$a_2 = a_3 = \frac{1}{2}\left(1 + \frac{1}{[1+(dx_{\alpha'})^2]^{1/2}}\right)^{1/2} = 2^{-1/2}\cos\rho'$

$E_3 = -\kappa' + [\beta'^2 + \frac{1}{4}\hbar^2(\omega_e + \omega_\mu)^2]^{1/2} = -\kappa' + \beta'(1 + x_{\beta'}^2)^{1/2}$

$a_1 = -a_4 = \frac{1}{2}\left(1 + \frac{1}{(1+x_{\beta'}^2)^{1/2}}\right)^{1/2} = 2^{-1/2}\cos\eta'$

$a_2 = -a_3 = -\frac{1}{2}\left(1 - \frac{1}{(1+x_{\beta'}^2)^{1/2}}\right)^{1/2} = -2^{-1/2}\sin\eta'$

$E_4 = -\kappa' - [\beta'^2 + \frac{1}{4}\hbar^2(\omega_e + \omega_\mu)^2]^{1/2} = -\kappa' - \beta'(1 + x_{\beta'}^2)^{1/2}$

$a_1 = -a_4 = \frac{1}{2}\left(1 - \frac{1}{(1+x_{\beta'}^2)^{1/2}}\right)^{1/2} = 2^{-1/2}\sin\eta'$

$a_2 = -a_3 = \frac{1}{2}\left(1 + \frac{1}{(1+x_{\beta'}^2)^{1/2}}\right)^{1/2} = 2^{-1/2}\cos\eta'$

222   *Muonium in matter*

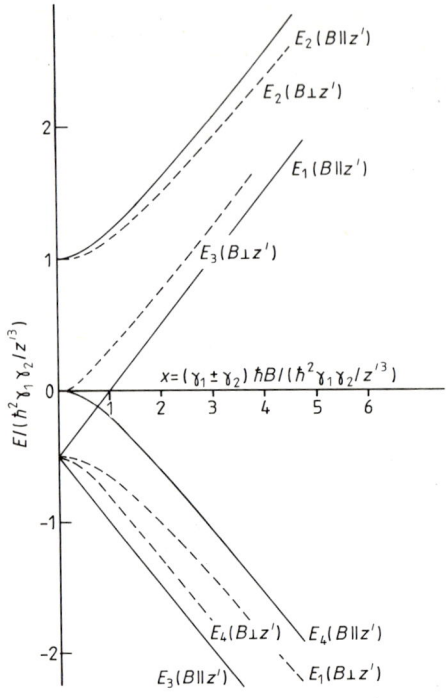

Figure 7.2 Breit–Rabi diagram of the energy eigenvalues if the hyperfine interaction is given by the dipole–dipole Hamiltonian, equation (7.18) and the external field is applied either along, or perpendicular to, the z' axis. $A_{x'x'} = A_{y'y'} = -\frac{1}{2}A_{z'z'} = \hbar^2 \gamma_1 \gamma_2 / z'^3$. Note that if $\gamma_1 = \gamma_2$ the field dependence of $E_1(B\|z')$, $E_3(B\|z')$, $E_1(B\perp z')$, $E_2(B\perp z')$ will disappear.

$$\begin{aligned}
\varphi_1 &= |\tfrac{1}{2}, \tfrac{1}{2}\rangle & &= |F=1, m_F=1\rangle \\
\varphi_2 &= 2^{-1/2}[|\tfrac{1}{2}, -\tfrac{1}{2}\rangle + |-\tfrac{1}{2}, \tfrac{1}{2}\rangle] = |F=1, m_F=0\rangle \\
\varphi_3 &= |-\tfrac{1}{2}, -\tfrac{1}{2}\rangle & &= |F=1, m_F=-1\rangle \quad (7.27) \\
\varphi_4 &= 2^{-1/2}[|\tfrac{1}{2}, -\tfrac{1}{2}\rangle - |-\tfrac{1}{2}, \tfrac{1}{2}\rangle] = |F=0, m_F=0\rangle.
\end{aligned}$$

That is, the eigenstates can be labelled in terms of the total angular momentum, $F = I + S$ and its projection on the quantisation axis.

In contrast, for zero external field and $\alpha \neq 0$ we obtain $(B\|z')$

$$\begin{aligned}
\varphi_1 &= 2^{-1/2}[|\tfrac{1}{2}, \tfrac{1}{2}\rangle + |-\tfrac{1}{2}, -\tfrac{1}{2}\rangle] = 2^{-1/2}[|F=1, m_F=+1\rangle + |F=1, m_F=-1\rangle] \\
\varphi_2 &= 2^{-1/2}[|-\tfrac{1}{2}, \tfrac{1}{2}\rangle + |\tfrac{1}{2}, -\tfrac{1}{2}\rangle] = |F=1, m_F=0\rangle \quad (7.28) \\
\varphi_3 &= 2^{-1/2}[|\tfrac{1}{2}, \tfrac{1}{2}\rangle - |-\tfrac{1}{2}, -\tfrac{1}{2}\rangle] = 2^{-1/2}[|F=1, m_F=1\rangle - |F=1, m_F=-1\rangle] \\
\varphi_4 &= 2^{-1/2}[|\tfrac{1}{2}, -\tfrac{1}{2}\rangle - |-\tfrac{1}{2}, \tfrac{1}{2}\rangle] = |F=0, m_F=0\rangle
\end{aligned}$$

and one sees that m_F is not providing a good quantum number anymore.

In treating the case **B** parallel to the x' axis, the z' axis was retained as the axis of quantisation with respect to which the basis χ_i was defined. Alternatively we can now define the x' axis as the new axis of quantisation with respect to which a new basis χ_i^0 is introduced. Let us label the new axis of quantisation z^0. The coordinate system as defined is obtained from the old one by a rotation around the y axis, which is the same in both coordinate systems. The rotation is represented by the matrix

$$\mathbf{R} = \begin{pmatrix} \cos\theta & 0 & \sin\theta \\ 0 & 1 & 0 \\ -\sin\theta & 0 & \cos\theta \end{pmatrix}. \tag{7.29}$$

where θ is the angle between the z^0 and z' axis. The tensor $\mathbf{A}^0$ in the new system is obtained from $\mathbf{A}'$ as follows

$$\mathbf{A}^0 = \mathbf{R}^{-1} \cdot \mathbf{A}' \cdot \mathbf{R} \tag{7.30}$$

and we find the new hyperfine constants

$$\begin{aligned}
\alpha_0 &= \tfrac{1}{4}(A_{x'x'}\cos^2\theta + A_{z'z'}\sin^2\theta - A_{y'y'}) \\
\beta_0 &= \tfrac{1}{4}(A_{x'x'}\cos^2\theta + A_{z'z'}\sin^2\theta + A_{y'y'}) \\
\gamma_0 &= \delta_0 = \tfrac{1}{2}(A_{x'x'} - A_{z'z'})\sin\theta\cos\theta \\
\kappa_0 &= A_{x'x'}\sin^2\theta + A_{z'z'}\cos^2\theta.
\end{aligned} \tag{7.31}$$

Inserting these constants with $\theta = 90°$ into equation (7.21), now derived with respect to the new coordinate system (x^0, y^0, z^0), we obtain the same energy eigenvalues as before (table 7.1(b)), as expected, since the energy eigenvalues are independent of the particular choice of the coordinate system. Inserting these parameters with $\theta = 90°$ into equation (7.25), we find the new eigenfunctions in terms of this new basis, χ_i^0.

As a simple example, let us wrote down the eigenfunctions for an axially symmetric hyperfine Hamiltonian with zero external magnetic field (i.e. $\alpha = 0$). In the original coordinate system the eigenfunctions were given by equation (7.27). In the new coordinate system rotated by $\theta = 90°$, instead we find

$$\begin{aligned}
\varphi_1^0 &= \sqrt{2}(|\tfrac{1}{2},\tfrac{1}{2}\rangle^0 + |-\tfrac{1}{2},-\tfrac{1}{2}\rangle^0) \\
\varphi_2^0 &= \sqrt{2}(|\tfrac{1}{2},-\tfrac{1}{2}\rangle^0 + |-\tfrac{1}{2},\tfrac{1}{2}\rangle^0) \\
\varphi_3^0 &= \sqrt{2}(|\tfrac{1}{2},\tfrac{1}{2}\rangle^0 - |-\tfrac{1}{2},-\tfrac{1}{2}\rangle^0) \\
\varphi_4^0 &= \sqrt{2}(|\tfrac{1}{2},-\tfrac{1}{2}\rangle^0 - |-\tfrac{1}{2},\tfrac{1}{2}\rangle^0).
\end{aligned} \tag{7.32}$$

If the applied field does not point along one of the principal directions of the tensor **A**, no analytic solutions for the eigenvalues can be obtained. This means, in particular, that at a given field strength the angular dependence of the eigenvalues with respect to the direction of the field cannot be written explicitly. However, an approximate solution can be obtained if the electronic Zeeman energy, $\hbar\gamma_e B$ is much larger than the hyperfine interaction energy (Slichter 1978, Hintermann et al 1980). The electron spin is then decoupled from the muon spin, and is well described by the quantum numbers $m_e = \pm\frac{1}{2}$. The z axis is defined by the direction of **B**. This implies that all terms in the Hamiltonian equation (7.20) which involve the operators S^+, S^- disappear. The truncated Hamiltonian thus assumes the form

$$\mathcal{H}_{tr} = \delta(I^+ S_z + I^- S_z)^{-1} \kappa S_z I_z - \gamma_\mu \hbar B I_z + \gamma_e \hbar B S_z. \quad (7.33)$$

The energy eigenvalues are easily obtained from equation (7.21)

$$E_{1/2} = \tfrac{1}{2}\hbar\omega_e \pm [(\tfrac{1}{2}\hbar\omega_\mu - \kappa)^2 + \delta^2]^{1/2}$$
$$E_{3/4} = -\tfrac{1}{2}\hbar\omega_e \pm [(\tfrac{1}{2}\hbar\omega_\mu + \kappa)^2 + \delta^2]^{1/2}. \quad (7.34)$$

We would now like to express κ and δ in terms of the $A_{x'x'}$, $A_{y'y'}$ and $A_{z'z'}$, as defined in the coordinate system where **A**′ is diagonal, and in terms of the direction cosines which relate the coordinate system x', y', z' to the system x, y, z, where $B \| z$. For simplicity, let us only consider the case in which the system (x, y, z) is obtained from the system (x', y', z') by a rotation around the $y = y'$ axis. Such a rotation has just been treated, and the desired relations for κ and δ are given by equations (7.31).

So far, we have assumed that the Zeeman interaction of the electron and the muon are of an isotropic nature, that is the g factors are assumed to be c numbers. It is however well known from ESR experiments in crystals, that the Zeeman interaction of the electron is often also represented by an operator of the form

$$\mathcal{H}_z = \mathbf{S} \cdot \mathbf{g} \cdot \mathbf{B} \quad (7.35)$$

where **g** is now a tensor. On the other hand, the nuclear Zeeman interaction is known to simply involve an isotropic g factor, so we will also assume that this holds true in general for the muon Zeeman interaction. The total Hamiltonian now reads

$$\mathcal{H} = \mathbf{I} \cdot \mathbf{A} \cdot \mathbf{S} + \hbar \mathbf{S} \cdot \mathbf{g} \cdot \mathbf{B} - g_\mu \hbar \mathbf{I} \cdot \mathbf{B}. \quad (7.36)$$

If the tensors **A** and **g** have different principal axes, we have quite a complex situation which has not, in fact, been treated in the μSR literature yet. Blazey et al (1981) have derived formulae for the energy eigenvalues in the much simpler case when both **A** and **g** are axially isotropic with the same axis of symmetry, and that the external field is applied either along the axis of symmetry or perpendicular to it. Since these formulae are a straightforward generalisation of the solutions of equation (7.21) they will not be reproduced here. For further details, the reader is referred to Belousov et al (1978a,b) and Hintermann et al (1980).

7.1.2 Evolution of μ^+ polarisation in muonium ground state

Muonium is formed when a μ^+ captures an electron (see also §7.2). The μ^+ are assumed to be spin polarised, say, along the z axis. The spin state of the μ^+ is therefore represented by the function

$$\chi_\mu^+ = |+\tfrac{1}{2}\rangle_\mu. \tag{7.37}$$

The electrons, on the other hand, are unpolarised. Their spin states are therefore represented, with respect to the chosen z axis, by

$$\chi_e^+ = |+\tfrac{1}{2}\rangle_e \qquad \chi_e^- = |-\tfrac{1}{2}\rangle_e \tag{7.38}$$

with a population of 50% in each state. Upon formation of muonium, the combined spin state is then given by

$$\chi_{\mu^+e^-}^+ = |\tfrac{1}{2}\rangle_e|\tfrac{1}{2}\rangle_\mu = |\tfrac{1}{2},\tfrac{1}{2}\rangle \qquad \chi_{\mu^+e^-}^- = |-\tfrac{1}{2}\rangle_e|\tfrac{1}{2}\rangle_\mu = |-\tfrac{1}{2},\tfrac{1}{2}\rangle. \tag{7.39}$$

Again, each state has a population of 50%. By comparing these expressions with equation (7.27) for $x=0$, we immediately realise that the second state is not an eigenstate of the hyperfine Hamiltonian equation (7.5), which is switched on as muonium formation takes place. The spin arrangement, $|-\tfrac{1}{2}\rangle_e|+\tfrac{1}{2}\rangle_\mu$, is hence unstable. Therefore the muon and electron spin start to precess in the magnetic field produced by the other particle, when in this arrangement. In other words, the spin polarisation vector of those μ^+ in this initial spin arrangement will become time dependent, and will evolve in time. However, the configuration $|+\tfrac{1}{2}\rangle_e|+\tfrac{1}{2}\rangle_\mu$ is an eigenstate of the hyperfine Hamiltonian, and consequently the moun polarisation in this state will not display any time dependence. If an external field is applied, as we shall see, a rather complex time dependence of the muon's spin polarisation $\boldsymbol{P}_\mu$ evolves.

In order to calculate the time evolution of $\boldsymbol{P}_\mu(t)$, we make use of the spin density matrix formalism. If the spin density matrix is called $\tilde{\rho}$, the polarisation of the μ^+, $\boldsymbol{P}_\mu = \langle\boldsymbol{\sigma}_\mu\rangle$, of the electron, $\boldsymbol{P}_e = \langle\boldsymbol{\sigma}_e\rangle$ and the mixed terms, $P_{ij} = \langle\sigma_\mu^i\sigma_e^j\rangle$ ($i,j = x, y, z$) can be obtained from the well known expressions

$$\boldsymbol{P}_\mu = \mathrm{Tr}\,(\tilde{\rho}\boldsymbol{\sigma}_\mu) \tag{7.40}$$

$$\boldsymbol{P}_e = \mathrm{Tr}\,(\tilde{\rho}\boldsymbol{\sigma}_e) \tag{7.41}$$

$$P_{ij} = \mathrm{Tr}\,(\tilde{\rho}\sigma_\mu^i\sigma_e^j). \tag{7.42}$$

The spin density matrix generally also has the property that

$$\mathrm{Tr}\,\rho = 1. \tag{7.43}$$

The expressions can be used to determine the explicit form of $\tilde{\rho}$ for muonium, which must be a 4×4 matrix (combination of two spin $\tfrac{1}{2}$ states)

$$\tilde{\rho} = \tfrac{1}{4}(\mathbf{1}_\mu \otimes \mathbf{1}_e + \boldsymbol{P}_\mu\cdot\boldsymbol{\sigma}_\mu \otimes \mathbf{1}_e + \boldsymbol{P}_e\cdot\mathbf{1}_\mu \otimes \boldsymbol{\sigma}_e + \sum_{i,j} P_{ij}\sigma_\mu^i \otimes \sigma_e^j. \tag{7.44}$$

σ_μ^i, σ_e^j are the traceless Pauli spin matrices.

Alternatively, one can introduce a spin density operator ρ, which is commonly written as (see e.g. Messiah 1965)

$$\rho = \sum_{ij} |i\rangle \rho_{ij} \langle j| \qquad (7.45)$$

where $|i\rangle$ is a system of basis states (e.g. the χ_i defined in equation (7.10)) and

$$\rho_{ij} = \langle i|\rho|j\rangle = \tilde{\rho}_{ij}. \qquad (7.46)$$

Hence

$$\rho = \tfrac{1}{4}(1 + \boldsymbol{P}_\mu \cdot \boldsymbol{\sigma}_\mu + \boldsymbol{P}_e \cdot \boldsymbol{\sigma}_e + \sum P_{ij}\sigma_\mu^i \sigma_e^j). \qquad (7.47)$$

The calculation of the time evolution of $\boldsymbol{P}_\mu(t)$ can be performed either in the Heisenberg representation, that is

$$\boldsymbol{P}_\mu(t) = \mathrm{Tr}\,[\rho(0)\boldsymbol{\sigma}_\mu(t)] \qquad (7.48)$$

or in the Schrödinger representation,

$$\boldsymbol{P}_\mu(t) = \mathrm{Tr}\,[\rho(t)\boldsymbol{\sigma}_\mu]. \qquad (7.49)$$

$\tilde{\rho}(0)$ represents the initial polarisation status of muonium at the instant of formation. $\boldsymbol{\sigma}_\mu(t)$ and $\rho(t)$ are obtained from the solution of the equations

$$-i\hbar\dot{\boldsymbol{\sigma}}_\mu(t) = [\mathcal{H}, \boldsymbol{\sigma}_\mu(t)] \qquad (7.50)$$

$$i\hbar\dot{\rho}(t) = [\mathcal{H}, \rho(t)]. \qquad (7.51)$$

$\mathcal{H}$ is the muonium hyperfine Hamiltonian. For the rest of this section we shall use the Heisenberg representation. An example for the application of the Schrödinger picture is given in §7.1.3.

The formal solution of equation (7.51) is

$$\boldsymbol{\sigma}_\mu(t) = \exp\,(\mathrm{i}\mathcal{H}t/\hbar)\boldsymbol{\sigma}_\mu \exp\,(-\mathrm{i}\mathcal{H}t/\hbar). \qquad (7.52)$$

The initial condition is characterised by a non-zero polarisation, $\boldsymbol{P}_\mu(0)$, of the μ^+ and zero electron polarisation $\boldsymbol{P}_e(0) = 0$ and $P_{ij}(0) = 0$, i.e.

$$\rho(0) = \tfrac{1}{4}(1 + \boldsymbol{P}_\mu(0)\boldsymbol{\sigma}_\mu). \qquad (7.53)$$

Inserting equations (7.52) and (7.53) into equation (7.49), and choosing the eigenfunctions of $\mathcal{H}$ as a basis, denoted by $|n\rangle$, we arrive at

$$\boldsymbol{P}_\mu(t) = \tfrac{1}{4} \sum_{k,n,m} \langle k|n\rangle\langle n|1 + \boldsymbol{P}_\mu(0)\boldsymbol{\sigma}_\mu|m\rangle\langle m|\boldsymbol{\sigma}_\mu(t)|k\rangle$$

$$= \tfrac{1}{4} \sum_{n,m} \langle n|1 + \boldsymbol{P}_\mu(0)\boldsymbol{\sigma}_\mu|m\rangle\langle m|\boldsymbol{\sigma}_\mu|n\rangle \exp\,\mathrm{i}\omega_{mn}t \qquad (7.54)$$

where

$$\omega_{mn} = (E_m - E_n)/\hbar. \qquad (7.55)$$

Magnetic properties in the 1S ground state

Taking into account that $\mathrm{Tr}\,\sigma_\mu^i = 0$ ($i = x, y, z$), equation (7.54) simplifies to

$$P_\mu(t) = \tfrac{1}{4} \sum_{n,m} \langle n|P_\mu(0)\sigma_\mu|m\rangle\langle m|\sigma_\mu|n\rangle \exp i\omega_{mn}t. \quad (7.56)$$

In practice, the eigenfunctions $|n\rangle$ will be given in terms of the basis states, χ_i, that is

$$|n\rangle = \sum_i a_{ni}|\chi_i\rangle \quad (7.57)$$

$$\langle n| = \sum_j \langle \chi_j|a_{jn}^{-1} = \sum_j \langle \chi_j|a_{nj}^* \quad (7.58)$$

where the a_{nj} are, for example, given in table 7.1(a) and (b). Note that the matrix (a_{ij}) represents a unitary transformation. We then obtain

$$P_\mu(t) = \tfrac{1}{4} \sum_{\substack{n,m,\\i,j,k,l}} a_{nj}^* a_{mi} a_{mk}^* a_{nl} \langle \chi_j|P_\mu(0)\sigma_\mu|\chi_i\rangle\langle \chi_k|\sigma_\mu|\chi_l\rangle \exp i\omega_{mn}t. \quad (7.59)$$

We are now in the position to write down the solution for $P_\mu(t)$ in a straightforward manner for various special situations. Let us first consider the case of an isotropic hyperfine Hamiltonian, plus an applied field along the z direction. The energy eigenvalues for this case are given by equations (7.22). Two initial conditions for the muon polarisation will be considered.

(1) $P_\mu(t=0) = P_z(t=0)\hat{z}_0 = 1\hat{z}_0$

Here the μ^+ is polarised in the z direction. Applied field and μ^+ polarisation are parallel. This condition is commonly referred to as the *longitudinal field case*. For this case the initial muonium states are

$$|\tfrac{1}{2}, \tfrac{1}{2}\rangle \qquad |-\tfrac{1}{2}, \tfrac{1}{2}\rangle$$

With the help of table 7.1(a), equation (7.59) is easily evaluated and yields

$$|P_\mu(t)| = P_z(t) = 1 - \tfrac{1}{2}[1 - \cos \omega_0(1+x^2)^{1/2}t]/(1+x^2) \quad (7.60)$$

with $\omega_0 = A/\hbar$. Note that the muon polarisation is also, in the course of time, restricted to the z direction. No components perpendicular to the field direction develop. The evaluation of $P_z(t)$ is characterised by a time independent, but field dependent, term (called the residual polarisation) and a time and field dependent term with periodicity

$$(\omega_0/2\pi)(1+x^2)^{1/2} \geqslant 4.46\,\mathrm{GHz}$$

for muonium in vacuum. Since in most experiments the time resolution is not sufficient to observe such a fast modulation, the time dependent term is, in effect, averaged out to zero. The residual polarisation is then given by

$$P_z = (1 + 2x^2)/(2 + 2x^2). \tag{7.61}$$

In zero field, P_z is effectively reduced to 50% of the initial polarisation (corresponding to the population in the initial $|\frac{1}{2},\frac{1}{2}\rangle$ state) and approaches 100% upon increasing the applied field to large values (see figure 7.3). This is a consequence of the decoupling of electron and muon spin in strong magnetic fields (Paschen–Back effect) and is usually referred to as the 'quenching of depolarisation'.

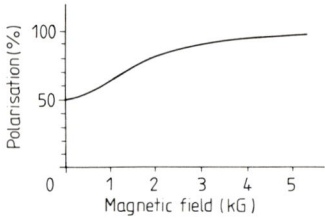

Figure 7.3 Residual (time independent) μ^+ polarisation in the Mu state in a longitudinal magnetic field. The repolarisation with increasing field reflects the decoupling of electron and μ^+ spin (Paschen–Back effect).

(2) $\boldsymbol{P}_\mu(t=0) = P_x(t=0)\hat{\boldsymbol{x}}_0 = 1\hat{\boldsymbol{x}}_0$.

Here the μ^+ are polarised in the x direction perpendicular to the applied field (the z direction). This is referred to as the *transverse field case*. We choose the direction of the initial polarisation as the axis of quantisation, that is the x axis. The initial situation is then again described by the two states $|\frac{1}{2},\frac{1}{2}\rangle, |-\frac{1}{2},\frac{1}{2}\rangle$ with populations of 50% each. The eigenvectors and eigenstates for this case are listed in table 7.1(b), setting $\alpha' = 0$. We only have to interchange the labelling of the x and z axes. The evaluation of equation (7.59) is lengthy but straightforward, and results in

$$P_x(t) = \tfrac{1}{4}[(1+\delta)\cos\omega_{12}t + (1-\delta)\cos\omega_{14}t \\ + (1+\delta)\cos\omega_{34}t + (1-\delta)\cos\omega_{23}t] \tag{7.62}$$

$$P_y(t) = \tfrac{1}{4}[-(1+\delta)\sin\omega_{12}t + (\delta-1)\sin\omega_{14}t \\ + (\delta-1)\sin\omega_{23}t + (1+\delta)\sin\omega_{34}t] \tag{7.63}$$

$$P_z(t) = 0 \tag{7.64}$$

where $\omega_{ij} = (E_i - E_j)/\hbar$ and $\delta = x(1+x^2)^{-1/2}$.

We notice that the time evolution of the muon's polarisation is characterised by four frequencies. These correspond to the possible transitions among the hyperfine states of muonium obeying the selection rules $\Delta F = 0, 1$, $\Delta m_F = 1$ at low fields and $\Delta m_e = 0, \Delta m_\mu = 1$ at very high fields. Note that at these fields the terms with ω_{14} and ω_{23} involving the electronic transitions $\Delta m_e = 1$ vanish due to the effective decoupling of μ^+ spin and electron spin in the Paschen–Back limit.

As a final example we consider the time evolution of the muons' polarisation in the presence of an anisotropic hyperfine Hamiltonian only (i.e. no external field). We are interested in particular in the dependence of $P_\mu(t)$ on the direction of the initial muon polarisation with respect to the principal axis (x', y', z') of the tensor $\mathbf{A}$. The z' axis is chosen as the axis of quantisation. The energy eigenvalues and the eigenfunctions are obtained from table 7.1(a) by setting $x=0$

$$E_{1,3} = \tfrac{1}{4}[A_{z'z'} \pm (A_{x'x'} - A_{y'y'})] \qquad \varphi_{1,3} = 2^{-1/2}(\chi_1 \pm \chi_4) \quad (7.65)$$

$$E_{2,4} = \tfrac{1}{4}[-A_{z'z'} \pm (A_{x'x'} + A_{y'y'})] \qquad \varphi_{2,4} = 2^{-1/2}(\chi_2 \pm \chi_3). \quad (7.66)$$

The initial μ^+ polarisation along the x', y' and z' axes are obtained from a projection of $P_\mu(0)$ onto these axes

$$P_{x'}(0) = P(0) \cos \alpha_{x'}$$
$$P_{y'}(0) = P(0) \cos \alpha_{y'} \qquad (7.67)$$
$$P_{z'}(0) = P(0) \cos \alpha_{z'}$$

where $\cos \alpha_i$ are the direction cosines.

The evaluation of equation (7.59) is straightforward and yields for the various components (Holzschuh 1982)

$$P_{x'}(t) = \tfrac{1}{2}P(0) \cos \alpha_{x'}(\cos \omega_{12}t + \cos \omega_{34}t)$$
$$P_{y'}(t) = \tfrac{1}{2}P(0) \cos \alpha_{y'}(\cos \omega_{14}t + \cos \omega_{23}t) \qquad (7.68)$$
$$P_{z'}(t) = \tfrac{1}{2}P(0) \cos \alpha_{z'}(\cos \omega_{13}t + \cos \omega_{24}t)$$

where

$$\omega_{12} = (A_{z'z'} - A_{y'y'})/2\hbar = \omega_{x'}^- \qquad \omega_{34} = (A_{z'z'} + A_{y'y'})/2\hbar = \omega_{x'}^+$$
$$\omega_{23} = (A_{x'x'} - A_{z'z'})/2\hbar = \omega_{y'}^- \qquad \omega_{14} = (A_{z'z'} + A_{x'x'})/2\hbar = \omega_{y'}^+ \quad (7.69)$$
$$\omega_{13} = (A_{x'x'} - A_{y'y'})/2\hbar = \omega_{z'}^- \qquad \omega_{24} = (A_{x'x'} + A_{y'y'})/2\hbar = \omega_{zi}^+.$$

If the polarisation is monitored in the direction of the initial polarisation, $P_\mu(0)$, as is usually done, $P_\mu(t)$ must be projected back onto this axis, that is

$$P(t) = P_\mu(t) \cdot P_\mu(0)/|P_\mu(0)|$$

$$= \sum_{i=x',y',z'} P_i(t) \cos \alpha_i \qquad (7.70)$$

$$= \tfrac{1}{2} P_\mu(0) \sum_{i=x',y',z'} \cos^2 \alpha_i (\cos \omega_i^+ t + \cos \omega_i^- t). \qquad (7.71)$$

Equation (7.71) reduces to the proper limits for an isotropic hyperfine Hamiltonian ($\omega_{12} = \omega_{13} = \omega_{23} = 0$, $\omega_{24} = \omega_{34} = \omega_{14}$) or an axially symmetric hyperfine one ($\omega_{13} = 0$). In the general case, one observes six frequencies. The spectral weights of these frequencies strongly depend on the initial direction of $P_\mu(0)$. Frequencies can be made to appear or disappear through a proper choice of $P_\mu(0)$.

7.1.3 Inclusion of electron relaxation and chemical lifetime

A muonium atom formed in condensed matter may be subject to various interactions and chemical reactions of muonium are one possibility. These can place the muon into a diamagnetic compound and thereby depopulate the muonium state. This leads to a corresponding reduction of the muonium related μSR signal. Another possibility is interactions involving the muonium electron spin which results in relaxation of the electron spin. In this section, we would like to explore how these effects are manifest in the μSR signal. If chemical reactions which place the μ^+ into a diamagnetic state are present, it may become possible to observe the corresponding μSR signal and this is much easier to measure than the muonium signal. As we will see, the amplitude and, in the transverse field case, the phase of this signal reflect much of what has happened in the muonium phase.

Our treatment again will be based on the spin density formalism. However, this time we follow a more general procedure in which the time dependence is carried by the spin density matrix $\rho(t)$. In this we are following Iakovleva (1958), Nosov and Yakovleva (1962), Ivanter and Smilga (1967, 1968, 1971, 1972), Ivanter (1968).

As is well known, the time evaluation of the spin density matrix obeys the differential equation

$$i\hbar d\rho(t)/dt = [\mathcal{H}, \rho(t)] \qquad (7.72)$$

which has to be solved in order to find the evolution of the muon's polarisation. The spin density matrix of muonium is a 4×4 matrix (there are four hyperfine states) with trace zero and is given by (see equation (7.44))

$$\tilde{\rho} = \tfrac{1}{4}(\mathbb{1}_\mu \otimes \mathbb{1}_e + \boldsymbol{P}_\mu \cdot \boldsymbol{\sigma}_\mu \otimes \mathbb{1}_e + \boldsymbol{P}_e \cdot \mathbb{1}_\mu \otimes \boldsymbol{\sigma}_e + \sum_{i,j} P_{ij}\sigma_\mu^i \otimes \sigma_e^j). \qquad (7.73)$$

For the initial state ($P_{ij}=0$, $\boldsymbol{P}_e=0$), this reduces to

$$\rho_0 = \tfrac{1}{4}(\mathbb{1}_\mu \otimes \mathbb{1}_e + \boldsymbol{P}_\mu(0) \cdot \boldsymbol{\sigma}_\mu \otimes \mathbb{1}_e). \qquad (7.74)$$

In the following, we will only consider the hyperfine Hamiltonian given in its isotropic form

$$\mathcal{H} = \hbar\omega_0 \boldsymbol{S} \cdot \boldsymbol{I} + \gamma_e \hbar \boldsymbol{S} \cdot \boldsymbol{B} - \gamma_\mu \hbar \boldsymbol{I} \cdot \boldsymbol{B}. \qquad (7.75)$$

Inserting equation (7.73) into equation (7.72), with $\mathcal{H}$ given by equation (7.75), evaluating the commutator and equating terms on both sides of equation (7.72) which involve the same operator $\sigma_\mu^i \sigma_e^j$, we finally arrive at the following coupled system of 15 differential equations $(i, j = 1, 2, 3)$

$$\dot{P}_{i0} = -\frac{\omega_0}{2} \sum_{j,k=1}^{3} P_{jk}\varepsilon_{ijk} - \sum_{l,k=1}^{3} \omega_l^\mu P_{l0}\varepsilon_{ilk}$$

$$\dot{P}_{0i} = \frac{\omega_0}{2} \sum_{j,k=1}^{3} P_{jk}\varepsilon_{ijk} + \sum_{l,k=1}^{3} \omega_l^e P_{0k}\varepsilon_{ilk} - 2v_i P_{0i} \quad (7.76)$$

$$\dot{P}_{ij} = \frac{\omega_0}{2} \sum_{l=1}^{3} (P_{l0}\varepsilon_{ijl} - P_{0l}\varepsilon_{ijl}) - \sum_{k,l=1}^{3} \omega_k^\mu P_{lj}\varepsilon_{ikl} + \sum_{k,l=1}^{3} \omega_k^e P_{il}\varepsilon_{jkl} - 2v_i P_{ij}$$

where $\omega_l^\mu = |\gamma_\mu B_l|$, $\omega_l^e = |\gamma_e B_l|$ and $\varepsilon_{ijk} =$ antisymmetric unit tensor.

In writing down these so-called Wangsness–Bloch equations (Wangsness and Bloch 1953), we have added an additional damping term with rate $2v_i$ to the right side of the equations describing the evolution of the pure and mixed electron polarisation components. In this purely phenomenological manner, the possibility of electron spin relaxation is taken into account. In its most general form, we allow for the possibility that each electron component has a different relaxation rate.

The system of equations (7.76) can be separated into two irreducible subsystems, one only involving the components of muon and electron polarisation $(P_\mu^\parallel, P_e^\parallel)$ along the magnetic field direction $\mathbf{B}$ (longitudinal subsystem), and the other only the components $(P_\mu^\perp, P_e^\perp)$ perpendicular to $\mathbf{B}$ (transverse field subsystem). We specify the z axis to be parallel to $\mathbf{B}$.

(1) Longitudinal subsystem

The equations of motion for those components of polarisation coupled to the muon polarisation along the field, P_{30}, are

$$\dot{P}_{30} = -\tfrac{1}{2}\omega_0(P_{12} - P_{21}) \qquad \dot{P}_{03} = \tfrac{1}{2}\omega_0(P_{12} - P_{21}) - 2v_\parallel P_{03}$$

$$\dot{P}_{11} = -\omega_e P_{12} + \omega_\mu P_{21} - 2v_\perp P_{11} \qquad \dot{P}_{22} = \omega_e P_{21} - \omega_\mu P_{12} - 2v_\perp P_{22}$$

$$\dot{P}_{12} = \tfrac{1}{2}\omega_0(P_{30} - P_{03}) + \omega_e P_{11} + \omega_\mu P_{22} - 2v_\perp P_{12} \quad (7.77)$$

$$\dot{P}_{21} = -\tfrac{1}{2}\omega_0(P_{30} - P_{03}) - \omega_e P_{22} - \omega_\mu P_{11} - 2v_\perp P_{21}.$$

where we have used $v_3 = v_\parallel$, $v_1 = v_2 = v_\perp$. $v_\parallel = 1/T_1$ is the longitudinal or spin–lattice relaxation rate, and $v_\perp = 1/T_2$ is the transverse relaxation rate of the electron spin.

The solution of these equations is rather involved in the general case. Here we shall only treat a few special cases.

(i) $v_\parallel = v_\perp = 0$. This is trivial, and the solution for $P_{30}(t) = P_z(t)$ is given by equation (7.60).

232   Muonium in matter

(ii) *Fast electron relaxation* $v = v_\perp = v_\parallel \gg \omega_0(1+x^2)^{1/2}$. ($x$ is defined by equation (7.23).) According to Nosov and Yakovleva (1962) one finds

$$P_z(t) = P_{30}(t) \simeq \exp(-t/\tau_1) \tag{7.78}$$

with

$$\tau_1 = 4v/\omega_0^2. \tag{7.79}$$

The polarisation is exponentially damped with a damping rate τ_1^{-1} which is independent of the applied field strength. We also notice that the oscillating term in equation (7.60) has disappeared. This property reflects the fact that at high electron spin-flip rates the hyperfine coupling of electron and muon spin is broken. Accordingly, with rising relaxation rate, v, of the electron, the muon is less affected by it and increasingly behaves as if it was 'free'.

(iii) *Slow electron relaxation* $v = v_\parallel = v_\perp < \omega_0(1+x^2)^{1/2}$. According to Nosov and Yakovleva (1962), one obtains a somewhat more complicated expression

$$P_z(t) = P_{30}(t) = \frac{1+2x^2}{2(1+x^2)} \exp\left(-\frac{v}{1+x^2}t\right) + \frac{1}{4(1+x^2)} \exp\left(-\frac{(3+4x^2)v}{2(1+x^2)}t\right)$$

$$\times \left(2\cos\omega_0(1+x^2)^{1/2}t + \frac{(5+8x^2)v}{\omega_0(1+x^2)^{3/2}} \sin\omega_0(1+x^2)^{1/2}t\right). \tag{7.80}$$

For $v \to 0$, this formula reduces to equation (7.60). Of particular interest is the non-oscillating term which is now exponentially damped with a damping rate proportional to the electron spin relaxation rate, v. For vanishing fields, that is $x \to 0$, the muon spin closely follows the electron spin. The decoupling of electron and muon spin (Paschen–Back effect) reduces the muon spin relaxation rate for larger applied fields.

(2) *Transverse field subsystem*

The magnetic field $\boldsymbol{B}$ is directed along the z axis and the initial muon polarisation is assumed to be along the x axis. Following Ivanter and Smilga (1967) we introduce the complex four component vector

$$\check{P} = \begin{Bmatrix} P_{10} + iP_{20} \\ P_{01} + iP_{02} \\ P_{13} + iP_{23} \\ P_{31} + iP_{32} \end{Bmatrix}. \tag{7.81}$$

The first component is the complex transverse muon polarisation, the quantity in which we are interested. The time derivative of $\check{P}$ can be written with the use of equations (7.76)

$$\dot{\check{P}} = i\mathbf{A}\check{P} \tag{7.82}$$

where **A** is a 4×4 complex matrix

$$\mathbf{A} = \frac{\omega_0}{2} \begin{pmatrix} -2\omega_\mu/\omega_0 & 0 & 0 & -1 \\ 0 & (i\gamma_\perp + 2\omega_e/\omega_0) & -1 & 1 \\ 1 & -1 & (i\gamma_\| - 2\omega_\mu/\omega_0) & 0 \\ -1 & 1 & 0 & (i\gamma_\perp + 2\omega_e/\omega_0) \end{pmatrix} \tag{7.83}$$

where

$$\gamma_{\|,\perp} = \frac{4v_{\|,\perp}}{\omega_0}.$$

The equation of motion (7.82) is solved by diagonalising matrix **A**. The orthogonal matrix **M** that diagonalises **A**,

$$\mathbf{M}^{-1}\mathbf{A}\mathbf{M} = \lambda \tag{7.84}$$

and the resulting eigenvalues λ can be found by standard, but tedious, manipulations. The solution for $\check{P}$ is then given by

$$\check{P}(t) = \sum_{k=1}^{4} \check{P}_k \exp{-\lambda_k t} \tag{7.85}$$

subject to the initial condition

$$\check{P}(0) = \begin{pmatrix} 1 \\ 0 \\ 0 \\ 0 \end{pmatrix}. \tag{7.86}$$

Again we shall only discuss a few special cases.

(i) $v_\| = v_\perp = 0$. **A** is now a real symmetric matrix. The solutions for $P_x(t) = P_{10}(t)$ and $P_y(t) = P_{20}(t)$ are given by equations (7.62) and (7.63).

(ii) Fast electron relaxation $v = v_\| = v_\perp \gg \omega_0(1 + x^2)^{1/2}$. According to Nosov and Yakoleva (1962), one finds

$$P_x(t) = P_{10}(t) \simeq \exp{(-\omega_0^2 t/4v)} \cos{\omega_\mu t}. \tag{7.87}$$

Similarly to the longitudinal field case the muon behaves as quasi-free and precesses with its own Larmor frequency $\omega_\mu = \gamma_\mu B$. The polarisation is exponentially damped with the same rate as in the longitudinal fields case.

(iii) Slow electron relaxation $v = v_\| = v_\perp < \omega_0(1+x^2)^{1/2}$. Gurevich et al (1971) have calculated the time dependence of $P_x(t)$ in the limit $v^2 \ll (\omega_0/2)^2 x^4$. Neglecting the terms with frequencies ω_{14} and ω_{24} (which usually cannot be resolved in an experiment), they arrive at the result

$$P_x(t) \simeq \tfrac{1}{2} \exp(-t/\tau)[(\cos \Omega_\gamma t + \Omega \sin \Omega_\gamma t / 3\tau \Omega_\gamma^2) \cos \omega_- t$$
$$+ (2\omega_+ \Omega^2 / \omega_0 \Omega_\gamma^2) \sin \Omega_\gamma t \sin \omega_- t] \qquad (7.88)$$

where

$$\omega_- = \tfrac{1}{2}(\omega_{12} + \omega_{23}) = \tfrac{1}{2}(|\omega_e| - |\omega_\mu|) \qquad (7.89)$$
$$\omega_+ = \tfrac{1}{2}(|\omega_e| + |\omega_\mu|) \qquad (7.90)$$
$$\Omega = \tfrac{1}{2}(\omega_{23} - \omega_{12}) = \tfrac{1}{2}\omega_0[(1+x^2)^{1/2} - 1] \qquad (7.91)$$

and the beat frequency

$$\Omega_\gamma = \Omega(1 - v^2/4\Omega^2)^{1/2}. \qquad (7.92)$$

The precession signal is exponentially damped with

$$\tau = 2/3v. \qquad (7.93)$$

We notice that the beat frequency also depends on the relaxation rate v.

A remark on the intermediate case $v \simeq \omega_0/2\pi$ and small fields seems appropriate here. In this case, neither μ^+ nor muonium precession is experimentally observable, since the muon polarisation is lost to the medium through the electron in times of the order of ω_0^{-1}. Under special circumstances (small ω_0, small external field) this depolarisation may become accessible to experimental observation.

We shall now deal with the chemical reactions of muonium (Mu) which place the μ^+ into a diamagnetic state. The disappearance of muonium by chemical reactions can be described by an exponential decay law (see e.g. Walker 1983)

$$[\text{Mu}] = [\text{Mu}]_0 \exp(-t/\tau_{ch}) \qquad (7.94)$$

with τ_{ch} as the average chemical lifetime of Mu and $[\text{Mu}]_0 = 1$. Correspondingly, the typical muonium precession signal (i.e. equation (7.62)) will disappear exponentially with the time constant, τ_{ch}. The exponential disapperarance of the muonium signal has been intensively studied in connection with the chemical kinetics of muonium (see e.g. Walker 1983). Such studies are possible, however, only in those cases where the time constant, τ_{ch}, is long enough to render the observation of the Muonium precession signal possible. If the chemical lifetime is too short, there is still another possibility of gaining information on the muonium state. This is by looking at the so-called residual polarisation which is carried by the μ^+ after their transfer into the diamagnetic state.

We shall first consider the transverse field case. The muon's polarisation in the Mu state is expressed, as before, in the complex notation

$$\check{P}_\mu(t) = P_{10}(t) + iP_{20}(t). \tag{7.95}$$

The fraction of Mu's that react and disappear, at time t', in the interval dt', is given by

$$dn_\mu = -\tau_{ch}^{-1} \exp(-t'/\tau_{ch}) \, dt'$$

as follows from equation (7.94). This fraction with polarisation $\check{P}_\mu(t')$, starts to precess with the muon Larmor frequency ω_μ. The residual polarisation at time t is then given by a superposition of the polarisations of all muons that have evolved from the muonium state before t, precessing ever since with frequency ω_μ, and the polarisation of those μ^+ still in the muonium state

$$\check{P}_{res}(t) = \int_0^t \check{P}_\mu(t') \exp[i\omega_\mu(t-t')] \exp(-t'/\tau_{ch}) \frac{dt'}{\tau_{ch}} + \check{P}(t)\exp(-t/\tau_{ch}).$$

For t going to infinity or, in practice, to values much greater than τ_{ch}, equation (7.96) reduces to

$$\check{P}_{res}(t) = \left(\frac{1}{\tau_{ch}} \int_0^\infty \check{P}_\mu(t') \exp[-(i\omega_\mu + \tau_{ch}^{-1})t'] \, dt'\right) \exp(i\omega_\mu t). \tag{7.97}$$

The amplitude of this precession signal with frequency ω_μ is now the residual polarisation $\check{P}_\perp(\infty)$.

As was first noted by Ivanter and Smilga (1967), $\check{P}_\perp(\infty)$ can be understood as the Laplace transform

$$\mathcal{L}(u) = \int_0^\infty \exp(-ut)\check{P}_\mu(t) \, dt \tag{7.98}$$

of the polarisation $\check{P}_\mu(t)$ with $u = (i\omega_\mu + \tau_{ch}^{-1})$.

Applying the same Laplace transformation to the system of differential equations (7.82), one derives a set of linear equations, which allows a straightforward evaluation of $\check{P}_\perp(\infty) = \mathcal{L}(u)$

$$(\mathbf{A} - u\mathbf{E})\mathcal{L}(u) = -\check{P}(t=0) = \begin{pmatrix} 1 \\ 0 \\ 0 \\ 0 \end{pmatrix} \tag{7.99}$$

where $\mathbf{A}$ is the 4×4 matrix defined in equation (7.83), and $\mathbf{E}$ is the unit 4×4 matrix and $\check{P}(t=0) = P_{10}(t=0)$ represents the initial muon polarisation (assumed to be along the x axis). The solution of this system of inhomogeneous linear equations is simply given by

$$\check{P}_\perp(\infty) = \tau_{ch}^{-1} \mathbf{M}(u)/|\mathbf{A} - u\mathbf{E}|. \tag{7.100}$$

$|\mathbf{A} - u\mathbf{E}|$ is the characteristic determinant of the system (7.99), and $\mathbf{M}(u)$ is the corresponding minor. The final result for the transverse field case is (Ivanter and Smilga 1967, Ivanter 1968)

$$\check{P}_\perp(\infty) = \left(1 + \tfrac{1}{2}i\tau_{ch}\omega_0 \frac{B(A+B)}{[AB^2 - (A+B)]}\right) \tag{7.101}$$

where

$$B = i\left(\frac{2}{\omega_0 \tau_{ch}} + \frac{4v_\perp}{\omega_0}\right) - 2x$$

$$A = i\left(\frac{2}{\omega_0 \tau_{ch}} + \frac{4v_\parallel}{\omega_0}\right)$$

and

$$x = \frac{\gamma_e - \gamma_\mu}{\omega_0} B.$$

Note that $\check{P}_\perp(\infty)$ is a complex quantity, from which we get the desired polarisations in the x and y directions

$$P_x(\infty) = \mathrm{Re}[\check{P}_\perp(\infty)]$$
$$P_y(\infty) = \mathrm{Im}[\check{P}_\perp(\infty)] \tag{7.102}$$

with phase

$$\tan\phi = P_y(\infty)/P_x(\infty). \tag{7.103}$$

As an example, figure 7.4 shows schematically the dependence of P_x and ϕ on the average chemical lifetime in a small field, assuming $v_\parallel = v_\perp = 0$.

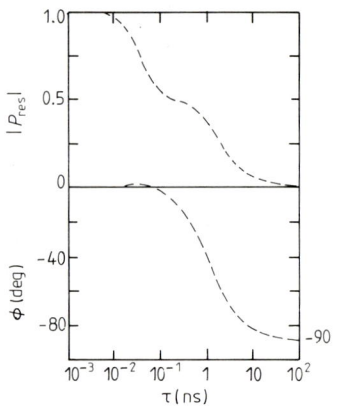

Figure 7.4 Amplitude, $|P_{res}|$, and phase, ϕ, of the residual μ^+ polarisation in a transverse field (100 G), as a function of the chemical lifetime, τ, of the muonium state (equations (7.102) and (7.103)).

One can also derive the result for the residual polarisation for the longitudinal field case in the same manner. The result is (Ivanter 1968)

$$\tilde{P}_\parallel(\infty) = P_z(\infty) = 1 - (\omega_0\tau_{ch})^2(\tfrac{1}{2} + v_\perp\tau_{ch})$$

$$\times \left[(1+2v_\perp\tau_{ch})^2 + (\omega_0\tau_{ch})^2\left(\frac{(1+2v_\perp\tau_{ch})(1+v_\parallel\tau_{ch})}{1+2v_\parallel\tau_{ch}} + x^2\right)\right]^{-1}.$$

(7.104)

We see that in both expressions (7.101) and (7.104), $v_\parallel$ as well as $v_\perp$ enter. These relaxation rates are to be identified with the longitudinal and transverse electron spin relaxation rates involving the relaxation times $T_1 = 1/v_\parallel$ and $T_2 = 1/v_\perp$. These rates may not necessarily be identical, depending on the origin of the electronic relaxation. In general both rates are also expected to depend on the applied field strength. It is obvious that measurements of the residual polarisation, both in the longitudinal and the transverse fields, provides a means of studying fast chemical reactions as well as the effects of electronic relaxation. While chemical reaction rates have been studied in this way (Brewer et al 1974), no corresponding investigations of the electronic relaxation are so far known.

7.1.4 Transitions between muonium states with different hyperfine Hamiltonians

In the preceding section, expressions were found for the μ^+ polarisation in the diamagnetic state which had evolved at random times from a precursor muonium state. The transition from the muonium state to the diamagnetic state was ascribed to a chemical reaction although other processes terminating the hyperfine coupling in the muonium state, may also be considered. The scheme of treating such transitions, as outlined in the previous section can be easily extended to the case when the terminal state is also a muonium state, but with a different hyperfine Hamiltonian (see Meier 1982). If the muonium electron in either the precursor or the terminal state does not relax, a very straightforward derivation of the μ^+ polarisation in the second state can be achieved (Percival and Fischer 1976). Since, in all the practical cases so far the muonium electron relaxation can be neglected, we will describe here just such a treatment. It is, in fact, very similar to the one in §7.1.2. We are interested specifically in the μ^+ polarisation $P_\mu(t, t')$ in the second state, characterised by a hyperfine Hamiltonian $\tilde{\mathcal{H}}$, where

$$\tilde{\mathcal{H}} = \tilde{\mathcal{H}}_{hf} + \text{Zeeman terms.} \qquad (7.105)$$

$\tilde{\mathcal{H}}_{hf}$ may be of the general form equation (7.7). This state is entered at time t' by a transition from the precursoe state governed by the Hamiltonian

$$\mathcal{H} = \mathcal{H}_{hf} + \text{Zeeman terms.} \qquad (7.106)$$

238   Muonium in matter

$\mathcal{H}_{hf}$ may also be of the general form equation (7.7). We assume that both the μ^+ polarisation and the electron polarisation, as well as the correlations between these, are conserved in the transition event. The starting conditions at time t' in the second state are therefore given by the spin density matrix $\rho(t')$ which has evolved from some initial condition at $t=0$ under the action of $\mathcal{H}$, that is

$$\rho(t') = \exp\left(-\frac{i}{\hbar}\mathcal{H}t'\right)\rho(0)\exp\left(\frac{i}{\hbar}\mathcal{H}t'\right). \tag{7.107}$$

Denoting the eigenstates of $\mathcal{H}$ by $|i\rangle, |j\rangle$ etc, the spin density operator $\rho(0)$ can be expressed as

$$\rho(0) = \sum_{i,j} |i\rangle\langle i|\rho(0)|j\rangle\langle j| \tag{7.108}$$

and

$$\rho(t') = \sum_{i,j} \exp\left(-\frac{i}{\hbar}(E_i - E_j)t'\right)|i\rangle\langle i|\rho(0)|j\rangle\langle j| \tag{7.109}$$

where E_i and E_j are the corresponding energy eigenvalues.

The polarisation $\boldsymbol{P}(t, t')$ with $t > t'$ follows from equation (7.39) and is given by

$$P_\mu(t, t') = \mathrm{Tr}[\tilde{\rho}(t - t')\sigma_\mu] \tag{7.110}$$

where

$$\tilde{\rho}(t - t') = \exp\left(-\frac{i}{\hbar}\tilde{\mathcal{H}}(t - t')\right)\rho(t')\exp\left(\frac{i}{\hbar}\tilde{\mathcal{H}}(t - t')\right). \tag{7.111}$$

It is convenient to introduce now, as a basis, the eigenfunctions of $\tilde{\mathcal{H}}$, denoted by $|\tilde{n}\rangle, |\tilde{m}\rangle$ etc. In this basis, the spin density operator $\rho(t')$ assumes the form

$$\rho(t') = \sum_{n,m} |\tilde{n}\rangle\langle \tilde{n}|\rho(t')|\tilde{m}\rangle\langle \tilde{m}|$$

$$= \sum_{n,m}\sum_{i,j} |\tilde{n}\rangle\langle \tilde{n}|i\rangle\langle i|\rho(0)|j\rangle\langle j|\tilde{m}\rangle\langle \tilde{m}| \exp\left(-\frac{1}{\hbar}(E_i - E_j)t'\right) \tag{7.112}$$

and

$$\tilde{\rho}(t - t') = \sum_{n,m}\sum_{i,j}\left[\exp\left(-\frac{i}{\hbar}(E_i - E_j)t'\right)\exp\left(-\frac{i}{\hbar}(\tilde{E}_n - \tilde{E}_m)(t - t')\right)\right.$$

$$\left.\times |\tilde{n}\rangle\langle \tilde{n}|i\rangle\langle i|\rho(0)|j\rangle\langle j|\tilde{m}\rangle\langle \tilde{m}|\right] \tag{7.113}$$

where $\tilde{E}_n, \tilde{E}_m$ are the corresponding energy eigenvalues of $\tilde{\mathcal{H}}$. The trace in equation (7.110) can now be evaluated and one obtains

$$P_\mu(t, t') = \sum_k \langle \bar{k}|\tilde{\rho}(t-t')\sigma_\mu|\bar{k}\rangle$$

$$= \sum_{k,m,i,j} \exp\left(-\frac{i}{\hbar}(E_i - E_j)t'\right) \exp\left(-\frac{i}{\hbar}(\tilde{E}_k - \tilde{E}_m)(t-t')\right)$$

$$\times \langle \bar{k}|i\rangle\langle i|\rho(0)|j\rangle\langle j|\tilde{m}\rangle\langle \tilde{m}|\sigma_\mu|\bar{k}\rangle. \qquad (7.114)$$

Finally we have to superimpose, at time t, the various contributions of $P_\mu(t, t')$, associated with different transition times, t'. If the first state disappears exponentially with rate λ, the number of μ^+ emerging from the first state at time t' in the time interval dt' is given by

$$dn = \lambda \exp(-\lambda t') dt' \qquad (7.115)$$

and the total polarisation $P(t)$ at time t is given by

$$P_\mu(t) = \lambda \int_0^t \exp(-\lambda t') P_\mu(t, t') dt'. \qquad (7.116)$$

For times $t \gg \lambda^{-1}$, this expression reduces to

$$P_\mu(t) = \sum_{i,j,k,m} \left(\frac{\lambda^2 \exp[-i(\tilde{\omega}_{km} t + \varphi_{ijkm})]}{(\omega_{ij} - \tilde{\omega}_{km})^2 + \lambda^2} \langle \bar{k}|i\rangle\langle i|\rho(0)|j\rangle\langle j|\tilde{m}\rangle\langle \tilde{m}|\sigma_\mu|\bar{k}\rangle \right)$$

$$(7.117)$$

where

$$\omega_{ij} = \frac{1}{\hbar}(E_i - E_j) \qquad \tilde{\omega}_{km} = \frac{1}{\hbar}(\tilde{E}_k - \tilde{E}_m) \qquad (7.118)$$

$$\tan \varphi_{ijkm} = (\omega_{ij} - \tilde{\omega}_{km})/\lambda.$$

Inspection of equation (7.117) directly reveals some of the important features of the μ^+ polarisation in the second state. Oscillating terms with $\tilde{\omega}_{km}$ will be small if $|\tilde{\omega}_{km} - \omega_{ij}| \gg \lambda$, as is to be expected because of dephasing, and will hardly be affected in magnitude for $|\tilde{\omega}_{km} - \omega_{ij}| \ll \lambda$. The dependence on the transition rate, λ, will be most pronounced in the vicinity of $|\omega_{ij} - \tilde{\omega}_{km}| \simeq \lambda$, concerning both the amplitude of the oscillating term with, $\tilde{\omega}_{km}$, and its phase.

According to §7.1.1, the eigenfunctions $m\rangle$ and $i\rangle$ are expressed in terms of basis vectors χ_i, which are defined (equation (7.10)) with respect to the selected axis of quantisation, the z axis. Examples are given in table 7.1(a) and (b). By selecting the axis of the initial muon polarisation as the axis of quantisation, the evaluation of $\langle i|\rho(0)|j\rangle$ becomes particularly simple.

A complete analytical evaluation of equations (7.114)–(7.117) is possible if $\mathcal{H}$ and $\tilde{\mathcal{H}}$ are isotropic. A special case would be if $\tilde{\mathcal{H}}$ reduced to the μ^+ Zeeman term only. This is the case treated by Ivanter and Smilga (1968). The transition from a muonium state to a free radical state containing a μ^+ is another example (Walker 1983). Meier (1982) has treated the transition from an isotropic hyperfine Hamiltonian $\mathcal{H}$ to an axially symmetric $\mathcal{H}'$. Analytic solutions are possible for the two cases when the applied field is either parallel or perpendicular to the axis of axial symmetry. A transition between an isotropic and an axially symmetric hyperfine Hamiltonian seems to be observed in diamond (see §7.4.5).

In the derivation of equation (7.114), it was tacitly assumed that in the transition $\mathcal{H} \to \tilde{\mathcal{H}}$ not only the μ^+ polarisation, but also the electron polarisation and all mixed terms are conserved. Situations can be imagined in which this assumption may not be true. For example, the transition may involve a change of electrons, whereby the new electron is, of course, unpolarised. In this case, the initial condition at the onset of the evolution under the action of $\tilde{\mathcal{H}}$ is given by the μ^+ polarisation $\boldsymbol{P}(t')$ at this instant t' combined with an unpolarised electron.

$\boldsymbol{P}(t')$ follows from equation (7.39) and reads in this case

$$\boldsymbol{P}_\mu(t') = \text{Tr}\, \rho(t')\sigma_\mu = \sum_{i,j} \exp\left(-\frac{1}{\hbar}(E_i - E_j)\right) \langle i|\rho(0)|j\rangle \langle j|\sigma_\mu|i\rangle. \quad (7.119)$$

According to equations (7.53) and (7.112), the spin density operator $\rho(t')$ describing the new initial condition at time t' can be written as

$$\rho(t') = \sum |\tilde{n}\rangle\langle \tilde{n}| \mathbb{1} + \boldsymbol{P}(t')\sigma_\mu|\tilde{m}\rangle\langle \tilde{m}|. \quad (7.120)$$

The remainder of the evaluation leading to equation (7.117) is unchanged.

7.1.5 Superhyperfine interaction

The treatment in the previous sections of this chapter was based on the assumption that the hyperfine coupling between the μ^+ and the (unpaired) electron was the only hyperfine coupling present. In composite systems, like a molecule containing a μ^+ and an unpaired electron — a free chemical radical — or a muonium atom embedded in a solid, the unpaired electron may also couple to other nuclei. This can be either by direct overlap, leading to an additional contact Fermi term, and/or via a dipole–dipole interaction. This additional hyperfine coupling is usually weak compared with the μ^+ electron coupling, and is therefore sometimes called superhyperfine interaction.

The superhyperfine Hamiltonian is of the form (Beck *et al* 1975)

$$\mathcal{H}_{\text{Shf}} = \sum_i \left[a_i \boldsymbol{I}_i \cdot \boldsymbol{S} + \hbar^2 \frac{\gamma_i \gamma_e}{r_i^3}\left(\boldsymbol{I}_i \cdot \boldsymbol{S} - \frac{3(\boldsymbol{I}_i \cdot \boldsymbol{r}_i)(\boldsymbol{S} \cdot \boldsymbol{r}_i)}{r_i^3}\right)\right] = \sum_i \boldsymbol{S}_i \cdot \boldsymbol{A}_i \cdot \boldsymbol{I}_i \quad (7.121)$$

where the sum runs over all nuclei involved, and $\boldsymbol{r}_i$ is the distance vector from the electron to the ith nucleus. The nuclei may belong to different chemical species. The shf tensors, $\boldsymbol{A}_i$ are generally different for different nuclei in different

Magnetic properties in the 1S ground state

positions. In particular, they cannot be diagonalised all together. The dipole–dipole interaction term has also to be averaged over the spatial extension of the electron wave function.

The total hyperfine Hamiltonian, including Zeeman terms caused by an external field is now given by

$$\mathcal{H} = A_\mu \mathbf{S} \cdot \mathbf{I}_\mu + \sum_i \mathbf{S} \cdot \mathbf{A}_i \cdot \mathbf{I}_i + \gamma_e \hbar \mathbf{S} \cdot \mathbf{B} - \gamma_\mu \hbar \mathbf{I}_\mu \cdot \mathbf{B} + \sum_i \gamma_i \hbar \mathbf{I}_i \cdot \mathbf{B} \quad (7.122)$$

where we have assumed an isotropic hf interaction between the μ^+ and the electron. $\mathbf{A}_i$ is understood to be the proper average over the spatial extension of the electron. This Hamiltonian involves a greatly increased number of energy levels, and it is therefore to be expected that the evolution of the μ^+ polarisation is also characterised by a multitude of frequencies. This latter circumstance may render the time dependent part of $\mathbf{P}_\mu(t)$ unobservable, since the initial polarisation is spread over many frequency components with a corresponding reduction in the individual amplitudes.

The evolution of the μ^+ polarisation in time under the action of $\mathcal{H}$, can be derived in the same manner as in §7.1.2. Following Beck et al (1975) and Roduner and Fischer (1981), we first introduce the product of the spin functions as a basis

$$\chi_i = \chi^\mu \chi^e \prod_{j=1}^N \chi_j^I = |m_\mu\rangle |m_e\rangle \prod_{j=1}^N |m_j^I\rangle = |\chi_i\rangle \quad (7.123)$$

corresponding to the possible combinations of μ^+ spin, electron spin and N nuclear spins, I. For the sake of simplicity, we assume that only one nuclear spin species is present. The number D of possible combinations is

$$D = (2I_\mu + 1)(2S + 1) \prod_{j=1}^N (2I + 1) = 4(2I + 1)^N. \quad (7.124)$$

This is also the number of Hamiltonian eigenstates (equation (7.122)). The eigenstates $|m\rangle$ with eigen energies, E_m, are evaluated in terms of the basis χ_i.

$$|m\rangle = \sum a_{mi} \chi_i. \quad (7.125)$$

The coefficients depend on the details of $\mathcal{H}$ and the chosen coordinate system, and may not be easy to obtain. To describe the initial condition ($\mathbf{P}_\mu(0) \neq 0$, $\mathbf{P}_e(0) = 0$, $\mathbf{P}_{I,j}(0) = 0$), we again use the spin density matrix formalism. In an obvious generalisation of equation (7.74) we find

$$\rho(0) = D^{-1}[1 + \mathbf{P}_\mu(0) \boldsymbol{\sigma}_\mu]. \quad (7.126)$$

Further by applying equations (7.48) and (7.52) the evolution of the muons polarisation in time is given by the expression (analogous to equation (7.54))

Muonium in matter

$$P_\mu(t) = \sum_{k,m,n} \langle k|m\rangle\langle m|\rho(0)|n\rangle\langle n|\sigma_\mu(t)|k\rangle$$

$$= \sum_{m,n} \langle m|\rho(0)|n\rangle\langle n|\sigma_\mu|m\rangle \exp\left(\frac{i}{\hbar}(E_n - E_m)\right)t$$

$$= \frac{1}{D} \sum_{m,n} \langle m|P_\mu(0)\sigma_\mu|n\rangle\langle n|\sigma_\mu|m\rangle \exp -i\omega_{nm}t. \quad (7.127)$$

Here we have made use of the fact that Tr $\sigma_\mu^i = 0$. Furthermore expressing the eigenstates $|n\rangle$ etc in terms of the $|\chi_i\rangle$, we have

$$P_\mu(t) = \frac{1}{D} \sum_{\substack{m,n \\ i,j}} a_{mi}^* a_{nj} a_{nj}^* \langle \chi_i|P_\mu(0)\sigma_\mu|\chi_j\rangle\langle \chi_j|\sigma_\mu|\chi_i\rangle \exp -i\omega_{nm}t. \quad (7.128)$$

Let us specify this expression for the two cases $P_\mu(0) \parallel B$ and $P_\mu(0) \perp B$. The field direction in both cases defines the z axis (Roduner and Fischer 1981).

(1) $P_\mu(0) \parallel B$

$$P_z(t) = \frac{P_\mu(0)}{D} \sum_m \left(\left|\sum_i a_{mi}^* a_{mi} 2m_\mu^i\right|^2 + 2 \sum_{n<m} \left|\sum_i a_{mi}^* a_{ni} 2m_\mu^i\right|^2 \cos\omega_{nm}t \right) \quad (7.129)$$

where

$$m_\mu^i = \tfrac{1}{2}\langle \chi_i|\sigma_\mu^z|\chi_i\rangle = \pm 1.$$

(2) $P_\mu(0) \perp B$

$$P_x(t) = \frac{2P_\mu(0)}{D} \sum_m \sum_{n<m} \left|\sum_{i,j} a_{mi}^* a_{nj} \langle \chi_i|\sigma_\mu^x|\chi_j\rangle\right|^2 \cos\omega_{nm}t$$

$$= \frac{2P_\mu(0)}{D} \sum_m \sum_{n<m} \left|\sum_{i,j} a_{mi}^* a_{nj} \langle \chi_i^\mu|\sigma_\mu^x|\chi_j^\mu\rangle \delta_{ij}^e \prod_{k=1}^N \delta_{ij}^k\right|^2 \cos\omega_{nm}t. \quad (7.130)$$

Equations (7.129) and (7.130) are similar to equations (7.60) and (7.62). $P_z(t)$ consists of a constant term and a number of oscillating terms, while $P_x(t)$ only consists of oscillating terms. The oscillating terms of $P_z(t)$ originate from transitions between states $|n\rangle$ and $|m\rangle$, which contain the same basis functions $|\chi_i\rangle$. The oscillating terms of $P_x(t)$ involve transitions between states $|n'\rangle$ and $|m'\rangle$ with basis functions $|\chi_i\rangle$ and $|\chi_j\rangle$ respectively which differ only in their muon part and not their electron and nuclear parts.

The allowed transitions are governed by selection rules which follow from the matrix elements

$$\langle n|\sigma_\mu|m\rangle = \sum_{i,j} a_{ni}^* a_{mj} \langle \chi_i|\sigma_\mu|\chi_j\rangle \neq 0.$$

Limiting ourselves to the case when both the hyperfine and superhyperfine interaction is isotropic, the Hamiltonian equation (7.122) only mixes basis functions, $|\chi_i\rangle$, in equation (7.125), which involve equal magnetic quantum numbers, $M = m_\mu + m_e + \sum_k m_k$. For $\boldsymbol{P}_\mu(0) \| \boldsymbol{B} \| z$ axis, it follows from equation (7.129)

$$\Delta M = 0 \tag{7.131}$$

and for the perpendicular case we find from equation (7.130)

$$\Delta M = \pm 1. \tag{7.132}$$

At zero or low $\boldsymbol{B}$, the eigenstates $|n\rangle$ etc are also eigenstates of the total angular momentum operators $(\boldsymbol{F})^2$ and F_z with

$$\boldsymbol{F} = \boldsymbol{S} + \boldsymbol{I}_\mu + \sum_{k=1}^N \boldsymbol{I}_k \tag{7.133}$$

that is

$$|n\rangle = |F, M\rangle \tag{7.134}$$

and

$$\langle n|\sigma_\mu|m\rangle = \langle F, M|\sigma_\mu|F', M'\rangle. \tag{7.135}$$

With help of the Wigner–Eckart theorem, one finds the additional selection rule

$$\Delta F = 0, \pm 1 \tag{7.136}$$

which applies to both field configurations.

At sufficiently high magnetic fields, that is, in the Paschen–Back regime, muon, electron and nuclear spins are effectively decoupled from one another. This implies that the eigenstates $|n\rangle$ reduce to the basis functions, $|\chi_i\rangle$. Thus the eigenstates are characterised by the set of magnetic quantum numbers m_μ, m_e, m_k $(k=1\ldots N)$.

It is easily seen from equation (7.129) that, in the case of strong $\boldsymbol{B}$, $P_z(t)$ reduces to

$$P_z(t) = P_\mu(0). \tag{7.137}$$

That is, the full muon polarisation is restored, as in the more simple case of isolated muonium. For the strong transverse field case, equation (7.130) yields the selection rules

$$\Delta m_\mu = \pm 1 \qquad \Delta m_e = 0 \qquad \Delta m_k = 0 \qquad k = 1, \ldots, N. \tag{7.138}$$

In order to identify the possible transition frequencies, we need to know the energy eigenvalues. Under the assumption that the electron Zeeman energy is much larger than A_μ and A_i $(i = 1, \ldots, N)$, first order perturbation theory yields

$$E_m = \hbar\omega_e m_e^m - \hbar\omega_\mu m_\mu^m - \sum_{k=1}^{N} \hbar\omega_I m_k^m + A_\mu m_\mu^m m_e^m + \sum_{k=1}^{N} A_k m_e^m m_k^m. \quad (7.139)$$

With the selection rules of equation (7.138), we find in strong transverse fields

$$\omega_{nm} = (E_n - E_m)/\hbar = \pm(\omega_\mu \pm \tfrac{1}{2} A_\mu). \quad (7.140)$$

In effect, only two precession frequencies can be observed, as in the isolated muonium case, while in zero or low applied fields the occurrence of many transition frequencies may prevent the actual observation of the oscillatory time depencence of $P_\mu(t)$. However, the decoupling effect in strong fields reduces the number of frequencies which share the full initial polarisation to two and hence the signal should be easily detectable.

The field dependence of the constant term in equation (7.129) is of interest, since it allows some details of the shf interaction to be measured. The corresponding dependence for isolated muonium was shown in figure 7.3. Figure 7.5 sketches schematically what is to be expected if two different shf interaction terms of rising strength are included. At first, the zero field value of the constant part of $P_z(t)$ is reduced considerably below the isolated muonium value. With increasing field strength one reaches the Paschen–Back region, with respect to the weakest SHF interaction term. A partial restoration of μ^+ polarisation will result, indicated by the first step in figure 7.5. With further rising field one will reach the Paschen–Back regions related to the more stronger shf terms consecutively, until finally the hf interaction between the μ^+ and the electron is also quenched. This will give rise to further increases in the constant term of $P_z(t)$ until the curve merges with the quenching curve of an

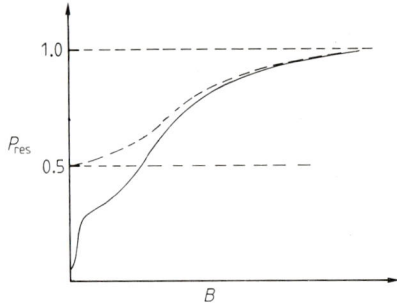

Figure 7.5 Residual (time independent) μ^+ polarisation, P_{res}, in the muonium state with (full curve) and without (broken curve) the additional superhyperfine interaction, as a function of applied longitudinal field, B. The schematic plot depicts the situation of two superhyperfine fields of different strength which are successively quenched by the applied field until eventually the pure muonium quenching curve is approached.

isolated muonium atom. Some more realistic curves of this kind were calculated by Beck *et al* (1975).

7.2 Muonium Formation

The presence of muonium in matter has been in evidence since the historical Garwin *et al* (1957) experiment. In various subsequent experiments on different substances, the amplitude of the μSR signal was much reduced or even zero, implying a partial or complete loss of polarisation in a very short time after implantation. Quenching experiments in longitudinal fields usually showed a restoration of the initial polarisation, suggesting strongly that muonium formation was involved (Orear *et al* 1957, Friedman and Telegdi 1957, Swanson 1958, Feher *et al* 1960, Breit and Hughes 1957, Ferrell and Chaos 1957, Ferrell *et al* 1960). Long lived muonium was directly detected first in Ar gas, via its typical precession frequency in the triplet state (Hughes *et al* 1960). Later experiments by Stambaugh *et al* (1974) revealed that no muonium was formed in pure He and Ne gas, but a small addition of Xe led to a muonium fraction greater than 70%. Still later experiments (Fleming *et al* 1982) showed that the gas pressure also had an influence on the visible fraction of formed muonium.

The first long-lived muonium signal in a solid was observed in quartz by Myasishcheva *et al* (1967). Long-lived muonium states have been detected subsequently in a few more substances, including the semiconductors Si, Ge (see §§7.3, 7.4). However, the evidence for muonium formation is largely of an indirect nature. It is interesting that in metals there is no indication for muonium formation, either directly or indirectly. In liquids the first long-lived muonium signal was detected by Percival *et al* (1976) in water. There the muonium fraction amounted to about 20%, whilst in ice it was of the order of 52%. Furthermore, in water the observed muonium fraction plus the free, diamagnetic μ^+ fraction do not account for all muons. There is a so-called 'missing fraction' in water and this is absent in ice (Percival *et al* 1978).

This short review shows that the formation of muonium in gases, liquids and solids is a rather complex phenomenon. Obviously it depends on the target substance, its state of matter, and other influential parameters too (pressure, temperature, crystalline state etc). Many questions arise: how is muonium formed, what determines the fraction of formation, what is the particular effect of a given environment etc? In addition, we have to consider the problem of μ^+ depolarisation associated with transient muonium states. From all experiments it seems to be clear that muonium must be formed early after implantation, say within 10 ns. Therefore, it seems very likely that the formation process is somehow associated with the slowing down and thermalisation of the μ^+ upon entering the target.

The same physics is studied experimentally with protons, where one measures the degree of neutralisation (H formation) after the protons have passed through a certain amount of matter (see e.g. Tawara and Russek 1973, Allison 1958). From these studies the following picture emerges for the early history of μ^+ in a target which it usually enters with energies above 4 MeV (Crowe et al 1972, Brewer et al 1975, Fleming et al 1982). Independent of the state of matter, three different regions of energy loss can be distinguished.

(1) In the high velocity regime, the μ^+ are slowed down by ionisation of the target atoms. The energy loss per unit time can be calculated on the basis of the Bethe–Bloch theory. The proton stopping power is well known in many substances. (For references see Fleming et al 1982.) This allows one to estimate typical slowing down times between an initial and a final energy. As soon as the μ^+ approaches a velocity regime which is compatible with the orbital velocity of the outer electrons of the target atoms, the second region of energy loss sets in. This happens around 35 keV and below. The time needed to reach this energy is of the order of 0.1–10 ns.

(2) This region is dominated by charge-exchange processes. The μ^+ is now able to capture from the target atoms an electron to form muonium. The binding energy is the same as in atomic hydrogen, that is 13.5 eV. Since there is enough energy available the electron is lost again and the μ^+, in fact, undergoes a cycle of rapid electron capture and loss events. This regime extends down to roughly 50 eV or less. Below this energy the available energy is not sufficient for further ionisation of target atoms, except in those substances in which the capture of an electron is an exothermic process. The status of the μ^+ at this energy now depends on the cross-sections for μ^+ neutralisation (i.e. muonium formation), σ_{01}, and muonium ionisation (i.e. loss of the electron) σ_{10}. In terms of these cross-sections, the fraction of neutralised and free μ^+ may be expressed as

$$f_{Mu} = \sigma_{01}/(\sigma_{01} + \sigma_{10}) \qquad f_{\mu^+} = \sigma_{10}/(\sigma_{01} + \sigma_{10}). \tag{7.141}$$

Table 7.2 shows a collection of measured relative fractions, f_{Mu} and f_{μ^+}, in various gases, which are compared with the expected neutral fraction, f_H from proton-exchange studies (Fleming et al 1982). We see that the μ^+ results can be understood quite well on the basis of the charge-exchange picture. Table 7.2 also lists the ionisation potentials of some of the target materials. With the exception of Xe, the capture process by the μ^+ is always epithermal. We notice that muonium formation is almost totally suppressed in those materials displaying the largest ionisation energies, while 100% muonium formation is observed in Xe. In view of the fact that the ionisation energy of atoms and molecules in the liquid and solid state is generally smaller than the muonium

Table 7.2 Relative fractions (in per cent) of muonium (f_{Mu}) and diamagnetic μ^+ (f_{μ^+}) as found in different gases (Fleming *et al* 1982). IP = ionisation potential.

Target gas	IP (eV)	Pressure/range of pressure (atm)	f_{μ^+}	f_{Mu}	f_H†
He	24.5	1.2–3.1	100 ± 1	0 ± 1	15
		50‡	99 ± 5	1 ± 5	
Ne	21.6	1.2	93 ± 5	7 ± 5	20
		26‡	100 ± 2	0 ± 2	
Ar	15.8	1.0–2.8	26 ± 4	74 ± 4	85
		30‡	35 ± 5	65 ± 5	
Kr	14.0	0.4–0.95	0 ± 5	100 ± 5	100
Xe	12.1	0.4–0.65	0 ± 4	100 ± 4	100
		4.4‡	10 ± 5	100	
H_2		3.0	39 ± 4	61 ± 4	95
N_2	15.6	1.0–2.4	16 ± 4	84 ± 4	90
NH_3		2.8	9 ± 4	91 ± 4	100
CH_4		1.2–3.0	13 ± 4	87 ± 4	100

† Expected neutral fraction from proton charge-exchange studies (Tawara and Russek 1973, Tawara 1978).
‡ Higher pressure values from earlier study of Stambaugh *et al* (1974).

ionisation energy, it is reasonable to expect muonium formation to prevail in most substances, independent of its state of matter.

The importance of the electron capture and loss regime is further emphasised by the pressure dependence of the observed absolute amplitudes of the μ^+/muonium signals in some noble and other gases. Table 7.3 gives the results of Fleming *et al* (1982). Listed are the combined μ^+ and muonium amplitudes, normalised to the μ^+ amplitude in aluminium: $A_{abs} = (A_{\mu^+} + 2A_{Mu})/A_{Al}$. The muonium amplitude, which results only from the observable precession of muonium in the triplet state, is multiplied by a factor of two to account for all muonium atoms (see discussion in §7.1.2). Note that the relative fractions, f_{Mu} and f_{μ^+}, listed in table 7.2, are pressure independent. The pressure dependent reduction in absolute amplitude is a result of depolarisation caused by the repeated muonium formation, which will lead to dephasing on account of the huge difference in μ^+ frequencies in the muonium (particularly in the singlet state) and the free state.

Just how much dephasing will occur depends on the number of capture and loss cycles, and the lifetime of the neutral state between two charge-exchange collisions. The number of cycles is estimated to be roughly 100. The total time spent in this second energy range may vary greatly, depending on the density of the target material. In condensed matter it is probably too short ($\ll 1/\omega_0$) to cause any noticeable depolarisation by dephasing. Not so in gases, particularly at low pressure, where the neutral state survives long enough (up

Table 7.3 Total muon (pressure dependent) amplitudes in various gases normalised to the μ^+ amplitude in aluminium (Fleming et al 1982) $A_{abs} = (A_{\mu^+} + 2A_{Mu})/A_{Al}$.

Gas	Pressure (atm)	A_{abs} (%)	Gas	Pressure (atm)	A_{abs} (%)
He	1.2	31	Xe	0.40	36
	2.7	48		0.60	48
	3.1	59		0.65	58
Ne	0.8	28	H_2	3.1	82
	1.2	41	N_2	1.0	92
	1.6	62		2.4	100
	2.0	82			
Ar	1.0	72	CH_4	1.2	63
	2.0	85		3.0	100
	2.4	90	NH_3	2.8	100
	2.8	96			
Kr	0.40	32			
	0.65	50			
	0.95	68			

to 0.3 ns in He at 1.2 atm) to produce a noticeable rotation of the μ^+ spin. The observed pressure dependence of the total amplitude in gases is compatible with a linear pressure dependence of neutral state's 'effective' lifetime, as one would expect for a collision dominated process.

Very surprising results are obtained when going from the gaseous to the liquid to the solid phase of the same substance (see table 7.4, Fleming et al 1982). For additional data see Fleming et al (1984) and Arseneau et al (1984). We notice drastic differences in the various fractions in the different phases. For liquids in particular, we find a 'missing' fraction, as mentioned before. It seems that the observed differences are not related to the charge-exchange regime and one has to search for further effects.

(3) This leads us to the third region of energy loss which starts below about 50 eV, and extends down to the final thermalisation of the μ^+. Around 50 eV we have a fraction of μ^+ in the neutral state muonium, and a fraction still in the free state. Except for substances with ionisation potentials much larger than 13.5 eV, we actually expect that most, if not all, μ^+ are in the muonium state. Energy loss in this region is facilitated by elastic collisions with the target atoms (the only mechanism in gases), and inelastic processes involving excitation of vibrational modes (particularly in solids) etc. These processes will not alter the fractions f_{μ^+} and f_{Mu}. The involved time constants may vary considerably from 10^{-12} s in solids to 10^{-7} s in gases.

Table 7.4 Absolute fractions f^A (%) for muonium formation in gases as compared with condensed media (Fleming et al 1982). f_L = missing fraction, f_H = neutral fraction of hydrogen.

Target	Medium	f_μ^A	f_{Mu}^A	f_L	f_H
He	Gas	100	0	0	15
	Liquid	90	2	8	
Ar	Gas	26 ± 4	74 ± 4	0	
	Liquid	1.6 ± 1.0	97 ± 30	3 ± 29	85
	Solid	0.8 ± 0.2	91 ± 9	8 ± 9	
Kr	Gas	0 ± 5	100 ± 5	0	
	Liquid	6.5 ± 0.1	57 ± 10	36 ± 10	100
	Solid	1.4 ± 1.8	100 ± 10	0 ± 10	
Xe	Gas	0 ± 4	100 ± 4	0	
	Liquid	3.3 ± 0.8	43 ± 9	54 ± 10	100
	Solid	5.0 ± 3.3	79 ± 25	16 ± 28	
H_2O	Gas	10 ± 5	90 ± 10	0	
	Liquid	62 ± 1	20 ± 1	18 ± 1	100
	Solid	48 ± 1	52 ± 2	0	

In this energy range, as was first conjectured by Williams (Crowe et al 1972), muonium might enter into so-called hot atom reactions. These are reactions that are not possible at thermal energies but are allowed in the epithermal range.

Through such an interaction, the μ^+ could be placed into a diamagnetic compound and would appear magnetically as a free particle. This could explain the appearance of the 'free' μ^+ fraction f_{μ^+} in many substances. Since this reaction is fast, no depolarisation caused by dephasing can result (see §7.1.3). Hence, the 'missing' fraction cannot be the result of hot atom reactions. It has also been found in ice, with no observable missing fraction, that the diamagnetic fraction f_{μ^+} decreases with temperature below 160 K, while f_{Mu} increases by the same proportion (figure 7.6, Percival et al 1978). Such a behaviour seems to suggest also that the diamagnetic fraction cannot result solely from hot atom reactions which are expected to be temperature independent. The increase of the muonium fraction, f_{Mu}, with decreasing temperature suggests that some additional muonium formation process besides the charge-exchange mechanism may have to be considered.

Percival et al (1978) (see also Percival 1981) have suggested that the interaction of thermalised μ^+ or muonium with radiolysis products in the terminal spur of the μ^+ may provide the answer. In water, the radiolysis

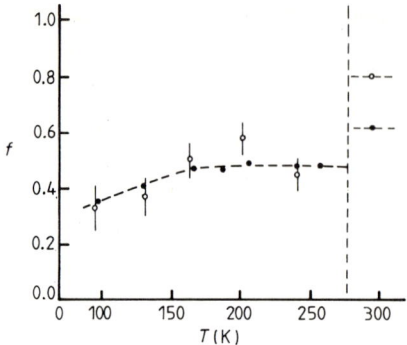

Figure 7.6 Diamagnetic and muonium fractions in ice (H_2O) and the liquid phase at ambient temperature: ●, f_{μ^+}; ○, $(1-f_{\mu^+})$ (Percival et al 1982).

products are presolvated electrons, e^- (which later become solvated electrons, e^-_{eq}) and the hydroxyl radical $^\bullet OH$. It is assumed that the μ^+ or muonium come to rest in the near vicinity (within a few nm) of the last spur products. Since the spur products only stay for a very short time (<1 ns) close to the muon, interactions with the former will be of a transient nature. Two possibilities can be discussed. First the μ^+ is thermalised as muonium. The following reactions may then occur

$$Mu + {}^\bullet OH \rightarrow MuOH$$

$$Mu + e^-_{eq} \xrightarrow{H_2O} MuH + OH^- \qquad (7.142)$$

$$(\mu^+ \uparrow e^- \uparrow) + e^-_{eq} \downarrow \rightarrow (\mu^+ \uparrow e^- \downarrow) + e^-_{eq} \uparrow.$$

The first two reactions produce a diamagnetic species in a very short time (<1 ns) with no loss of polarisation, and now contributing to the diamagnetic fraction f_{μ^+}. The third reaction is a spin-exchange reaction which leads to depolarisation and may account for the missing fraction. Those muonium atoms not taking part in any of these reactions comprise the fraction f_{Mu}. Secondly, the μ^+ is thermalised as a free particle. Possible reactions are now

$$\begin{array}{l} \mu^+ + e^- = Mu \rightarrow f_{Mu} \\ \left\{ \begin{array}{l} \mu^+ + H_2O \rightarrow MuH_2O^+ \\ MuH_2O^+ + H_2O \rightarrow H_3O^+ + MuHO \end{array} \right\} \rightarrow f_{\mu^+}. \end{array} \qquad (7.143)$$

The missing fraction may again result from spin-exchange reactions with solvated electrons near the terminal spur of the μ^+ track. The latter model ascribes muonium formation to the capture of presolvated electrons by free μ^+ after thermalisation. This model, known as the spur model, is believed to describe positronium formation in liquids adequately.

The spur model for muonium formation has been tested in various ways by scavanging the presolvated electrons. The results are much under dispute and no consensus on its validity has yet been reached (Walker 1981, Percival 1984b).

Certainly, if valid, the spur model should also be relevant to solids. It is not quite clear how it could explain the results in ice, particularly the absence of the missing fraction. It is very likely that spur processes and epithermal (hot) reactions can exist side by side. The gas phase results definitely indicate that muonium formation at the end of the charge-exchange regime is of great importance. There is no reason to doubt that the same processes do also take place in the condensed phases of matter. In summary, at present (1983) one is still far away from a consistent understanding of the observed differences in muonium formation for the various states of matter. In discussing the effects of muonium formation and associated phenomena one has to be aware that the invisibility of long-lived muonium in many solids and liquids (also a kind of missing fraction) is not caused by a lack of muonium formation, but by other processes long after thermalisation. Most important are chemical reactions (Walker 1983) and electron spin relaxation (see §§7.3, 7.4). In metals, muonium cannot persist because of the presence of many conduction electrons which cooperatively combine to screen the positive charge of the μ^+. A paramagnetic state with an unpaired electron cannot survive under these conditions (see §4.2).

7.3 Muonium in Quartz and Other Insulators

Long lived muonium has been identified by its characteristic triplet state precession frequency in quartz (SiO_2) (Myasishcheva et al 1967, Gurevich et al 1968, 1969, 1971, Brewer et al 1979, 1981a,b, Brown et al 1979, 1980 and Holzschuh et al 1981), in solid CO_2 (Myasisheva et al 1967), in ice (H_2O, D_2O) (Myasisheva et al 1967, Gurevich et al 1971, 1973, Percival et al 1978, 1982, 1984a), in solid neopentane (2.2–dimethylpropane) (Ito et al 1982) and in the oxides CaO, BeO (Brewer 1982), MgO and hexagonal GeO (Spencer et al 1984). It has also been possible to observe the precession of a muonic radical (2,3,5,6–tetramethyl cyclohexadienyl where muonium is added to the aromatic ring) in an anorgomic single crystal of durene (1,2,4,5–tetramethyl benzene) (Rudoner 1981).

Longitudinal field studies in quartz SiO_2 and Al_2O_3 (Minaichev et al 1970) and in KCl (Ivanter et al 1972) showed the existence of muonium by the Paschen–Back effect.

As already pointed out in §7.2, evidence for muonium or muonium-like states in many other compounds can be inferred from the observation of missing μ^+ polarisation (see e.g. table 3 in Brewer et al 1975) and repolarisation phenomena in longitudinal fields (see e.g. Eisenstein et al 1966, Myasishcheva et al 1969, Buhler et al 1965). In the following we will review a

7.3.1 Longitudinal field studies

Figure 7.7 shows results for the μ^+ polarisation in a single crystal of quartz, as a function of the longitudinal field strength (Minaichev et al 1970). The polarisation is derived from the forward–backward positron rate along the initial polarisation direction. The measurements were performed at room temperature. The field dependence of these data is excellently described by equation (7.61) with $A/(\gamma_e+\gamma_\mu)\hbar = 1636 \pm 78$ G. For muonium in vacuum, this value becomes 1585 G. At zero field we notice a slightly larger P_z than 0.5, which is caused by a small background signal contribution.

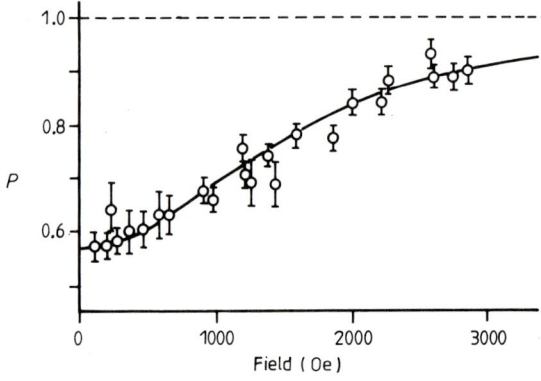

Figure 7.7 Residual μ^+ polarisation against longitudinal field strength in quartz (Minaichev et al 1970).

The same authors have also measured the repolarisation curve in a single crystal of corundum (Al_2O_3). The results are shown in figure 7.8. Again the field dependence of the data is well described by equation (7.61) above about 1000 G with $A/\hbar(\gamma_e+\gamma_\mu) = 1650 \pm 120$ G. Below about 1000 G we see a marked deviation when the polarisation drops to a very small value as the applied field is decreased to zero. This effect can be understood if one considers that the Al atoms in Al_2O_3 carry large nuclear magnetic moments, while in SiO_2 only the isotope ^{29}Si, with its low natural abundance of 4.7%, possesses a small nuclear magnetic moment. In Al_2O_3 one therefore expects that a superhyperfine interaction may couple the muonium electron spin with the Al spins. According to equation (7.129), this leads to a very small, time independent polarisation in zero external field. This coupling is expected to be relatively weak in comparison with the μ^+ spin–e^- spin coupling. It will already have

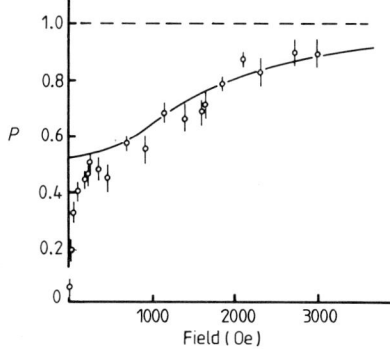

Figure 7.8 Resdiual μ^+ polarisation against longitudinal field strength in Al_2O_3 (corundum) (Minaichev et al 1970).

been suppressed in correspondingly smaller magnetic fields, and this explains the steep rise in P_z with rising applied field.

7.3.2 Transverse and zero field studies: hyperfine properties

In the pioneering experiment of Myasishcheva et al (1967) muonium precession was observed in a field of 7.2 G. Since the time resolution in this experiment was limited, only the slow precession components were actually visible, that is, the observable part of the time evolution of $P_x(t)$ (equation (7.62)) is given by

$$\langle P_x(t) \rangle = \tfrac{1}{4}[(1+\delta)\cos\omega_{12}t + (1-\delta)\cos\omega_{23}t] \tag{7.144}$$

where

$$\delta = x(1-x^2)^{-1/2} \qquad x = \frac{\hbar(\gamma_\mu + \gamma_e)}{A} \qquad B = \frac{\gamma_\mu + \gamma_e}{\omega_0} B.$$

The small applied field of 7.2 G is much smaller than $A/(\gamma_e + \gamma_\mu)\hbar = 1585$ G. Therefore $\delta \simeq 0$ and (equation (7.22))

$$\omega_{12} = \hbar^{-1}(E_1 - E_2) \simeq \omega_{23} = \hbar^{-1}(E_2 - E_3) \simeq \tfrac{1}{2}(\gamma_e + \gamma_\mu)B$$

$$\simeq \tfrac{1}{2}\gamma_e B. \tag{7.145}$$

At such a small field, we are in the linear Zeeman regime. The observable muonium precession is caused by the muonium atoms in the triplet state $|F=1, m_F=0, \pm 1\rangle$. The precession frequency is essentially that of a particle with spin one and the magnetic moment of an electron. The observed precession frequency is in agreement with equation (7.145). We realise that it is not possible to obtain any information on the hyperfine coupling constant, A, from the measurement of the observable part of $P_x(t)$ as long as we are in this linear Zeeman region.

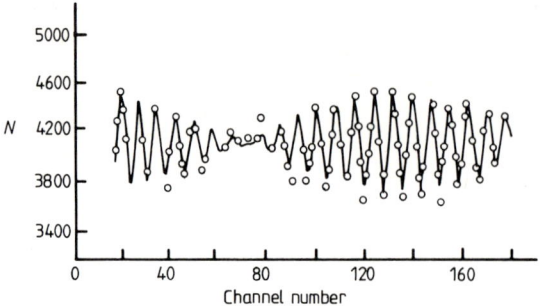

Figure 7.9 Two-frequency Mu precession signal in single crystal quartz in a transverse field of 95 G at room temperature (Gurevich et al 1971).

In a very famous experiment by Gurevich et al (1969, 1971), the applied field was raised to 95 G. The observed precession signal is displayed in figure 7.9. It shows a pronounced beating behaviour (two-frequency precessions), and this is easily understood on the basis of equation (7.144) if the condition $\omega_{12} \simeq \omega_{23}$ no longer holds. In this case, equation (7.144) can be rewritten

$$\langle P_x(t) \rangle = \tfrac{1}{2}\{\cos[\tfrac{1}{2}(\omega_{12}+\omega_{23})t]\cos[\tfrac{1}{2}(\omega_{12}-\omega_{23})t]$$
$$- \delta \sin[\tfrac{1}{2}(\omega_{12}+\omega_{23})t]\sin[\tfrac{1}{2}(\omega_{12}-\omega_{23})t]\}. \qquad (7.146)$$

With the help of equation (7.22) we find

$$\tfrac{1}{2}(\omega_{12}+\omega_{23}) = \omega^+ = \tfrac{1}{2}(\gamma_e + \gamma_\mu)B \qquad (7.147)$$

$$\tfrac{1}{2}(\omega_{12}-\omega_{23}) = \omega^- = \tfrac{1}{2}\omega_0\{1-[1+(\gamma_e-\gamma_\mu)^2 B^2/\omega_0^2]^{1/2}\}. \qquad (7.148)$$

For fields still sufficiently small $[(\gamma_e-\gamma_\mu)^2 B^2/\omega_0^2 \ll 1]$, and neglecting $\gamma_\mu \ll \gamma_e$ the latter expression reduces to

$$\omega^- \simeq -\tfrac{1}{4}\omega_0^{-1}\gamma_e^2 B^2 \simeq \omega^{+2}/\omega_0^2. \qquad (7.149)$$

The beat frequency is therefore inversely proportional to the hyperfine coupling constant, $A = \hbar\omega_0$, and proportional to the square of the external field. Since $\delta \simeq 0$ for the fields considered here, the second term in equation (7.146) can be neglected.

The data in figure 7.9 are described well by equation (7.146), yielding for $\omega_0/2\pi$ the value 4474 ± 130 MHz, again in good agreement with the vacuum value of 4463 MHz.

From these measurements it appeared that the muonium state in quartz was characterised by an isotropic hyperfine Hamiltonian, as in the atomic state in vacuum. However, careful studies in single crystal samples a few years later also revealed a beating behaviour at very small fields (see figure 7.10, Brewer et al 1979). The beating frequency turned out to be field independent, but strongly dependent on the crystal orientation with respect to the external field

Quartz and other insulators

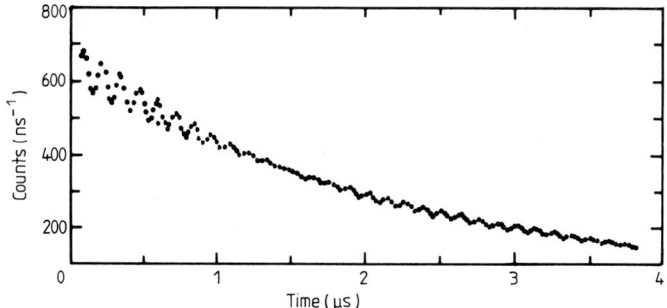

Figure 7.10 Low field (6.3 G) two-frequency pattern of the triplet Mu precession signal in single crystal quartz at room temperature (Brewer et al 1979).

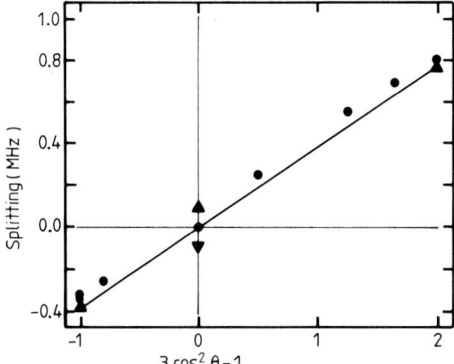

Figure 7.11 Angular dependence of the splitting between the two frequencies for Mu in single crystal quartz in low fields. θ is the angle between the c axis of the single crystal and the applied field (Brewer et al 1979).

direction (see figure 7.11). These findings were confirmed by Brown et al (1980) and Holzschuh (1982).

These results can be understood if the hyperfine Hamiltonian is assumed to be axially symmetric, that is

$$\begin{aligned}
\mathcal{H}_{hf} &= \mathbf{S} \begin{pmatrix} A_\perp & 0 & 0 \\ 0 & A_\perp & 0 \\ 0 & 0 & A_\parallel \end{pmatrix} \mathbf{I} + \text{Zeeman terms} \\
&= A_\parallel S_z I_z + A_\perp (S_x I_x + S_y I_y) + \text{Zeeman terms} \\
&= A_\perp \mathbf{S} \cdot \mathbf{I} + (A_\parallel - A_\perp) S_z I_z + \text{Zeeman terms}.
\end{aligned} \quad (7.150)$$

Here the z axis, the axis of axial symmetry is directed along the crystallographic c axis. From the precision measurements of Holzschuh *et al* (1982), one finds

$$h^{-1}A_\perp = (4496.2 \pm 0.4) \text{ MHz}$$

$$h^{-1}(A_\| - A_\perp) = -(0.412 \pm 0.004) \text{ MHz}$$

and these yield an isotropic value of $h^{-1}A_0 = \frac{1}{3}h^{-1}\text{Tr}\mathbf{A} = 4496.1(0.4)$ MHz.

The observed orientation dependence of the beating frequency can be derived as follows. We assume the z axis to coincide with the applied field. Since the hyperfine Hamiltonian is axially symmetric, we can, without loss of generality, assume the z axis to rotate in the plane $x'z'$ of the coordinate system in which the hyperfine tensor is diagonal. This allows us to make use of equations (7.29)–(7.31). Since $(A_\| - A_\perp)$ is very small in comparison to $A_\perp$, we neglect all terms in the energy matrix equation (7.21) carrying the factor $(A_\| - A_\perp)$, this means $\alpha \simeq \delta \simeq \gamma \simeq 0$. The energy matrix now has exactly the form for $\mathbf{B}$ applied parallel to a principal axis, such as the z' axis, of the hyperfine tensor. The solutions for the energy eigenvalues in this case are given by the equations in table 7.1(a). The strength of the applied field further implies that

$$(A_\| - A_\perp) \ll \hbar\gamma_e B \ll A_\perp$$

That is, in the further calculation we can again neglect terms $(A_\| - A_\perp)$ against $\hbar\gamma_e B$ and terms $\hbar\gamma_e B$ against $A_\perp \simeq A_\|$.

The energy eigenvalues are then given by

$$E_1 = \tfrac{1}{4}(A_\perp \sin^2\theta + A_\| \cos^2\theta) + \tfrac{1}{2}\hbar\omega_e$$
$$E_2 = \tfrac{1}{2}A_\perp \cos^2\theta + \tfrac{1}{4}A_\|(\sin^2\theta - \cos^2\theta)$$
$$E_3 = \tfrac{1}{4}(A_\perp \sin^2\theta + A_\| \cos^2\theta) - \tfrac{1}{2}\hbar\omega_e \quad (7.151)$$
$$E_4 = -\tfrac{1}{2}(A_\perp + \tfrac{1}{2}A_\|).$$

The precession of this atom, with a nearly isotropic hyperfine coupling at small magnetic fields (linear Zeeman regime), will be almost like the precession of ordinary muonium. Hence, the possible frequencies are $\omega_{12}, \omega_{23}, \omega_{14}, \omega_{34}$, of which only the first two will be resolvable in a standard μSR spectrometer. Hence the observed beat frequency is given by

$$\omega^- = \tfrac{1}{2}(\omega_{12} - \omega_{23}) = \tfrac{1}{2}h^{-1}[(E_1 - E_2) - (E_2 - E_3)]$$
$$= \tfrac{1}{4}(A_\perp - A_\|)(1 - 3\cos^2\theta) \quad (7.152)$$

in accordance with the experimentally observed angular dependence.

However, this is still not the whole story. At temperatures below 80 K in zero external field, using a single crystal quartz sample, Brewer *et al* (1981a,b) found a rather complicated precession pattern which contains three distinct frequencies with values 1.7(1) MHz, 6.2(1) MHz and 7.9(1) MHz. The amplitudes of these frequencies depend strongly on the orientation of the single crystal with respect to the initial polarisation (see figure 7.12). More recent zero field measurements by Holzschuh *et al* (1982) (see also Holzschuh 1982), using a special µSR spectrometer with an extremely high time resolution, in which the initial μ^+ polarisation vector was oriented perpendicular to the *c* axis of their single crystal sample, confirmed two of the frequencies with much improved precision (1.73(2) MHz, 6.22(2) MHz). In addition it was also possible to resolve two high-frequency components corresponding to triplet-singlet transitions: 4508 MHz and 4500 MHz.

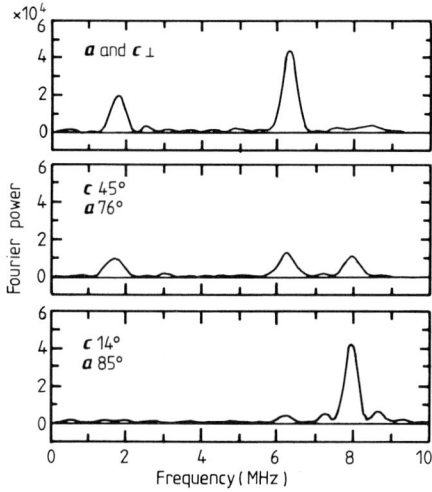

Figure 7.12 Fourier spectra of the Mu precession signal below 80 K in zero field in single crystal quartz. The three spectra correspond to three different orientations of the crystal with respect to $P_\mu(0)$ (Brewer *et al* 1981a,b).

The observed frequency spectrum, and the dependence of the frequency amplitudes on the relative orientation of $P_\mu(0)$ and the crystal imply a fully anisotropic hyperfine Hamiltonian with principal values $A_{x'x'}, A_{y'y'}, A_{z'z'}$. The possible transition frequencies and the orientation dependence were already derived in §7.1.2, yielding equations (7.68) and (7.69). Assigning of the measured frequencies to one of the ω_{ij} frequencies, and determining the position of the hyperfine tensor's principal axes relative to the crystal axes exploit the orientation dependence of the measured amplitudes (see equation (7.68)). Note also that $\omega_{12} + \omega_{23} = \omega_{13}$. The comparison of equation (7.68) with the data has to take into account that there may be several equivalent or non-equivalent muonium sites in the lattice. A systematic study of the amplitudes as a function of crystal orientation is, as yet, lacking. A preliminary analysis of the TRIUMF data by Brewer et al (1981a,b) yield tentative angles of $94\pm5°$ and $111\pm5°$ for the x' and z' axis relative to the crystallographic c axis.

Fortunately some guidance is available from electron paramagnetic resonance (EPR) measurements on atomic hydrogen in α quartz below 120 K (Perlson and Weil 1974, Weil 1981). An anisotropic hyperfine Hamiltonian was also observed here. According to the analysis of Perlson et al (1979) and Weil (1981), the hydrogen atom sites are associated with the three equivalent two-fold symmetry axes perpendicular to the three-fold c axis. The principal z' axis of the hyperfine tensor is found to be normal to the two-fold symmetry axis, and makes an angle of $114.3°$ with the c axis. The hydrogen is located between two Si atoms where the Si–Si direction makes an angle of $111.8°$ with the c axis.

Preliminary analysis of μSR data in quartz is sufficiently close to the EPR results that it seems reasonable to assume the muonium atom occupies the same site as the hydrogen atom. With this assumption, and the available information on the frequency amplitudes, an unambiguous identification of the observed frequencies becomes possible, with the result

$|v_{12}|=6.22(2)$ MHz† $\qquad v_{34}=4508$ MHz†

$|v_{23}|=1.73(2)$ MHz† $\qquad v_{14}=4500$ MHz†

$|v_{13}|=7.7(1)$ MHz‡ $\qquad v_{24}=$ not measured.

Since $\omega_{12}+\omega_{34}=\omega_{14}-\omega_{23}$ (see equation (7.69)), it follows that $v_{12}<0$, $v_{23}<0$ and $v_{13}<0$. The hyperfine tensor can now be easily determined with the help of equation (7.69).

However, we note, that the low frequencies v_{12} and v_{23} are much more accurately determined than the others. To make maximum use of these, we apply a trick suggested by Holzschuh (1982). We split the hyperfine tensor into an isotropic part

$$\mathbf{A}^{iso}=\tfrac{1}{3}\operatorname{Tr}\mathbf{A}\cdot\mathbf{1}=\tfrac{1}{3}(A_{x'x'}+A_{y'y'}+A_{z'z'})\cdot\mathbf{1} \qquad (7.153)$$

and an anisotropic part

$$\mathbf{A}^{an}=\mathbf{A}-\mathbf{A}^{iso} \qquad (7.154)$$

$$=\tfrac{1}{3}\begin{pmatrix} 2A_{x'x'}-A_{y'y'}-A_{z'z'} & 0 & 0 \\ 0 & 2A_{y'y'}-A_{x'x'}-A_{z'z'} & 0 \\ 0 & 0 & 2A_{z'z'}-A_{x'x'}-A_{y'y'} \end{pmatrix}$$

with $\operatorname{Tr}\mathbf{A}^{an}=0$.

† Holzschuh et al (1982).
‡ Brewer et al (1981a,b).

The elements of $\mathbf{A}^{\mathrm{an}}$ are easily expressed in terms of v_{12} and v_{23} only

$$(1/2\pi\hbar)A^{\mathrm{an}}_{x'x'} = \tfrac{2}{3}(v_{12}+2v_{23}) = -6.45(2) \text{ MHz}$$
$$(1/2\pi\hbar)A^{\mathrm{an}}_{y'y'} = -\tfrac{2}{3}(2v_{12}+v_{23}) = +9.44(2) \text{ MHz} \qquad (7.155)$$
$$(1/2\pi\hbar)A^{\mathrm{an}}_{z'z'} = \tfrac{2}{3}(v_{12}-v_{23}) = -2.99(2) \text{ MHz}.$$

The isotropic term, on the other hand, can be written as

$$(1/2\pi\hbar)\tfrac{1}{3}\operatorname{Sp}\mathbf{A} = \tfrac{1}{3}(2v_{14}-v_{12}+v_{34}) = 4505.0(5) \text{ MHz}. \qquad (7.156)$$

All these data refer to a temperature of 30 K.

Holzschuh *et al* (1982) also studied the precession frequencies at a few other temperatures. Figure 7.13 displays the behaviour of the high (triplet–singlet transition) frequencies as a function of temperature. Two frequencies were also clearly seen at a temperature of 69 K. Within the limits of accuracy they are the same as at 30 K. From 190 K up to room temperature only one frequency shows up corresponding to the axially symmetric hyperfine Hamiltonian. This frequency declines slightly with rising temperature (by 5.9 MHz from 190 K to 296 K).

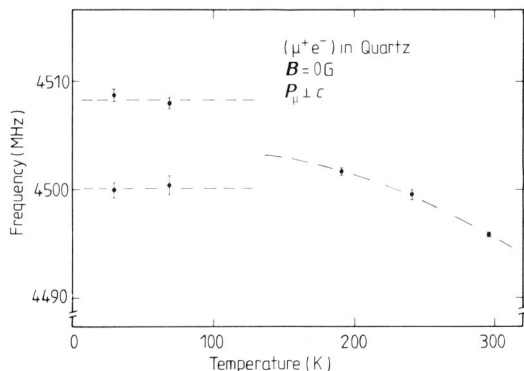

Figure 7.13 Temperature dependence of the triplet–singlet muonium transition frequencies, v_{14} and v_{34} of quartz in zero field (Holzschuh *et al* 1981).

The change from an anisotropic to an axially symmetric hyperfine Hamiltonian, somewhere between 80 K and 190 K, is ascribed to the onset of muonium diffusion between the three equivalent sites suggested by the atomic hydrogen EPR results (Brewer 1981). Each site is associated with the same anisotropic hyperfine Hamiltonian, but the principal axes of the hyperfine tensor have different orientations with respect to the initial polarisation. Averaging the hyperfine Hamiltonian over the three equivalent sites one does indeed arrive at an axially symmetric expression (Holzschuh 1982)

$$A_\parallel - A_\perp = \tfrac{3}{2}[A^{an}_{z'z'} + (A^{an}_{x'x'} - A^{an}_{z'z'})\sin^2\alpha] \tag{7.157}$$

where α is the angle of the principal axis, z', with respect to the c axis. With $\alpha = 24.2°$ one obtains

$$(A_\parallel - A_\perp) = -5.4 \text{ MHz}.$$

This value has the correct sign, but is much larger than the measured value. It seems to be clear that diffusion is indeed responsible for the change of the effective $\mathcal{H}_{hf}$. This supersedes earlier attempts to explain the axially symmetric hyperfine Hamiltonian in terms of an intrinsic quadrupole moment of muonium (Brewer et al 1979, Baryshevsky and Kuten 1977, 1978, Beder 1978).

7.3.3 Zero field studies: relaxation behaviour

The relaxation behaviour of the muonium precession signal in quartz has been studied by Brewer (1981). At low temperatures up to 80 K a temperature independent relaxation rate of $\lambda = 0.2 \ \mu s^{-1}$ is observed. This relaxation is presumably caused by the interaction with nuclear dipole fields originating from randomly placed ^{29}Si nuclei (natural abundance of 4.7%). Above 80 K, the relaxation rate increases rapidly (see figure 7.14(a)) until the signal becomes unobservable around 140 K. This fast increase is attributed to the onset of muonium diffusion between equivalent sites, each with different orientations of the hyperfine tensor's principal axes. This implies that, upon each jump, the amplitudes of the three low frequency components will change in concurrence with the phase change of each frequency signal. This constitutes an effective depolarisation mechanism. Plotting the logarithm of the damping rate against the inverse temperature (Arrhenius plot) one finds that all data points fall on a straight line. Thus, a temperature activated process is indicated with an activation energy of $E_a = 719(\pm 68)$ K (see figure 7.17(b)).

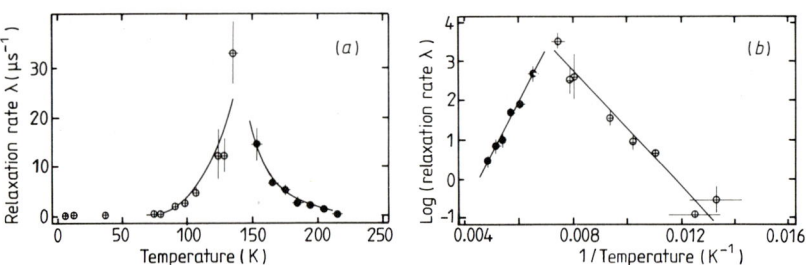

Figure 7.14 (a) Temperature dependence of the relaxation rate of the Mu signal in quartz in zero applied field: ○, low temperature points as deduced from the decay of the oscillating signal; ●, high temperature points as deduced from the decay of the longitudinal forward–backward asymmetry (Brewer 1981). (b) Arrhenius plot of the relaxation rates from figure 7.14(a) (Brewer 1981).

Measurements of the forward–backward positron rates above 150 K reveal a 'longitudinal' relaxation function which is well fitted by a simple exponential decay function. In contrast to the zero field precession studies below 150 K, the damping constant now decreases with increasing temperature (see figure 7.14(a)). This is interpreted in terms of motional averaging over the hyperfine coupling at the differently orientated equivalent sites (Brewer 1981). The Arrhenius plot of these data again reveal a temperature activated process with an activation energy of $E_a(>150\text{ K}) = 1336(\pm 40)$ K. A quantitative analysis of the data in terms of hopping between different, yet crystallographically equivalent sites is not yet available. In particular it is not possible to give a physical interpretation for the change in the apparent activation energy.

7.3.4 Double electron muon magnetic resonance

A new way of studying the behaviour of muonium was introduced by Brown *et al* (1979) by exposing it to a strong radio frequency field, with the RF frequency close to one of the muonium precession frequencies. Otherwise, the experimental arrangement was similar to a normal transverse field experiment. A transverse field of 125 G was applied perpendicular to the initial μ^+ polarisation and the positron rate was monitored as a function of elapsed μ^+ lifetime in the direction of the initial μ^+ polarisation.

The experiment is essentially an electron paramagnetic resonance (EPR) experiment. Since the RF field is large and relaxation is assumed to be negligible (as for muonium at room temperature in quartz), the coupled electron–muon spin system will follow the RF field coherently. These coherence effects produce a characteristic structure in the μSR spectrum which serves to locate the EPR transitions. In this respect the method is similar to electron–nuclear double resonance (ENDOR) as well as other double resonance techniques, and hence the name double electron muon magnetic resonance (DEMUR) has been given to it.

The basic phenomenon can be understood by considering a spin $\frac{1}{2}$ system coupled to a strong radio frequency field $\tilde{H}_1(t)$, perpendicular to a much larger static field H_0. This situation has already been considered in §2.2.2(2). We found there that the evolution of the μ^+ polarisation was characterised by three frequencies, namely $\omega_0 = \gamma_\mu H_0$, $\omega^\pm = \gamma_\mu(H_0 \pm \tilde{H}_1)$. Similarly, it is found that if the RF frequency is close to a particular allowed transition frequency of the muonium spectrum, this frequency will split into three components in the precession pattern. Other transition frequencies which share an energy level with that transition frequency nearly in resonance with the RF, show a splitting into two components.

A complete description of the theory of the DEMUR technique has been given by Estle and Vanderwater (1983). Here we shall only present a simplified deviation appropriate for the case of muonium in quartz. The small anisotropy of the hyperfine Hamiltonian will be ignored. In this we follow Brown *et al*

(1983). The hyperfine Hamiltonian, including the static Zeeman terms, is thus given by

$$\mathcal{H}_0 = A\mathbf{I}\cdot\mathbf{S} + \gamma_e\hbar\mathbf{S}\cdot\mathbf{H}_0 - \gamma_\mu\hbar\mathbf{I}\cdot\mathbf{H}_0. \tag{7.158}$$

In addition, we have the interaction with an RF field, supposedly circularly polarised in the xy plane perpendicular to $\mathbf{H}_0 = H_0\hat{z}_0$

$$\mathbf{H}_1(t) = \tilde{H}_1[\cos(\omega t + \phi)\hat{x}_0 + \sin(\omega t + \phi)\hat{y}_0]. \tag{7.159}$$

$\hat{x}_0, \hat{y}_0, \hat{z}_0$ are unit vectors. The corresponding Hamiltonian reads

$$\mathcal{H}_1(t) = \gamma_e\hbar\mathbf{S}\cdot\mathbf{H}_1(t) = \tfrac{1}{2}\gamma_e\hbar\tilde{H}_1[S^+\exp-i(\omega t + \phi) + S^-\exp i(\omega t + \phi)] \tag{7.160}$$

neglecting the much smaller μ^+ related contribution. The total Hamiltonian to be considered is

$$\mathcal{H}_{tot} = \mathcal{H}_0 + \mathcal{H}_1(t). \tag{7.161}$$

The time dependent solutions of the Schrödinger equation

$$i\hbar\dot{\psi}(t) = \mathcal{H}_{tot}\psi \tag{7.162}$$

can be expressed in terms of the four eigenfunctions, ψ_n, of the static $\mathcal{H}_0$ with energy eigenvalues $E_n = \hbar\omega_n$ (see §7.1), that is

$$\psi(t) = \sum_{n=1}^{4} a_n(t)\psi_n \exp -i\omega_n t. \tag{7.163}$$

The time dependent coefficients $a_n(t)$ have to obey the following set of coupled differential equations $(n = 1 \ldots 4)$

$$\dot{a}_n(t) = -i\sum_m a_m(t)[\![\Omega_{nm}^+ \exp\{-i[(\omega_{nm} + \omega)t + \phi]\}$$
$$+ \Omega_{nm}^- \exp\{-i[(\omega_{nm} - \omega)t + \phi]\}]\!] \tag{7.164}$$

where

$$\omega_{nm} = \omega_n - \omega_m \tag{7.165}$$

and

$$\hbar\Omega_{nm}^\pm = \tfrac{1}{2}\langle\psi_n|\gamma_e\hbar\tilde{H}_1 S^\pm|\psi_m\rangle = \hbar\Omega_{mn}^{\mp *}. \tag{7.166}$$

Note that $\Omega_{nn}^\pm = 0$ for the chosen geometrical arrangement. The initial conditions are specified by

$$a_n(0) = \langle\psi_n|\psi(0)\rangle. \tag{7.167}$$

$\psi(0)$ is specified by the initial μ^+ polarisation when muonium is formed with unpolarised electrons.

Once the coefficients $a_n(t)$, and therefore $\psi(t)$, are known, the time evolution of the μ^+ polarisation can be obtained from the expectation value of σ_μ, that is

$$\mathbf{P}_\mu(t) = \langle\psi(t)|\boldsymbol{\sigma}_\mu|\psi(t)\rangle. \tag{7.168}$$

An analytic evaluation of $a_n(t)$ is in general not possible. Several circumstances appropriate for the experiment in quartz, however allow equation (7.164) to be simplified in such a manner that analytical solutions for the $a_n(t)$ can be obtained. First a moderate static field of $H_0 = 125$ G leads to four clearly separated frequencies ω_{ij} of which the two high frequency components ω_{14} and ω_{34} are not resolved in the experiment. Secondly the RF frequency, ω, is chosen to be close to the transition frequency, ω_{12}. Close means that the energy separation, $\hbar(\omega - \omega_{12})$ is small compared with the RF transition energy $\hbar\Omega_{12}$, and that at the same time $|\omega + \omega_{12}| \gg \Omega_{12}$ and $|\omega \pm \omega_{ij}| \gg \Omega_{ij}$ ($i, j = 1, 2$). Equation (7.164) therefore contains the slowly modulated (near secular) and rapidly oscillating terms. The effect of the latter terms will be essentially averaged out to zero, on the timescale of the slowly modulated terms. Therefore, retaining only the terms which are nearly secular in equation (7.164), we arrive at

$$\dot{a}_1(t) = -\tfrac{1}{2}i\Omega_{12}^- a_2(t) \exp\{i[(\omega - \omega_{12})t + \phi]\}$$

$$\dot{a}_2(t) = -\tfrac{1}{2}i\Omega_{12}^{-*} a_1(t) \exp\{-i[(\omega - \omega_{12})t + \phi]\} \qquad (7.169)$$

$$\dot{a}_3(t) = \dot{a}_4(t) = 0.$$

These equations can be analytically solved (Brown et al 1983). We are particularly interested in the partial polarisations $P_{12}(t)$ and $P_{23}(t)$, that is, the amplitudes associated with the observable precession frequencies ω_{12} and ω_{23}. Generally

$$P_{mn}(t) = 2 \operatorname{Re} [a_m^*(t) a_n(t) \exp(i\omega_{mn} t) \langle \psi_m | \sigma_\mu | \psi_n \rangle]. \qquad (7.170)$$

Defining a reduced frequency displacement

$$z = (\omega - \omega_{12})/|\Omega_{12}| \qquad (7.171)$$

the solutions for $P_{12}(t)$ and $P_{23}(t)$ read (Brown et al 1983)

$$P_{12}(t) = \tfrac{1}{2} \operatorname{Re} \Bigl\{ \langle \psi_1 | \sigma_\mu | \psi_2 \rangle a_1^*(0) a_2(0)$$

$$\times \Bigl[\Bigl(1 + \frac{z}{(z^2+1)^{1/2}}\Bigr)^2 \exp\{i[\omega - (z^2+1)^{1/2}|\Omega_{12}|]t\}$$

$$+ \frac{2}{z^2+1} \exp(i\omega t) + \Bigl(1 + \frac{z}{(z^2+1)^{1/2}}\Bigr)^2 \exp\{i[\omega + (z^2+1)^{1/2}|\Omega_{12}|]t\} \Bigr] \Bigr\}$$

$$P_{23}(t) = \operatorname{Re} \Bigl\{ \langle \psi_2 | \sigma_\mu | \psi_3 \rangle a_2^*(0) a_3(0) \Bigl[\Bigl(1 - \frac{z}{(z^2+1)^{1/2}}\Bigr) \exp[i(\omega_{23} - \omega_2^+) t]$$

$$+ \Bigl(1 + \frac{z}{(z^2+1)^{1/2}}\Bigr) \exp[i(\omega_{23} - \omega_2^-) t] \Bigr] \Bigr\} \qquad (7.173)$$

where

$$\omega_2^\pm = \tfrac{1}{2}[z \pm (z^2+1)^{1/2}]|\Omega_{12}|. \qquad (7.174)$$

We indeed find as indicated above, that the precession term associated with the transition ω_{12} close to the RF frequency is split into three frequencies ω and $[\omega \pm (z^2+1)^{1/2}|\Omega_{12}|]$, and the term associated with the transition ω_{23} into two frequencies $\{\omega_{23} - [z \pm (z^2+1)^{1/2}]|\Omega_{12}|\}$. The splittings are all equal and proportional to the RF amplitude $\tilde{H}_1$ through $|\Omega_{12}|$.

These predictions were experimentally verified and figure 7.15 shows Fourier spectra of the μSR signal obtained under various conditions. Figure 7.15(a) was obtained with $\omega \simeq \omega_{12}$. Three lines are indeed observed in the vicinity of ω_{12} and two lines in the vicinity of ω_{23}. Figure 7.15(c) was obtained by setting $\omega \simeq \omega_{23}$. This time the upper frequency signal is split into three lines and the lower into two. Figure 7.15(b) only shows the usual frequencies ω_{12} and ω_{23}. Here the RF frequency was close to $\tfrac{1}{2}(\omega_{12} + \omega_{23})$ and far away from either ω_{12} and ω_{23} (Brown et al 1983).

The DEMUR technique will be particularly useful for investigations of muonium and muonium-like centres in strong external fields, that is in the Paschen–Back regime where the electron and muon spin are largely decoupled. As can be seen from equations (7.62) and (7.63), the frequencies corresponding to magnetic dipole transitions between the Zeeman states of the electron ($\Delta m_e = \pm 1$), vanish from the μSR spectrum. Therefore it is impossible to obtain any precise information on the Zeeman interaction of the electron i.e. its g factor or **g** tensor. The DEMUR technique would allow EPR transitions between the electronic Zeeman states to be induced, for example, by choosing $\omega \simeq \omega_{14}$ or $\omega \simeq \omega_{23}$, and the transitions through the μSR spectrum at ω_{12} and ω_{34} to be monitored.

7.4 Muonium in Ge, Si and Diamond

The considerable attention μ^+ behaviour in group IV semiconductors has attracted has held steady since the early days of μSR. Feher et al (1960) were the first to find evidence of muonium's presence in Si and Ge from the temperature dependence of the diamagnetic μ^+ precession signal's amplitude plus its dependence on doping concentration (p type and n type). Subsequent work by Eisenstein et al (1966) concentrated on longitudinal field repolarisation studies, which all implied the presence of some kind of muonium. By and large however the experimental results were rather inconclusive at that time. The first successful direct observation of muonium in Ge, from its characteristic triplet precession frequency, was by Adrianov et al (1969) at a temperature of 77 K. The same group (Adrianov et al 1970) succeeded in determining the

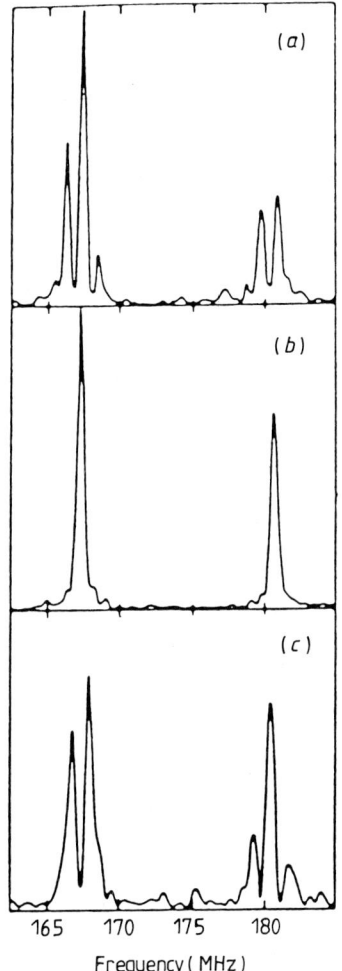

Figure 7.15 Fourier spectrum of the DEMUR signal in quartz in the presence of an RF field with (a) $\omega \simeq \omega_{12}$, (b) $\omega \simeq \frac{1}{2}(\omega_{12} + \omega_{23})$ and (c) $\omega \simeq \omega_{23}$ (Brown et al 1983).

hyperfine coupling constant, A, of muonium in silicon from a longitudinal field repolarisation measurement with the result $A(\text{Si}) = 0.42(3) A(\text{vac})$.

The reduction of the coupling constant in Si compared with the vacuum value, was ascribed to muonium's increased dimensions in Si, leading to a reduced electron density at the μ^+ (see equation 7.6). Considerable progress was achieved when Gurevich et al (1971) were able to detect the two-frequency precession of muonium in Ge, yielding $A(\text{Ge}) = 0.56(1) A(\text{vac})$. Two years later Brewer et al (1973) found the two-frequency precession signal in Si too, with the result $A(\text{Si}) = 0.45(2) A(\text{vac})$, and this agrees well with the value of Adrianov et al (1970). More importantly Brewer et al (1973) found a second muonium state in Si, later named Mu*, with a much reduced hyperfine coupling constant, $A^*(\text{Si}) \simeq 0.02 A(\text{vac})$, which depended on the orientation of the single crystal sample with respect to the field. This 'anomalous' muonium state was also detected later in Ge (Holzschuh et al 1979) and diamond (Holzschuh et al 1982), indicating that the appearance of the two muonium states in the isostructural group IV semiconductors may be based on the same mechanisms. In the wake of the detection of Mu and Mu* in silicon, much experimental work by various groups ensued, the main results of which will be discussed in the following sections.

Naturally, muonium states have been looked for in other semiconductors. Recent measurements indicate that muonium might have been found in GaAs (Barsov et al 1984a) and GaP (Blazey et al 1984). The search in other compounds might well unearth further interesting materials (see also Gurevich et al 1979).

7.4.1 Hyperfine properties of muonium states

The rich phenomenology of the μSR pattern in the group IV semiconductors is perhaps best illustrated by comparing it with quartz results. Figure 7.16 shows the frequency spectra in quartz and Si obtained in an applied transverse field of 100 G. These spectra were obtained by a Fourier transformation from the original μSR signal. We see two lines in the quartz spectrum, showing the two-frequency precession discussed before. The splitting of the two lines was given

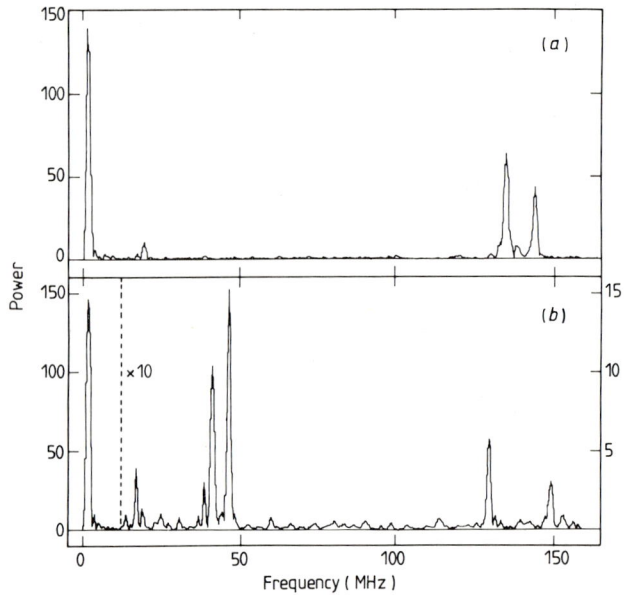

Figure 7.16 Comparison of the Fourier spectra of the Mu signals in (a) quartz and (b) silicon in a transverse field of 100 G. The latter data were taken at a temperature of 77 K (Brewer et al 1973).

by equation (7.149)

$$\omega^- \simeq -\tfrac{1}{4}\omega_0^{-1}\gamma_e^2 B^2. \qquad (7.175)$$

from which the hyperfine frequency, ω_0, could be determined. In Si we find two pairs of lines. One pair has the same average frequency as in quartz, but the splitting is clearly larger. Obviously, in view of equation (7.175) this must indicate a correspondingly reduced hyperfine frequency ω_0 (Brewer *et al* 1973). The second pair appears at frequencies reduced roughly by a factor of three in comparison to the first pair. These frequencies cannot be described by equations (7.147) and (7.149), since their dependence on the applied field looks quite different as can be seen from figure 7.17. These are more recent data (Patterson *et al* 1978) which show an additional set of two frequencies. The fifth one that is visible in figure 7.17 corresponds to the free μ^+ signal. These frequencies can be observed over a wide range of fields, quite in contrast to the high-frequency pair. The orientation dependence of the frequencies' position as the single crystal is rotated with respect to the applied field (see figure 7.18) supplies further information too. This clearly indicates that the detected 'anomalous' muonium state (Mu*) cannot be described by an isotropic hyperfine Hamiltonian (equation (7.5)). In contrast, the first pair of frequencies shows no angular dependence, and the field dependence is in accord with

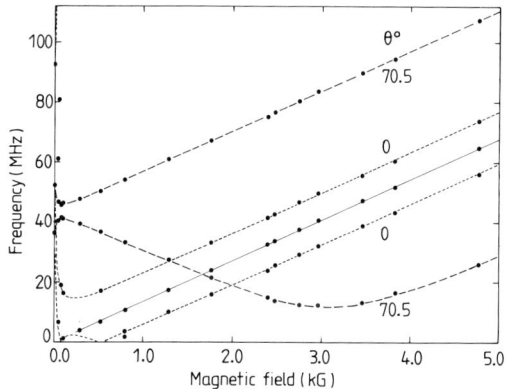

Figure 7.17 Field dependence of the precession frequencies in monocrystalline Si. The field was applied along one of the $\langle 111 \rangle$ directions. The data corresponding to the high-frequency pair in figure 7.16 are not included. Two pairs of frequencies can be identified, one ($\theta = 70.5°$) which corresponds to the low-frequency pair in figure 7.16. The two pairs correspond to angles of 0° and 70.5° between the axis of axial symmetry of the hf tensor and the applied field. The associated new anomalous muonium state is named Mu*. The single frequency (full curve) starting at the origin corresponds to a 'free' μ^+ signal (Patterson *et al* 1978).

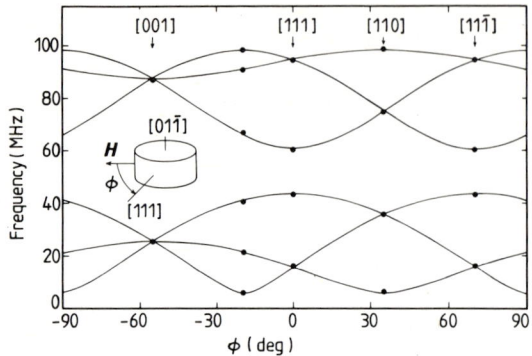

Figure 7.18 Orientation dependence of the Mu* precession frequencies in monocrystalline Si in a transverse field of 3.83 kG. The angle, ϕ, defines the direction of the field in the $(01\bar{1})$ plane of the crystal (Patterson *et al* 1978).

equations (7.175) and (7.147). Hence this state is labelled 'normal' muonium (Mu), although the hyperfine coupling has obviously been reduced from its vacuum value.

The frequency spectrum of the anomalous state, its field and orientation dependence are well accommodated by an axially symmetric hyperfine Hamiltonian

$$\mathcal{H}_{\text{Mu}^*} = A_\perp (I_x S_x + I_y S_y) + A_\parallel I_z S_z + \gamma_e \hbar \mathbf{S} \cdot \mathbf{B} - \gamma_\mu \hbar \mathbf{I} \cdot \mathbf{B}. \tag{7.176}$$

In terms of our earlier notation (equation (7.7)) we have

$$A_\perp = A_{x'x'} = A_{y'y'} \qquad A_\parallel = A_{z'z'}. \tag{7.177}$$

If the field is applied parallel to the principal axis, z', the energy eigenvalues of $\mathcal{H}_{\text{Mu}^*}$ follow from table 7.1(a)

$$\begin{aligned} E_1 &= \tfrac{1}{4} A_\parallel + \tfrac{1}{2}\hbar(\omega_e - \omega_\mu) \\ E_2 &= -\tfrac{1}{4} A_\parallel + [4A_\perp^2 + \tfrac{1}{4}\hbar^2(\omega_e + \omega_\mu)^2] \\ E_3 &= \tfrac{1}{4} A_\parallel - \tfrac{1}{2}\hbar(\omega_e - \omega_\mu) \\ E_4 &= -\tfrac{1}{4} A_\parallel - [4A_\perp^2 + \tfrac{1}{4}\hbar^2(\omega_e + \omega_\mu)^2]. \end{aligned} \tag{7.178}$$

The corresponding Breit–Rabi diagram is shown in figure 7.19. The possible precession frequencies are the same as in the isotropic case that is $\omega_{12}, \omega_{23}, \omega_{14}$ and ω_{34}. This can be easily verified by repeating the calculation leading to equations (7.62) and (7.63) with $A_{x'x'} = A_{y'y'} \neq A_{z'z'}$.

For $\hbar(\omega_e \pm \omega_\mu) \gg A_\parallel, A_\perp$ (Paschen–Back regime) only the two frequencies ω_{12} and ω_{34} will remain observable, with approximate values of

$$\begin{aligned} \omega_{12} &\simeq |\omega_\mu - \tfrac{1}{2} A_\parallel/\hbar| \\ \omega_{34} &\simeq |\omega_\mu + \tfrac{1}{2} A_\parallel/\hbar|. \end{aligned} \tag{7.179}$$

Ge, Si and diamond 269

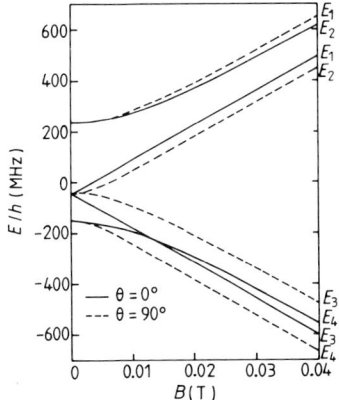

Figure 7.19 Breit–Rabi diagram for the anomalous Mu* state in diamond, with the applied field parallel ($\theta=0°$) or perpendicular ($\theta=90°$) to the axis of axial symmetry of the hyperfine Hamiltonian equation (7.176) (Holzschuh et al 1982).

These frequencies have been displaced from the free μ^+ Larmor frequency, ω_μ, by $\pm A_\parallel/\hbar$.

Figure 7.17 shows that the second pair of frequencies associated with Mu* does indeed display a behaviour described by equations (7.178) and (7.179) respectively. Since the external field here was applied parallel to one of the $\langle 111 \rangle$ directions of the Si crystal, (this has a diamond structure) it can immediately be deduced that the hyperfine tensor's principal z' axis is parallel to the $\langle 111 \rangle$ crystalline direction. The crystal structure of Si (also Ge, and diamond) is pictured in figure 7.20 and as this shows, there are four $\langle 111 \rangle$ crystalline directions. If the applied field is along one of these, the three other will make an angle of 70.5° with the applied field. Therefore one would expect there to be three times as many Mu* states for which the principal z' axis makes an angle of 70.5° with the external field. Indeed, the first pair of frequencies associated with Mu* can be identified with the transition frequencies ω_{12} and

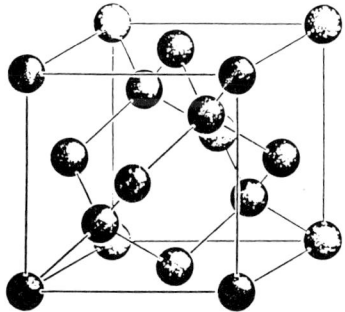

Figure 7.20 The diamond-type crystal structure.

ω_{34} for the said configuration, by a numerical evaluation of the field dependence of the energy eigenvalues. More μ^+ contribute to threse two frequencies and hence this was the only pair plainly visible in the original measurements of Brewer et al (1973) (figure 7.16).

The orientation dependence of the frequencies ω_{12} and ω_{34}, shown in figure 7.18, was observed in a relatively high external magnetic field (Paschen–Back regime), where the energy eigenvalues can be analytically evaluated quite generally. This case was treated in §7.1.1 and led to the expressions equation (7.34) for the energy eigenvalues. For the present case with $|A_\parallel - A_\perp| \ll A_\parallel$, we find as an approximate expression for ω_{12} and ω_{34}

$$\left.\begin{matrix}\omega_{12}\\ \omega_{34}\end{matrix}\right\} = \left\{\left[\omega_\mu \mp \tfrac{1}{2}\tfrac{1}{\hbar}(A_\perp \sin^2\theta + A_\parallel \cos^2\theta)\right]^2 + \left[\tfrac{1}{4}\tfrac{1}{\hbar}(A_\parallel - A_\perp)\sin(2\theta)\right]^2\right\}^{1/2} \qquad (7.180)$$

$$\simeq \omega_\mu \mp \tfrac{1}{2}\tfrac{1}{\hbar}[A_\perp + (A_\parallel - A_\perp)\cos^2\theta]$$

where we have made use of equations (7.30) and (7.31). θ is the angle between the principal z' axis, the axis of axial symmetry, and the external field. The angular dependence of ω_{12} and ω_{34} in figure 7.18 was obtained by rotating the applied field in the [011] plane. The two $\langle 111 \rangle$ directions not in this plane always make the same angle with the applied field, and are therefore associated with the same frequency pair ω_{12}, ω_{34}. The two directions $\langle 111 \rangle$ within the plane [011] have different angles with the applied field and hence lead to two distinct pairs. These only coalesce if the field is oriented midway between these $\langle 111 \rangle$ directions. Consequently we generally find three pairs ω_{12}, ω_{34} as is indeed evident from figure 7.19. All three pairs coalesce into one pair if the field is applied along the $\langle 100 \rangle$ or any equivalent crystalline directions.

From the field and orientation dependence of the observed frequencies, the hyperfine parameters $|A_\parallel|$ and $|A_\perp|$ and their relative sign were determined to a high degree of accuracy. The results extrapolated to 0 K are collected in table 7.5. However, it must be noted that the absolute sign of $A_\parallel$ and $A_\perp$ cannot be determined from the measured frequency values.

As mentioned before, normal (Mu) and anomalous (Mu*) muonium states were also found in germanium (Holzschuh et al 1979) and in diamond (Holzschuh et al 1982). The various hyperfine parameters obtained can be found in table 7.5.

Measurements in zero field which allow the hyperfine splitting to be determined directly are of particular interest. This has been done for the normal muonium state in Si and Ge (Holzschuh 1983) and for the anomalous muonium state in diamond (Holzschuh et al 1982). As an example, figure 7.21 shows the zero field spectrum of Mu* in diamond at temperatures of 636 K

Ge, Si and diamond

Table 7.5 Compilation of the hyperfine coupling constants of Mu and Mu* at 0 K $A_\parallel$ and $A_\perp$ are determined up to an unknown common sign. The sign of v_0 is assumed to be positive in agreement with the vacuum value, which theoretically carries a positive sign. Also listed are the fitted parameters (C) following from the Debye model of the temperature dependence of the hf parameters. The last line contains the literature values on the Debye temperatures (Holzschuh 1983, Blazey et al 1983).

	Parameter	Diamond	Silicon	Germanium
Mu	$v_0(0)$ (MHz)	3711(21)	2006(2)	2359.5(2)
	C		0.68(5)	0.27(7)
	θ_D (K)		655(25)	273(26)
Mu*	$A_\perp(0)/h$ (MHz)	392.59(6)	92.59(5)	131.037(9)
	$A_\parallel(0)/h$ (MHz)	−167.98(6)	16.79(1)	27.279(13)
	$C_\perp$	0.38(2)	0.371(148)	0.113(20)
	$C_\parallel$	0.73(4)	0.160(34)	−0.149(24)
	θ_D (K)	1902(51)	425(66)	152(14)
	θ_D (K)	1860	625	360

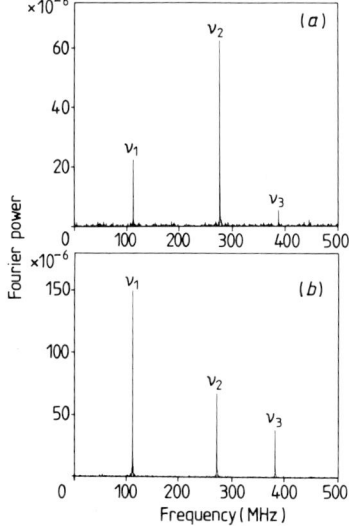

Figure 7.21 Fourier spectrum of the Mu* signal in zero field in diamond at (a) 636 K and (b) 925 K (Holzschuh et al 1982).

and 925 K. As expected for an axially symmetric hyperfine Hamiltonian, three frequencies show up corresponding to (see equations (7.69))

$$\omega_{23} = -\omega_{12} = \frac{1}{2h}(A_\perp - A_\parallel) \quad \omega_{34} = \omega_{14} = \frac{1}{2h}(A_\parallel + A_\perp) \quad \omega_{24} = \frac{1}{h}A_\perp. \quad (7.181)$$

The difference in the spectral weights of the frequencies at 623 K and 925 K is discussed in §7.4.5.

7.4.2 Temperature and pressure dependence of hyperfine parameters

The temperature dependence of the hyperfine parameter $A_0 = \hbar\omega_0$ of normal muonium (Mu) in Si and Ge has been directly measured from the zero field triplet–singlet splitting frequency by Holzschuh (1983). The temperature dependence of the hyperfine parameters, $A_\parallel$ and $A_\perp$, of anomalous muonium (Mu*) in Si and Ge were studied by Blazey et al (1981, 1983) using a transverse field geometry. Zero field measurements in diamond were used to determine the temperature dependence of $A_\parallel$ and $A_\perp$ of Mu* from the position of the three zero field frequencies (equation (7.181)) (Holzschuh et al 1982). The data on Mu are displayed in figure 7.22 and on Mu* in figure 7.23.

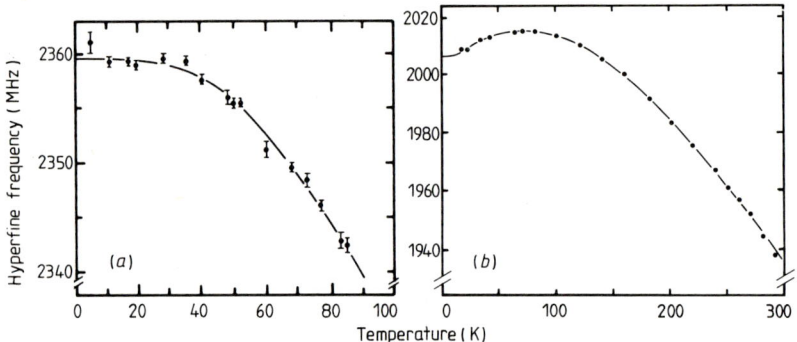

Figure 7.22 Temperature dependence of the hyperfine frequency, v_0, of normal Mu in (a) Ge and (b) Si. The full curve represents a fit of the Debye model to the data (see text for Si case) (Holzschuh 1983).

It turns out that all the data (except for Mu in Si) can be well described by the following expression

$$A(T) = A(0)\left[1 - C\left(\frac{T}{\theta_D}\right)^4 \int \frac{x^3 \, dx}{e^x - 1}\right]. \tag{7.182}$$

Here θ_D is the Debye temperature. The full curves in the figures represent fits of equation (7.182) to the data. The fit parameters are $A(T=0)$, C and θ_D. The results for this fit for both normal and anomalous states are collected in table 7.5.

The temperature dependence of $A(T)$ in Si can be described in the same fashion, if it is additionally assumed that the muonium atom diffuses between

Ge, Si and diamond   273

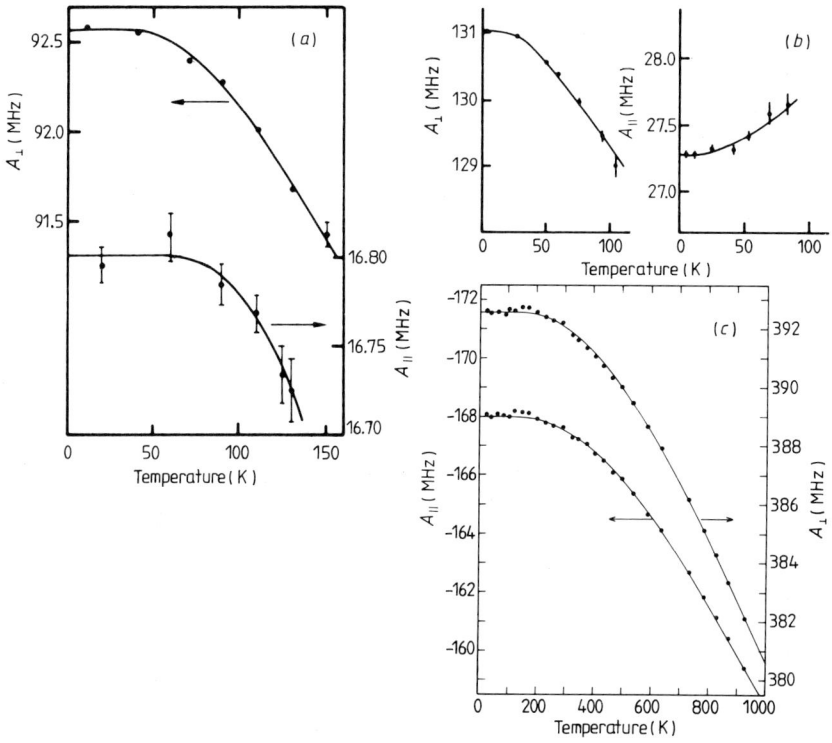

Figure 7.23 Temperature dependence of the hyperfine constants $A_\parallel$, $A_\perp$ of Mu* in (a) Ge, (b) Si and (c) diamond (Blazey et al 1981, 1983, Holzschuh 1983).

different sites with slightly different $A_0 = 2\pi\hbar v_0$, causing a spread δv_0 in the triplet–singlet transition frequency v_0. This idea is inferred from the observation that the zero field signal disappears abruptly below 18 K. In contrast, the frequencies v_{12} and v_{23} in transverse fields up to 220 G remained observable down to 4.2 K with no apparent change in amplitude or relaxation rate. This can be understood if one recalls that a change of v_0 by δv_0 causes a change to the first order of

$$\delta v_{12} = -\delta v_{23} = \tfrac{1}{4}x^2 \, \delta v_0 \qquad (7.183)$$

in the intra-triplet transition frequencies (see equation (7.22)). Hence a spread of a few MHz in the frequency v_0, sufficient to cause a rapid damping of the zero field signal by dephasing, will have only a tiny negligible effect on the intra-triplet frequency signals. For simplicity it was assumed that there are only two different sites (Holzschuh 1983) which are occupied according to the Boltzmann factor, and that the transition rate Γ is fast enough to produce a single observable zero field signal. With these assumptions the measured frequency is an average of the two frequencies associated with the two different sites weighted by the Boltzmann factor,

$$v_0(T) = \left(v_0(0) + \frac{\delta v_0}{\exp(\Delta E/k_B T) + 1}\right)\left(1 - CT^4 \int \frac{x^3 \, dx}{e^x - 1}\right) \quad (7.184)$$

where ΔE is the energy difference between the two sites. Note that it is assumed that each individual transition frequency has a temperature dependence according to equation (7.182). Equation (7.184) fits the data on Mu in Si excellently. The obtained $v_0(0)$ (table 7.5) is in good agreement with the result obtained by the two-frequency method. The frequency spread δv_0 and the energy difference ΔE are determined to be

$$\delta v_0 = 31 \pm 3 \text{ MHz} \qquad \Delta E/k_B = 55 \pm 14 \text{ K}.$$

Equation (7.182) was originally derived by Simanek and Orbach (1966) in order to explain the temperature dependence of the hyperfine field of Mn^{3+} in MgO. They assumed a phonon-induced admixture of s electron states in the Mn^{3+} ground state which cause a shift of the hyperfine field with reference to the direction of the magnetisation to more positive values. The net change of the hyperfine field is thereby proportional to the mean square strain $\langle \varepsilon^2 \rangle$. Assuming a Debye spectrum for the phonons the time average of ε^2, that is $\langle \varepsilon^2 \rangle$, is given by the second term in equation (7.182). Blazey et al (1981) first proposed that a similar mechanism might also be responsible for the temperature dependence of the muonium hyperfine splitting and applied equation (7.182) successfully to their results on Mu* in Ge. Comparing the fitted values for the Debye temperature in table 7.5 with values taken from the literature (de Launay 1956) we find good agreement for Mu in Si and Ge and Mu* in diamond, and much worse correspondence for Mu* in Si and Ge. A more in-depth analysis of these results is as yet missing and the physical significance of these findings, therefore, remains unclear.

Another source for the temperature dependence of the hyperfine parameters could be the lattice constant change with temperature, and hence the difference in volume available to the muonium states. To test this possibility, a measurement of the pressure dependence of the hyperfine splitting of Mu in Si was performed at 273 K temperature (Holzschuh 1983). A linear fit yields

$$\partial v_0/\partial p|_{273 \text{ K}} = -0.12 \pm 0.13 \text{ MHz kbar}^{-1}.$$

Using the bulk modulus of Si ($B = 980$ kbar) the above result may be expressed as a derivative with respect to the lattice constant, a:

$$\frac{1}{v_0}\frac{\partial v_0}{\partial a} = 0.078 \pm 0.084 \text{ Å}^{-1}$$

essentially giving a null result for the dependence of the hyperfine splitting on the lattice constant a. Since the change of a produced by the maximum pressure of 1.5 kbar is about twice as large as the change of a between 120 and 300 K, it is obvious that the static change of the lattice constant cannot be the source for the observed temperature dependence.

7.4.3 Determination of electronic g factor

In non-zero external fields, the electronic Zeeman energy makes a major contribution to the total energy splitting. This provides the possibility of measuring not only the hyperfine parameters, but also the Landé or g factor of the muonium electron. It would be particularly interesting to find out whether the g factor could also be anisotropic, since this would imply some kind of spin–orbit coupling to the host, which could aid in understanding the nature of the particular muonium state.

In the case of normal muonium, the electronic g factor can be obtained from the intra-triplet frequencies v_{12} and v_{23} in the linear and non-linear Zeeman region with help from the following expression

$$g_e = -(m_e/m_\mu)g_\mu[(v_{12}+v_{23})/v_\mu+1] \tag{7.185}$$

where v_μ is the free muon precession frequency.

This expression is easily derived from equation (7.22). Under the assumption that the muonic g factor, g_μ is given by the vacuum value (table 2.1), g_e can be calculated simply from experimentally determined quantities. This also applies to v_μ, which can be separately measured in the same field in, for example, a metallic target, or which is present as a background signal or occurs naturally as a 'diamagnetic' fraction in the investigated sample.

Calculating, as above, g_e for Mu in diamond, Holzschuh et al (1982) have obtained a value of

$$g_e = -2.0034 \pm 0.0017$$

which may be compared with the vacuum value of

$$g_e = -2.002319\ldots$$

Corresponding numbers for normal Mu in Si and Ge have not been quoted in the literature.

With respect to the anomalous muonium state, a particularly interesting possibility for detecting an anisotropic g_e factor arises from measurements at the so-called 'magic' field B_M. Equation (7.180) expresses the angular dependence of v_{12} and v_{34} on the relative orientation of the principal z' axis and external field in the Paschen–Back regime. If the external field is set to

$$B \simeq (A_\perp + A_\parallel)/(2\pi h \gamma_\mu) = B_M \tag{7.186}$$

v_{12} turns out to be independent of the orientation

$$v_{12} \simeq \tfrac{1}{4}(A_\perp - A_\parallel)/2\pi h. \tag{7.187}$$

At this field all frequencies, v_{12}, from the various Mu* states with their axes of axial symmetry along one of the $\langle 111 \rangle$ crystalline directions coincide. This implies that, at this field, a precession signal should also be observable in a polycrystalline sample. This has indeed been verified by Holzschuh *et al* (1982) in diamond. A precise calculation going beyond the strong field approximation shows, however, that this orientational degeneracy only occurs exactly if the electronic g factor is isotropic. This can be tested by measuring with some precision the field dependence of the various v_{12} frequencies which occur in a single crystal sample for a particular orientation. This has been done for Mu* in Si and Ge (Blazey *et al* 1984). Figure 7.24 displays the results. While the data in Si clearly have a crossover point at the magic field, B_M (within the bounds of accuracy) for all three frequencies, Mu* data obtained in Ge show no such behaviour. Assuming an axially isotropic g tensor, with the axis of symmetry also along the $\langle 111 \rangle$ crystalline directions, the data in Ge can be fitted excellently with the result

$$(g_\| - g_\perp) = 0.033 \pm 0.0012.$$

A similar fit to the data in Si yields

$$(g_\| - g_\perp) = 0.0003 \pm 0.0008.$$

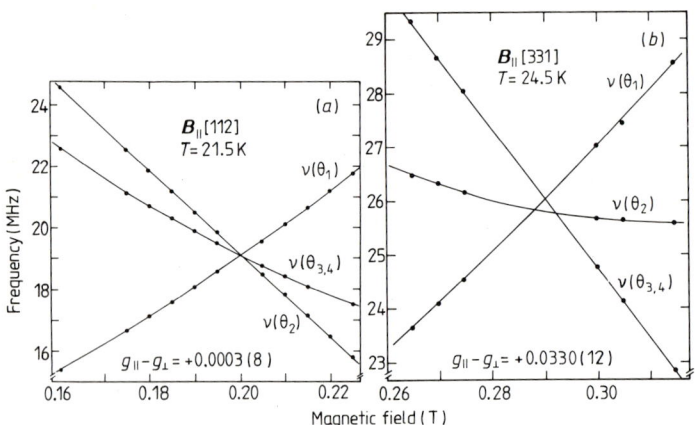

Figure 7.24 Field dependence of the transition frequencies v_{12} of Mu* at three non-equivalent sites (*a*) in Si and (*b*) Ge in the vicinity of the 'magic field' (Blazey *et al* 1984b).

7.4.4 Relaxation

The relaxation of the Mu and Mu* signal in Si and Ge has been the subject of a number of studies. The relaxation in longitudinal fields was investigated in Ge by Gurevich *et al* (1975), and in Si by Ivanter *et al* (1978) and Barsov *et al*

(1980). Transverse field studies were performed by Blazey *et al* (1981), Boekema *et al* (1981), Bucci *et al* (1981), Albert *et al* (1984), Patterson *et al* (1984a) and Weidinger *et al* (1983) in Si and by Balzer *et al* (1981), Clawson *et al* (1981), Weidinger *et al* (1981, 1983), Blazey *et al* (1983), Döring *et al* (1983), Estle and Vanderwater (1983), Döring *et al* (1984) in Ge.

We will limit the discussion here to some principal features of relaxation apparent in transverse field data. Figure 7.25 shows results on the relaxation (damping) rate of the Mu signal ($\omega_{12} \simeq \omega_{23}$) in an external field of 6 G in Ge (from Weidinger *et al* 1981). The μ^+ were stopped in three different samples:

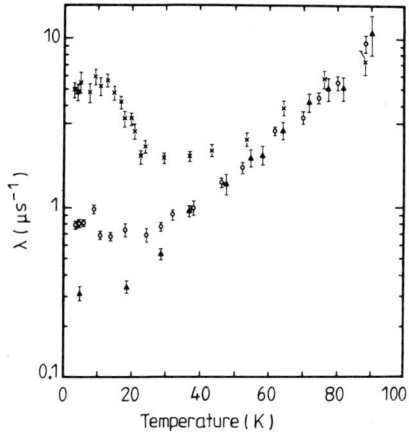

Figure 7.25 Temperature dependence of the relaxation rate of the normal Mu signal in a transverse field of 6 G in differently doped Ge samples (Weidinger *et al* 1981). ×, n type, 10^{14} cm^{-3}; ○, p type, 10^{14} cm^{-3}; ▲, undoped $<1.5 \times 10^{10}$ cm^3.

one was pure undoped Ge, the two others were n-type and p-type Ge. The dependence of the relaxation rate on the dopant concentration was measured at 4.3 K and 30 K (corresponding to the minimum of λ in the n type sample) (see figure 7.26). The dependence on concentration c can be fitted to a power law

$$\lambda \propto c^{0.7 \pm 0.2} \quad (4.3 \text{ K}) \qquad \lambda \propto c^{0.95 \pm 0.20} \quad (30 \text{ K}).$$

To summarise, we find the relaxation rate has a strong dependence on dopant concentration at low temperatures, and a vanishing one at higher temperatures (>60 K). The overall temperature dependence (figure 7.25) suggests three different regions in which different mechanisms are responsible for relaxation (Weidinger *et al* 1981).

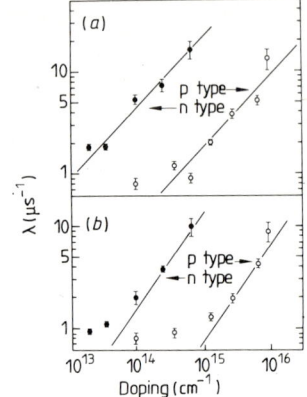

Figure 7.26 Dependence of the normal Mu relaxation rate on the type and concentration of dopants in Ge at (a) 4.3 K and (b) 30 K (Weidinger et al 1981).

(1) $T < 12$ K.

There is no temperature dependence but a strong dependence on the type and concentration of dopants. This behaviour is interpreted in terms of a static exchange interaction of the Mu electron with the dopants in their paramagnetic state at low temperatures. This causes a static spread in the Mu precession frequencies, leading to relaxation by dephasing (T_2 process).

(2) $20 < T < 50$ K.

In this temperature region, practically all dopants are ionised in Ge, and the number of free charge carriers is essentially constant and equal to the concentration of dopants. Since the spin-flip rate of the free electrons is quite rapid, the static depolarisation of region (1) is much reduced by exchange narrowing. However spin exchange scattering of the conduction electrons by the Mu electron could lead to a different kind of relaxation. This is known as the Korringa mechanism in NMR (see for example Slichter 1978). Here depolarisation is the result of an induced spin flipping of the Mu electron, and should be also observable in a longitudinal field (T_1 process). For low electron density systems (e.g. semiconductors), where the electrons obey Boltzmann statistics, the Korringa relaxation rate follows the relation

$$T_1^{-1} \propto n T^{1/2} \tag{7.188}$$

where n is the electron concentration. Note that in a metal $T_1^{-1} \propto T$. Indeed, the observed concentration dependence at 30 K is consistent with equation (7.188). The data are not precise enough to test also the $T^{1/2}$ dependence.

(3) $T > 50$ K (> 30 K for undoped Ge).

The relaxation rates are independent of type and concentration of dopants. The damping of the Mu signal may indicate a disappearance of Mu because of chemical reactions or transitions into a different state (see next section), or may be the result of spin–lattice relaxation caused by the Raman process. The latter mechanisms (T_1 process) leads to a temperature dependence of the relaxation rate of

$$\lambda = T_1^{-1} \propto T^9 \qquad \text{(Raman process)} \qquad (7.189)$$

while the former implies

$$\lambda \propto \exp(-E/k_B T) \qquad \text{(activated process)}. \qquad (7.190)$$

The data do not allow any distinction to be made between these two models.

The relaxation of Mu in Si was also studied by Weidinger et al (1983). Figure 7.27 displays results obtained for three differently doped n type samples in an external field of 10 G. In contrast to the Ge data, the relaxation rates in the purest sample show a pronounced minimum and, for all doping concentrations studied, a more similar low temperature plateau, indicating that the dopant concentration has little influence below 80 K. The data around and above 150 K are similar in their temperature and doping dependence to the corresponding data in Ge. Longitudinal field studies performed in parallel, indicate that only the relaxation data in the vicinity of the minimum of λ are of dynamical origin (T_1 process). This observation seems to rule out the presence of a Raman process in the high temperature region and favour some sort of a chemical process (Weidinger et al 1983).

The relaxation of the anomalous Mu* signal shows quite different properties. Figure 7.28 shows data obtained in Ge in external fields of 3 kG

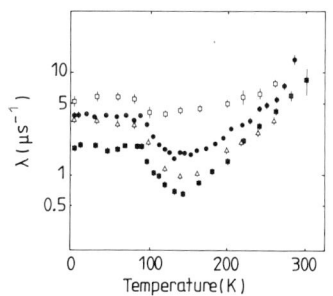

Figure 7.27 Temperature dependence of the relaxation rate of the normal Mu signal in n type Si in a transverse field of 10 G for various dopant concentrations (cm^{-3}): □, 4×10^{14}; ●, 11×10^{14}; △, 4×10^{13}; ■, 1.2×10^{13} (Weidinger et al 1984).

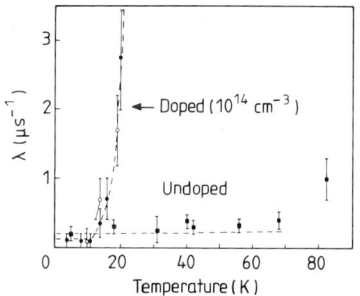

Figure 7.28 Temperature dependence of the relaxation rate of the anomalous Mu* signal in Ge. The transversely applied field was 0.4T below 40 K, and 0.3T above 40 K (Weidinger et al 1981).

and 4 kG (frequencies ω_{12} and ω_{34} are observed) (Weidinger et al 1981). At low temperatures (<10 K) the relaxation rate λ^* is extremely small and independent of the Ge sample type used. In pure Ge, λ^* remains small with a very weak temperature dependence up to 70 K, while λ^* in the two doped samples increases rapidly above 10 K, rendering the signals unobservable above 20 K. The Mu* signal in pure Ge becomes very weak above 70 K and no reliable information on the further trends on λ^* can be extracted.

The relaxation rates of the Mu* signal in Si display very interesting behaviour. This experiment was performed by Albert et al (1984) applying a field of 2 kG along one of the $\langle 111 \rangle$ crystalline axes. Three frequencies are then observed at 19 MHz, 35 MHz and 70 MHz, with weights of 4:1:3 (see §7.4.1). The relaxation of the 19 MHz component is displayed in figure 7.29 for a number of differently doped n type Si crystals. The data are characterised by a very small relaxation rate at low temperatures, and a sudden onset of a rapidly increasing relaxation rate around 30 K, independent of the doping concentration. At higher temperatures, one finds a levelling off of λ^* where the height of the λ^* plateau increases with increasing dopant concentration. At still higher temperatures, a second fast increase of λ^* is observed which is the same in all samples (broken curve in figure 7.29).

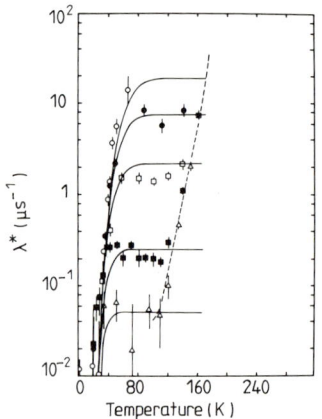

Figure 7.29 Temperature dependence of the relaxation rate of the anomalous Mu* signal (19 MHz component) in n type Si with different dopant concentrations, in a transverse field of 0.2T (Albert et al, 1984). ○, 1×10^{14}; ●, 4×10^{13}; □, 1.2×10^{13}; ■, 1.2×10^{12}; △, $<2 \times 10^{11}$.

The fast increase of λ^* around 30 K coincides with the temperature at which the dopants (phosphorus) become ionised. The full curves in figure 7.29 show the temperature dependence of the conduction electron density in the different samples, adjusted to the μSR data by the same normalisation constant. Obviously the relaxation rate, λ^*, is strictly proportional to the conduction electron density up to the onset of the high temperature relaxation regime. This observation again suggests an exchange scattering mechanism of the conduction electrons with the Mu* electron. According to this model, the relaxation rate λ^* can be expressed as (Albert et al 1984)

$$\lambda^* \propto J_{ex}^2 c(T) T^{1/2} \qquad (7.191)$$

where J_{ex} is the exchange scattering strength, and $c(T)$ the temperature dependent conduction electron concentration. The data show that J_{ex} is independent of the doping concentration in an n type Si sample, but is an order of magnitude larger in a p type Si sample. Although there is, as yet, no microscopic explanation of J_{ex} available, these data may very well be the starting point for the study of intrinsic semiconductor properties by the μSR technique. Then the Mu* or Mu system would be used as a spectator only.

The relaxation phenomena discussed so far manifest themselves as a damping of the Mu* and Mu signals (i.e. of the various precession components). The possibility that the observed damping is related to a spin-flip rate, v, of the muonium electron was already treated in §7.1.3. There the electron spin-flip rate was introduced phenomenologically as a damping term into the Wangsness–Bloch equations (7.76). Explicit solutions were mentioned for the two extreme cases of fast ($v \gg \omega_0 (1+x^2)^{1/2}$, equation (7.87)) and slow ($v < \omega_0(1+x^2)^{1/2}$ equation (7.88)) electron relaxation. In the extreme case of fast relaxation, the μSR signal in transverse fields displays a frequency which is given by the free μ^+ Larmor frequency, plus an exponential damping, independent of the strength of the applied field. For slow electron relaxation, the typical Mu or Mu* frequencies are obtained likewise showing an exponential damping with time constant $\tau = \frac{2}{3}v$ (for $x \gg 1$).

The fact that the typical Mu or Mu* signal was also observed in the temperature range where electron relaxation was assumed to be the source of depolarisation, indicates that the electron spin-flip rate can only have a moderate value. This is indeed found from the data analysis.

7.4.5 Formation probabilities and transitions among different states

The relative fraction of μ^+, Mu and Mu* in the group IV semiconductors and their dependence on temperature, doping and conditions of crystal growth have been investigated by various groups. In Si, Boekema et al (1981) investigated the temperature and doping dependence of the amplitudes of all three states. The results are shown in figure 7.30. The temperature dependence was obtained from a p type sample (450 Ωcm) and the effects of doping were investigated at a temperature of 20 K. One finds that the relative fraction ($\simeq 12.5\%$) of Mu is rather independent of temperature up to 220 K where the signal becomes unobservable because of rapid relaxation. The free μ^+ fraction stays constant up to about 150 K, where it starts to rise from $\simeq 13\%$ to $\simeq 45\%$

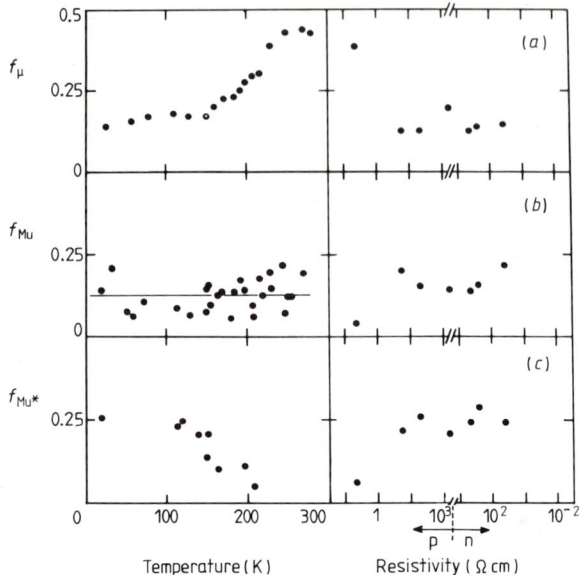

Figure 7.30 Temperature and doping (at 20 K) dependence of the fractions of (a) μ^+, (b) Mu and (c) Mu* in p type Si (Boekema et al 1981).

at 250 K. The Mu* fraction shows the opposite behaviour by starting to decline at 150 K, until it becomes undetectably small above 200 K.

The sum of all three fractions is roughly constant in the temperature interval investigated. However, the sum $f_{\mu^+} + f_{Mu} + f_{Mu^*}$ accounts for only about 50% of all stopped μ^+, so there is a unexplained sizable missing fraction. The doping concentration has only little effect on f_{μ^+}, f_{Mu} and f_{Mu^*}, except for the highest p doped sample where f_{Mu} and f_{Mu^*} are significantly reduced. The rise of f_{μ^+} with temperature, in conjunction with the relaxation behaviour of the Mu component suggests a transition Mu→μ^+, for example, by chemical reactions. This view is supported by a field dependent phase shift of the μ^+ signal in a transverse field (Kudinov et al 1975). The chemical reaction might be preceded by diffusion of the Mu system towards a reaction partner.

In this respect it is very interesting to note that Barsov et al (1984b) have indeed found clear evidence for a diffusion of the Mu system, from measurements on a neutron irradiated single crystal sample. Diffusion of the Mu* system can be excluded, since the anisotropy of the hyperfine interaction is unaffected by the temperature. The decline of f_{Mu^*} above 150 K is still unexplained.

Measurements in Ge concentrated on the study of the possible influence of electronically neutral impurities on the formation probabilities of various states. Döring et al (1983) investigated in particular the effect of Si impurities. It was found that the μ^+ fraction increases with temperature in a similar fashion to that in Si, except for a pronounced step-like increase around 15 K. A slight increase of $f_{\mu+}$ with Si concentration (5×10^{12}–10^{17} cm^{-3}) was indicated by the data. No influence of temperature and Si impurity concentration on both f_{Mu} and f_{Mu^*} was detected. The appearance of a temperature dependent phase of the μ^+ precession signal is again suggestive of a transition from the Mu state to a diamagnetic state ('bare' μ^+ or $e^-\mu^+e^-$ ion) (Döring et al 1984).

Clawson et al (1984) studied the influence of other electronically neutral impurities (C, N, O, H) on the behaviour of the various states. No significant differences from sample to sample could be detected.

Very interesting results are available from measurements in powdered diamond. Holzschuh et al (1982) found a temperature independent Mu fraction ($\simeq 20\%$) up to 150 K, above which the signal became obscured because of very rapid damping. The Mu* signal on the other hand could be observed up to $\simeq 1050$ K, showing a pronounced increase in amplitude (about five-fold) between 600 and 800 K. In this zero field measurement, all three possible frequencies $v_{12} = |v_{23}|$, $v_{14} = v_{34}$, v_{24} (see §7.4.1) were observed (figure 7.21).

Their unnormalised amplitudes are shown as a function of temperature in figure 7.31. A closer look reveals that the amplitude increase does not occur for all three frequencies at the same temperature, rather distinct differences are visible. This surprising behaviour can be easily understood in terms of a

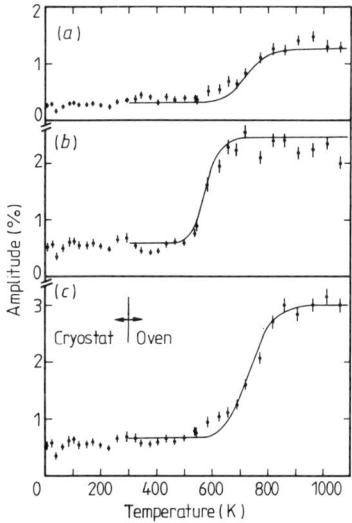

Figure 7.31 Temperature dependence of the amplitudes of: (a) the v_{24}, (b) the v_{12}, (c) the v_{14} components of anomalous Mu* (unnormalised) in diamond in zero field. The full curves represent fits of equation (7.193) (see text) (Holzschuh et al 1982).

transition Mu→Mu*. The polarisation, $P_\mu(t)$ in the second state was derived in §7.1.4 and when electron polarisation is conserved, is given by equation (7.117). For zero external field equation (7.117) reduces to the simple expression (in x direction = direction of initial $P_\mu(0)$).

$$P_x(t) = \tfrac{1}{6}[1 + 2a_1 \cos(\omega_{12}t + \phi_1) + 2a_2 \cos(\omega_{14}t + \phi_2) + a_3 \cos(\omega_{24}t + \phi_3)]$$
(7.192)

where

$$a_1 = [1 + (\omega_0 - \omega_{12})^2/\lambda^2]^{-1/2}$$
$$a_2 = [1 + \omega_{14}^2/\lambda^2]^{-1/2} \quad (7.193)$$
$$a_3 = [1 + (\omega_0 - \omega_{24})^2/\lambda^2]^{-1/2}$$

$$\phi_1 = \arctan[(\omega_0 - \omega_{12})/\lambda]$$
$$\phi_2 = \arctan[\omega_{14}/\lambda] \quad (7.194)$$
$$\phi_3 = \arctan[(\omega_0 - \omega_{24})/\lambda].$$

λ is the transition rate Mu→Mu*, ω_0 is the hyperfine splitting frequency of (isotropic) Mu and ω_{ij} are the splitting frequencies of (anisotropic) Mu*. For $\lambda = \infty$ we have, of course, $a_1 = a_2 = a_3 = 1$ and $\phi_1 = \phi_2 = \phi_3 = 0$. Using the known values for ω_0 and ω_{ij} one can calculate a_i and ϕ_i as a function of transition rate λ. The result is shown in figure 7.32. We notice in particular that a_2 rises appreciably for transition rates at which a_1 and a_3 are still very small. This is a consequence of ω_0 being an order of magnitude larger than ω_{ij}, and not appearing in the expressions for a_2 and ϕ_2. This earlier rise of a_2 is also observed in the data. The full curves in figure 7.31 are fits of equation (7.193) to the data, assuming that 10% of all stopped μ^+ instantaneously form Mu* and

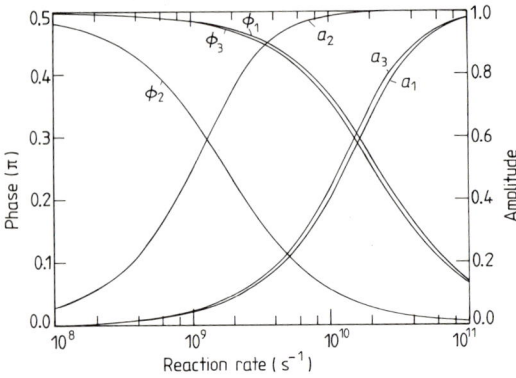

Figure 7.32 Amplitudes a_i, and phases ϕ_i, of the three Mu* signals in zero field as a function of the transition rate Mu → Mu* (Holzschuh *et al* 1982).

that λ can be described by an Arrhenius law

$$\lambda = \lambda_0 \exp(-U/k_B T). \tag{7.195}$$

The fit yields

$$\lambda_0 \simeq 4 \times 10^{14} \text{ Hz} \qquad U/k_B \simeq 7000 \text{ K}.$$

The values obtained have not been explained in terms of a physical model.

Another good demonstration of the transition Mu→Mu* in diamond has recently been achieved by Patterson et al (1984b) using measurements in non-zero transverse fields. Recalling equation (7.117), we immediately recognise that the amplitude of a particular frequency $\tilde{\omega}_{ij}$ shows a resonance-like behaviour for $(\tilde{\omega}_{ij} - \omega_{km})/\lambda \ll 1$, that is the amplitude has a maximum if the transition frequency in the final state coincides with a transition frequency of the precursor state. Patterson et al (1984b) verified this effect experimentally in a single crystal, with the $\langle 110 \rangle$ axis oriented parallel to the applied field. The results are shown in figure 7.33.

Finally, table 7.6 collects the results on the fractions f_{μ^+}, f_{Mu} and f_{Mu^*} in various diamond samples for $T < 296$ K (Holzschuh et al 1982). We see that an appreciable missing fraction exists. The five-fold increase of the Mu* fraction

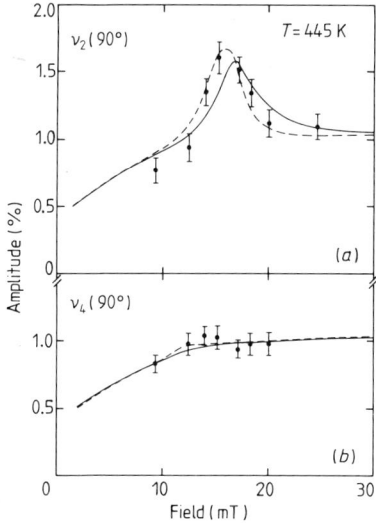

Figure 7.33 Experimental demonstration in single crystal diamond of the resonance-like behaviour of a Mu* precession amplitude, as a function of applied field at a temperature of 445 K. (a) shows the signal when frequencies of the precursor Mu state and the final Mu* state are degenerate. (b) shows the signal when this degeneracy condition is not fulfilled. The external field is applied along the $\langle 110 \rangle$ crystalline axis. For a more in-depth discussion see Patterson et al (1984).

Table 7.6 Formation probabilities for muon states in diamond. The values are corrected for background effects from cryostat, etc, and normalised with the precession amplitude in a Cu sample (Holzschuh *et al* 1982).

Physical state	T (K)	Mu (%)	Mu* (%)	μ^+ (%)	Missing (%)
Powder	4.2–90	18.5 ± 0.9	9.9 ± 0.7	<10	$\gtrsim 60$
Powder, magic field	296	—	14.5 ± 1.3	<5	$\gtrsim 80$
Single crystal	4.2	20 ± 4	11.9 ± 0.9	<10	$\gtrsim 60$

above 600 K indicates that this state cannot be formed from just the Mu fraction visible below 150 K. Rather, it must be fed from some part of the missing fraction as well. A sizable missing fraction still persists, even at the highest temperatures used.

Contradictory results were seemingly found by Spencer *et al* (1984) in a single crystal sample of diamond probably of higher purity than the one used by Holzschuh *et al* (1982) (the latter contained about 0.1 atomic % nitrogen). A Mu signal was visible up to room temperature accounting for about 30% of all stopped μ^+ (lower limit). However, no Mu* signal could be detected. It is an open question whether the different nitrogen content could explain the conflicting observations, particularly in view that the very pure powder sample of Holzschuh *et al* (1982) yielded results essentially in agreement with their single crystal data.

7.4.6 Nature of muonium states

Although the two different paramagnetic muon states in diamond, Si and Ge (also in GaAs and GaP) (Barsov *et al* 1984b) are very well identified experimentally, through their distinctive hyperfine parameters, their true nature and origin is still somewhat of a mystery. Even more so since no paramagnetic hydrogen centres in the group IV semiconductors could be observed unambiguously by ESR investigations, although it is known otherwise that hydrogen is a major impurity in Si and Ge. In view of the fact that the muonium states are observed within a few μs after μ^+ implantation, one might conjecture that the observed states are of transient nature and have not yet reached thermal equilibrium (Mainwood and Stoneham 1983). Another obstacle to a sufficient theoretical model is the lack of knowledge about the crystalline site of the two paramagnetic centres. However, there is hope, that a site determination by μ^+ and e^+ channelling experiments will soon be available (see e.g. Patterson 1984, Meier 1984). Although the 'anomalous' Mu* state has only been investigated theoretically very recently (Sahoo *et al* 1985) various theoretical calculations have, nevertheless, attempted to explain the hyperfine properties of the 'normal' muonium state (Mu). There is an apparent absence of Mu* diffusion in diamond (Holzschuh et al 1982) as shown by the anisotropy of the hyperfine interaction, regardless of temperature (in contrast to Mu behaviour). This, and the observed transition Mu → Mu*, has led to the suggestion that the Mu* state may be understood as a bound chemical complex resembling a chemical radical (e.g. Fischer 1984).

Any attempt to explain the properties of the normal Mu state must start with the observation that this state possesses a deep bound level which forbids the application of the so-called effective mass theory (EMT) (for a short review see Pantelides 1979). The EMT approach is very successful in explaining the properties of so-called shallow donors or acceptors. These possess binding energies of several tens of meV, and effective Bohr radii of several times $10a_B$, while the Bohr radius of normal Mu in diamond, Si and Ge is just 6–30% larger than the Bohr radius, a_B.

Wang and Kittel (1973) were the first to try and explain the properties of the normal Mu state. The Schrödinger equation of Mu is written as

$$\left(\frac{\hbar^2}{2m^*(r)}\nabla^2 + U(r)\right)\psi(r) = E\psi(r). \quad (7.196)$$

$m^*(r)$ is the position dependent effective mass, and $U(r)$ is the effective potential of the μ^+

$$U(r) = -e^2/\varepsilon_0(r)r \quad (7.197)$$

where $\varepsilon_0(r)$ is the position dependent dielectric function. Two models were considered.

(i) the so-called cavity model with the following approximations

$$U(r) = -(e^2/r) + e^2(1 - 1/\varepsilon_0)/R \qquad \text{for } r < R$$

$$U(r) = -e^2/\varepsilon_0 r \qquad \text{for } r > R \quad (7.198)$$

where ε_0 is the static dielectric constant (C: $\varepsilon_0 = 5.7$, Si: $\varepsilon_0 = 11.9$, Ge: $\varepsilon_0 = 16.0$), R is the cavity radius and

$$m(r) = m_0 \qquad \text{for } r < R$$

$$m(r) = m^* \qquad \text{for } r > R \quad (7.199)$$

where m^* is the bulk effective electron mass. Alternatively, $m(r)$ was fixed to m_0 everywhere.

The cavity radius was determined from the radius of the interstitial sphere that would fit inside the touching hard spheres of the lattice atoms. Two interstitial sites with tetrahedral (T) and hexagonal (H) local symmetry were considered as likely sites for the Mu system (C: $R_T = 1.45a_B$, $R_H = 1.33a_B$; Si: $R_T = 2.22a_B$, $R_H = 2.03a_B$; Ge: $R_T = 2.31a_B$, $R_H = 2.12a_B$).

(ii) the dielectric function approach. $\varepsilon_0(r)$ is expressed by a more sophisticated relation. Wang and Kittel (1973) used the following expression, which dates back to Hermanson (1966),

$$\varepsilon_0(r)^{-1} = \varepsilon_0^{-1} + (1-\varepsilon_0^{-1})\exp-Q_D r. \quad (7.200)$$

The parameter Q_D is obtained from a fit to band structure calculations of the momentum dependent dielectric function $\varepsilon(q)$. Holzschuh et al (1982) have, in addition, considered a different expression for $\varepsilon_0(r)$ (Resta 1977) which is obtained by solving the linearised Thomas–Fermi equation

$$\varepsilon_0(r) = \begin{cases} \varepsilon_0 \kappa R\{\sinh[\kappa(R-r)+\kappa r]\}^{-1} & \text{for } r \leqslant R \\ \varepsilon_0 & \text{for } r > R \end{cases} \quad (7.201)$$

where $\kappa = (4k_F/\pi)^{1/2}$ and $R = 2.76a_B$, $4.28a_B$ and $4.71a_B$ for diamond, Si and Ge, respectively. (For more recent calculations of $\varepsilon_0(r)$ see Cornolti and Resta (1978), Csavinzky and Braunstein (1981).)

Using the approximations for $U(r)$ and $m(r)$, the Schrödinger equation can be solved. Results for $f = |\psi(0)|^2/|\psi(0)_{\text{vac}}|^2$ for both models are collected in table 7.7. For the cavity model, only the results for $m(r) = m_0$ are quoted as these show the best agreement with the data. Q_D was determined with reference to band structure calculations of Vinsome et al (1971) (C: $Q_D = 1.15a_B^{-1}$, Si: $Q_D = 0.92a_B^{-1}$, Ge: $Q_D = 0.82a_B^{-1}$). All calculations yield the sequence $f_C < f_{Si} < f_{Ge}$ while the data clearly indicate that $f_C > f_{Ge} > f_{Si}$. Obviously these calculations are unable to reproduce the systematic trend of the data correctly, although the order of magnitude of f is not too much in disagreement with the data.

Table 7.7 Compilation of theoretical calculations of the normalised hyperfine field, $f = |\psi(0)|^2/|\psi(0)_{\text{vac}}|^2$, using the cavity model (equation (7.198)) or the dielectric function approach ((a) equations (7.200) and (b) (7.201)). The cavity model using $m = m_0$ for all r is evaluated both for Mu at tetrahedral (T) and hexagonal (H) sites (Holzschuh et al 1982).

Host	Model				Experiment
	Cavity		Dielectric function		
	T	H	(a)	(b)	
C	0.419	0.313	0.381	0.114	0.829
Si	0.834	0.754	0.428	0.170	0.450
Ge	0.856	0.786	0.487	0.175	0.529
Sequence	$f_C < f_{Si} < f_{Ge}$	$f_C < f_{Si} < f_{Ge}$	$f_C < f_{Si} < f_{Ge}$	$f_C < f_{Si} < f_{Ge}$	$f_C > f_{Ge} > f_{Si}$

Manninen and Meier (1982) have developed another simple theory, similar in spirit to that of Wang and Kittel. This theory is based on the density functional formalism and a pseudo-potential description of the semiconductor. Assuming that there is a well localised impurity state with charge density $n_1(r)$ the corresponding Schrödinger equation involves an effective potential of the form

$$W_1(r) = -\frac{1}{r} + \sum_m \tilde{V}_{ps}(r - R_m) + \int dr' \frac{n_p(r')}{|r - r'|}. \qquad (7.202)$$

The first term represents the Coulomb potential of the μ^+, the second term contains the screened pseudo-potentials of the host lattice atoms at site R_m and the third term stems from the charge density, $n_p(r)$ caused by the polarisation of the lattice due to the muonium impurity. Within linear response theory $n_p(r)$ can be calculated from $n_1(r)$ and the dielectric function $\varepsilon_0(r)$ with the result that (Manninen and Meier 1982)

$$W_1(r) = -\frac{1}{r\varepsilon_0(r)} + \int dr' \left(\frac{1}{\varepsilon(r-r')} - 1 \right) \frac{n_1(r')}{|r-r'|} + \sum_m \tilde{V}_{ps}(r - R_m) \qquad (7.203)$$

and

$$n_1(r) = |\psi_1(r)|^2.$$

This potential not only contains the term used in equation (7.197), but also the lattice potential (third term) which now allows the actual site dependence of the impurity state $\psi_1(r)$ to be explicitly included. The second term guarantees that no overscreening of the μ^+ charge can occur.

The actual self-consistent solution of the Schrödinger equation is simplified by replacing the sum over the lattice pseudo-potentials by a spherical average around the μ^+ site (spherical solid model). The dielectric function $\varepsilon_0(r)$ took the form of equation (7.101). In addition, Manninen and Meier (1982) have also taken the influence of the zero-point motion of the muonium atom on the hyperfine field into account by forming the average

$$\langle |\psi_1(0)|^2 \rangle = \int dr_\mu P_\mu(r_\mu) |\psi_1(r_\mu)|^2 \qquad (7.204)$$

where $P_\mu(r)$ is the probability that the muonium is at the point r_μ and $\psi_1(r_\mu)$ is calculated for a fixed μ^+ at the site r_μ. $P_\mu(r_\mu)$ is assumed to have a Gaussian form

$$P_\mu(R_\mu - r_\mu) = (2\pi\sigma^2)^{-3/2} \exp(-|R_\mu - r_\mu|/2\sigma^2) \qquad (7.205)$$

centred at the interstitial site R_μ. The results for the normalised hyperfine field

$$f = \langle |\psi_1(0)|^2 \rangle / |\psi(0)_{vac}|^2 \qquad (7.206)$$

as a function of the Gaussian width, σ, are shown in figure 7.34 for the tetrahedral and hexagonal interstitial site. We notice first that the zero-point motion has a pronounced influence on f. Without zero-point motion (fixed Mu at the centre of the interstitial site) the calculated f are much smaller than the experimental ones. Only zero-point amplitudes of 1.0 to 1.5 au lead to effective hyperfine fields close to the measured numbers. Secondly, there is no strong site dependence and thirdly f in diamond is always larger than in Ge and Si for comparable σ, also basically in agreement with experiment.

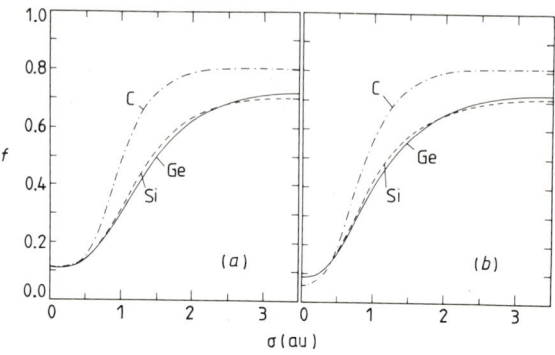

Figure 7.34 Normalised hyperfine field, $f = |\psi(0)|^2/|\psi(0)|^2_{\text{vac}}$ for Mu in diamond, Si and Ge at (a) tetrahedral and (b) octahedral sites. f is plotted as a function of the gaussian width σ of the μ^+ distribution following from the zero-point motion (Manninen and Meier 1982).

Whether the role of zero-point motion is really that important remains doubtful in view of rather detailed calculations of f in Si and diamond. In Si Katayama-Yoshida et al (1983) calculated f in Si for a *fixed* Mu at the centre of the tetrahedral interstitial site using the local spin density functional formalism and the LCAO (linear combination of atomic orbitals) Green's function method. Their result is $f_{\text{Si}} = 0.406$ in excellent agreement with the experimental value. In diamond Sahoo et al (1983) applied an unrestricted Hartree–Fock self-consistent-field cluster approach. Their calculation favours a tetrahedral site occupancy for Mu. For a fixed Mu at the centre of the tetrahedral interstitial site they obtained $f_C = 0.90$. Taking the spatial distribution of the μ^+ wave function at the tetrahedral site into account also, an averaged value of $\langle f_C \rangle = 0.75$ is calculated. Both values agree reasonably well with the experimental value of $f_C = 0.829$. Compared with the calculations of Manninen and Meier (1982), there is a sizable discrepancy for the centre of site values, and also for the trends in f when zero-point motion is included. Further theoretical studies are clearly necessary before a consistent understanding of both Mu and Mu* in group IV semiconductors can be reached. Moreover, we are also confronted with the problem of the mechanisms leading to the simultaneous formation of both Mu and Mu*.

7.5 Surface Interactions of Muonium in Oxide Powders

It is well known from positron annihilation studies on fine oxide powder samples, such as MgO and SiO_2, that ortho-positronium is present in the empty inter-granular volume. It is believed that the positronium atoms are formed inside the grains, then diffuse rapidly to the grain surfaces, and out into the vacuum because of a negative work function (Brandt and Paulin 1968). This behaviour has served as a motivation to look for similar effects when μ^+ are brought to rest in an oxide powder sample. Indeed, Marshall et al (1978) were able to observe a muonium precession signal in a SiO_2 powder sample consisting of grains with an average diameter of 70 Å. The observed precession amplitude indicated that about 45% of all stopped μ^+ formed muonium. It was further observed that the signal displayed a damping which increased linearly with the amount of oxygen (O_2) concentration outside the grains. The analysis showed that the relaxation rate was the same as obtained in gas phase studies, using an argon–oxygen target with the same oxygen concentration as in the SiO_2 powder target (Fleming et al 1980). This was taken as strong evidence that the interaction of muonium with the paramagnetic O_2 molecules must have taken place outside the grain particles. It was soon found that muonium is also formed in the intergranular volume of Al_2O_3 and MgO powders (Kiefl et al 1979). The experiments, previously performed at room temperature only, were then extended down to low temperatures (Kiefl et al 1982). In these low temperature studies, the powder was immersed in either a He or a Ne atmosphere.

Table 7.8 compares the room temperature μSR results in evacuated powders with the low temperature results in a He atmosphere. One notices that the

Table 7.8 Comparison of room temperature (295 K) μSR results for evacuated powders with low temperature results (6 K) in a He atmosphere (Kiefl et al 1982).

Target	Particle size (Å)	Temperature (K)	Muonium fraction (atomic %)	Muonium relaxation rate (μs^{-1})
SiO_2	70	6	49 ± 3	0.46 ± 0.03
SiO_2	70	295	61 ± 3	0.18 ± 0.03
SiO_2	140	6	$35 \pm 5(a)$†	$4.1 \pm 0.7(a)$†
			$35 \pm 5(b)$	$0.16 \pm 0.05(b)$
SiO_2	140	295	60 ± 3	0.18 ± 0.03
Al_2O_3	75	6	29 ± 3	0.35 ± 0.05
Al_2O_3	75	295	35 ± 14	11.3 ± 4.4
MgO	300	6	12 ± 3	0.22 ± 0.03
MgO	300	295	15 ± 3	1.9 ± 0.5

† The (a) and (b) refer to the two components resolved in the fit.

fraction of observed muonium atoms is nearly the same at both temperatures. Interesting differences show up in the relaxation rates for Al_2O_3 and MgO powders. Figure 7.35 shows the actual temperature dependence of the relaxation rate in Al_2O_3 powder between 4.5 and 16 K. A drastic almost step-like change of the relaxation rate is observed above 12 K.

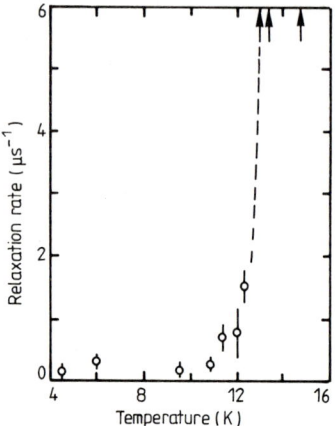

Figure 7.35 Temperature dependence of the Mu spin relaxation rate in 75 Å Al_2O_3 powder in a He atmosphere. The arrows indicate points that are off scale (Kiefl *et al* 1982).

The data indicate that muonium is also expelled from the grains at temperatures as low as 6 K, because the observed relaxation rates are much too low to be compatible with intra-grain or grain surface Mu states. In Al_2O_3, the superhyperfine interaction with the ^{27}Al nuclei should result in a very rapid relaxation rendering the Mu precession unobservable. Also, in bulk quartz the Mu relaxation rate at 6 K is 3.3 ± 0.5 μs^{-1} which is much faster than the rate observed in the powder samples.

It is known that He readily adsorbs at these low temperatures on such oxide surfaces forming a protective layer which prevents the Mu atom from approaching the bare surface of the grains too closely. The He surface film develops vacancies as it evaporates with increasing temperature, thus exposing the bare surface to approaching Mu atoms. This explains the sharp increase in the relaxation rate in the Al_2O_3 powder above 12 K, at which temperature the Mu atom starts to make direct contact with the bare surface and the superhyperfine interaction comes into play. This mechanism was studied in more detail by a series of measurements in which a controlled amount of He or Ne was adsorbed on the powder surfaces at a constant temperature. It was found that the relaxation rate was a decreasing linear function of the adsorbed gas density below monolayer coverage (Kiefl *et al* 1982).

At low enough temperatures, the Mu atom will become adsorbed on the oxide surface itself. This should give rise to a variety of interesting phenomena which could establish the μ^+ as a useful probe also for surface physics. One of the interesting possibilities is the observation of two-dimensional diffusion of the Mu atom on the surface, and its interactions with other adsorbed impurity atoms and molecules. A study of this kind has indeed recently been reported using a silica (amorphous SiO_2) powder with controlled amounts of hydroxyl (OH^-) groups on the surface (Harshman *et al* 1984). In this study the depolarisation rate was measured as a function of temperature and surface hydroxyl concentration. The hydroxyl concentration was varied by baking the powder at different temperatures. Normally, the concentration of surface hydroxyls is about 0.5 per surface Si atom. The results are displayed in figure 7.36. The low temperature plateau is identified with the frozen-in atoms at host adsorption sites. The decrease at rising temperatures is caused by the onset of diffusion leading to motional narrowing. At higher temperatures, trapping causes the peak in the relaxation rate. At still higher temperatures the decrease of the relaxation is the result of de-trapping and eventually desorption from the surface. According to the analysis of Harshman *et al* (1984) the relaxation in the host adsorption sites is caused by dipolar interaction with the protons in the hydroxyl groups. The origin of relaxation in the trap site, as well as the nature of the trap site, could not be determined. A three-state (host-adsorption

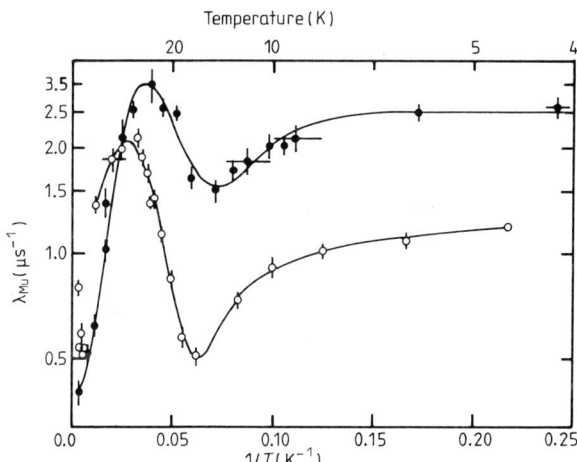

Figure 7.36 Temperature dependence of the Mu spin relaxation rate in silica powder (mean grain diameter 70 Å) for samples prepared at: ●, 110°C; ○, 600°C. The full curve is a fit of the three-state model mentioned in the text (Harshman *et al* 1984).

site, surface-trap site, desorbed state) non-equilibrium model was fitted to the data and yielded the following activation energies for the three states

$$63^{+10}_{-8}\text{ K} \qquad 118^{+25}_{-17}\text{ K} \qquad \text{and} \qquad 212^{+108}_{-43}\text{ K}$$

Recently the hyperfine properties of muonium in evacuated SiO_2 powder (grain diameter ~ 70 Å) were also studied in a 300 G field by the two-frequency method (Kiefl et al 1984a). Above 100 K the hyperfine splitting frequency is compatible with the vacuum value, indicating that the Mu atoms are free through moving in the void regions between the powder grains. Below 100 K. however, the splitting frequency drops rapidly to a value of 4446.5 ± 3.0 MHz, corresponding to a shift of -16.8 ± 3.0 MHz relative to the vacuum value. This sharp decrease is attributed to the adsorption of the Mu atoms on the SiO_2 grain surfaces. It is interesting to note that the relative frequency shift for Mu in bulk quartz is $+32.9$ MHz. Analysis of the data indicated a binding energy of Mu to the SiO_2 grain surface of ~ 300 K, essentially in agreement with the value of Harshman et al (1984).

References

Adrianov D G, Minaichev E V, Myasishcheva G G, Obukhov Yu V, Roganov V S, Savelev G I, Firsov V G and Fistul V I 1970 *Zh. Eksp. Teor. Fiz.* **58** 1896 (Engl. transl. 1970 *Sov. Phys.–JETP* **31** 1019)

Adrianov D G, Myasishcheva G G, Obukhov Yu V, Roganov V S, Firsov V G and Fistul V I 1969 *Zh. Eksp. Teor. Fiz.* **56** 643 (Engl. transl. 1969 *Sov. Phys.–JETP* **29** 643)

Albert E, Möslang A, Recknagel E and Weidinger A 1984 *Hyperfine Interactions* **17–19** 611

Allison S K 1958 *Rev. Mod. Phys.* **30** 1137

Arseneau D J, Garner D M, Senba M and Fleming D G 1984 *J. Chem. Phys.* (in press)

Balzer G, Graf H, Recknagel E, Weidinger A and Wichert Th 1981 *Hyperfine Interactions* **8** 393

Barsov S G, Firsov V G, Getalov A L, Gordeev V A, Kruglov S P, Kudinov V I, Kuzmin L A, Mikirtichyants S M, Minaichev E V, Myasishcheva G G, Obukhov Yu V, Savelev G I and Scherbakov G V 1984a *Hyperfine Interactions* **17–19** 635

Barsov S G, Getalov A L, Gordeev V A, Evseev V A, Konopleva R F, Kruglov S P, Kudinov V I, Kuzmin L A, Mikirtychants S M, Minaichev E V, Myasishcheva G G, Obukhov Yu V, Savelev G I, Firsov V G and Sherbakov G V 1984b *Hyperfine Interactions* **17–19** 551

Barsov S G, Getalov A L, Gordeev V A, Konopleva R F, Kruglov S P, Kudinov V I, Kuzmin L A, Mikritychants S M, Minaichev E V, Myasishcheva G G, Obukhov Yu V, Savelev G I, Firsov V G and Shcerbakov G V 1980 *Zh. Eksp. Teor. Fiz.* **79** 1461 (Engl. transl. 1980 *Sov. Phys.–JETP* **52** 738)

Baryshevsky V G and Kuten S A 1977 *Phys. Lett.* **64A** 238
—— 1978 *Phys. Lett.* **67A** 355
Beck R, Meier P F and Schenck A 1975 *Z. Phys.* **B22** 109
Beder D 1978 *Nucl. Phys.* A **305** 411
Belousov Yu M, Gorelkin V N and Smilga V P 1978a *Zh. Eks0. Teor. Fiz.* **74** 629 (Engl. transl. 1978 *Sov. Phys.–JETP* **47** 331)
—— 1978b *Zh. Eksp. Teor. Fiz.* **75** 1999 (Engl. transl. 1978 *Sov. Phys.–JETP* **48** 1007)
Blazey K W, Brown J A, Cooke D W, Dodds S A, Estle T L, Heffner R H, Leon M and Vanderwater D A 1981 *Phys. Rev.* B **23** 5316
Blazey K W, Estle T L, Felber J, Holzschuh E, Keller H, Kiefl R F, Kündig W, Meier P F, Odermatt W, Patterson B D, Rudaz S L and Schneider J W 1984a *SIN Newsletter No 16* (Swiss Institute for Nuclear Research, Villigen, Switzerland)
Blazey K W, Estle T L, Holzschuh E, Meier P F, Patterson B D and Richner M 1984b *Hyperfine Interactions* **17–19** 595
Blazey K W, Estle T L, Holzschuh E, Odermatt W and Patterson B D 1983 *Phys. Rev.* B **27** 15
Boekema C, Holzschuh E, Kündig W, Meier P F, Patterson B D, Reichart W and Rüegg K 1981 *Hyperfine Interactions* **8** 401
Brandt W and Paulin R 1968 *Phys. Rev. Lett.* **21** 193
Breit G and Hughes V W 1957 *Phys. Rev.* **106** 1293
Brewer J H 1981 *Hyperfine Interactions* **8** 375
—— 1982 cited by A Schenck in *Proc. Workshop Muon Science Facilities, Los Alamos* compiled by R H Heffner (Los Alamos, LA-9582-C conference)
Brewer J H, Beder D S and Spencer D P 1979 *Phys. Rev. Lett.* **42** 808
Brewer J H, Crowe K M, Gygax F N, Johnson R F, Fleming D G and Schenck A 1974 *Phys. Rev.* A **9** 495
Brewer J H, Crowe K M, Gygax F N, Johnson R F, Patterson B D, Fleming D G and Schenck A 1973 *Phys. Rev. Lett.* **31** 143
Brewer J H, Crowe K M, Gygax F N and Schenck A 1975 *Muon Physics* vol. 3 ed. V W Hughes and C S Wu (New York: Academic)
Brewer J H, Fleming D G and Spencer D P 1981b *Nuclear and Electron Resonance Spectroscopies Applied to Materials Science* ed. Kaufmann and Shenoy (Amsterdam: North-Holland)
Brewer J H, Spencer D P, Fleming D G and Coope J A R 1981a *Hyperfine Interactions* **8** 405
Brown J A, Dodds S A, Estle T L, Heffner R H, Leon M and Vanderwater D A 1980 *Solid State Commun.* **33** 613
Brown J A, Heffner R H, Leon M, Dodds S A, Estle T L and Vanderwater D A 1983 *Phys. Rev.* B **27** 3980
Brown J A, Heffner R H, Leon M, Dodds S A, Vanderwater D A and Estle T L 1979 *Phys. Rev. Lett.* **43** 1751
Bucci C, de Renzi R, Guidi G, Podini P, Tedeschi R and Norlin L O 1981 *Hyperfine Interactions* **8** 385
Buhler A, Massam T, Muller T, Schneegans M and Zichichi A 1965 *Nuovo Cimento* **29** 812

Clawson C W, Crowe K M, Haller E E and Rosenblum S S 1984 *Hyperfine Interactions* **17–19** 603
Clawson C W, Haller E E, Crowe K M, Rosenblum S S and Brewer J H 1981 *Hyperfine Interactions* **8** 417
Cornolti F and Resta R 1978 *Phys. Rev.* B **17** 3239
Crowe K M, Hague J F, Rothberg J E, Schenck A, Williams D L, Williams R W and Young K K 1972 *Phys. Rev.* D **5** 2145
Csavinszky P and Brownstein K R 1981 *Phys. Rev.* B **24** 4566
Döring K-P, Arnold K-P, Gladisch M, Haas N, Haller E E, Herlach D, Jacobs W, Krause M, Krauth M, Orth H and Seeger A 1984 *Hyperfine Interactions* **17–19** 629
Döring K-P, Haas N, Haller E E, Herlach D, Jacobs W, Krauth M, Orth H, Rosenkranz J, Seeger A, Vetter J, Arnold K-P, Aurenz Th and Bossy H 1983 *Physica* **116B** 354
Eisenstein B, Prepost R and Sachs A M 1966 *Phys. Rev.* **142** 217
Estle T L, Rudaz S L, Holzschuh E, Kiefl R F, Patterson B D, Kündig W and Blazey K W 1984 *Hyperfine Interactions* **17–19** 623
Estle T L and Vanderwater D A 1983 *Phys. Rev.* B **27** 3962
Feher G, Prepost R and Sachs A M 1960 *Phys. Rev. Lett.* **5** 515
Ferrell R A and Chaos F 1957 *Phys. Rev.* **106** 1322
Ferrell R A, Lee Y C and Pal K M 1960 *Phys. Rev.* **118** 317
Fischer H 1984 *Hyperfine Interactions* **17–19** 751
Fleming D G, Arseneau D J, Garner D M, Senba M and Mikula R J 1984 *Hyperfine Interactions* **17–19** 655
Fleming D G, Mikula R J and Garner D M 1980 *J. Chem. Phys.* **73** 2751
—— 1982 *Phys. Rev.* A **26** 2527
Friedman J I and Telegdi V L 1957 *Phys. Rev.* **106** 1290
Garwin R L, Ledermann L M and Weinrich M 1957 *Phys. Rev.* **105** 1415
Gurevich I I, Ivanter I G, Makariyna L A, Meleshko E A, Nikolskii B A, Roganov V S, Selivanov V I, Smilga V P, Sokolov B V, Shestakov V D and Jakovleva I V 1969 *Phys. Lett.* **29B** 387
Gurevich I I, Ivanter I G, Meleshko E A, Nikolskii B A, Roganov V S, Selivanov V I, Smilga V P, Sokolov B V and Shestakov V D 1971 *Zh. Eksp. Teor. Fiz.* **60** 471 (Engl. transl. 1971 *Sov. Phys.–JETP* **33** 253)
Gurevich I I, Makaryna L A, Meleshko E A, Nikolskii B A, Roganoy V S, Selivanov V I and Sokolov B V 1968 *Zh. Eksp. Teor. Fiz.* **54** 432 (Engl. transl. 1968 *Sov. Phys.–JETP* **27** 235)
Gurevich I I, Meleshko E A, Muratova I A, Nikolskii B A, Roganov V S and Selivanov V I 1973 *Zh. Eksp. Teor. Fiz. Pis. Red.* **18** 608 (Engl. transl. 1973 *Sov. Phys.–JETP Lett.* **18** 357)
Gurevich I I, Nikolskii B A, Ponomarev A N, Selivanov V I and Suetin V A 1979 *Zh. Eksp. Teor. Fiz.* **76** 340 (Engl. transl. 1979 *Sov. Phys.—JETP* **49** 174)
Gurevich I I, Nikolskii B A, Selivanov V I and Sokolov B V 1975 *Zh. Eksp. Teor. Fiz.* **68** 806 (Engl. transl. 1975 *Sov. Phys.–JETP* **41** 401)
Harshman D R, Keitel R, Senba M, Kiefl R F, Ansaldo E J and Brewer J H 1983 *Phys. Lett.* **104A** 472
Hermanson J 1966 *Phys. Rev.* **150** 660

Hintermann A, Meier P F and Patterson B D 1980 *Am. J. Phys.* **48** 956
Holzschuh E 1982 *Doctoral thesis* University of Zürich
—— 1983 *Phys. Rev.* **27** 102
Holzschuh E, Graf H, Recknagel E, Weidinger A, Wichert Th and Meier P F 1979 *Phys. Rev.* B **20** 4391
Holzschuh E, Kündig W, Meier P F, Patterson B D, Sellshop J P F, Stemmet M C and Appel H 1982 *Phys. Rev.* **B25** 1272
Holzschuh E, Kündig W and Patterson B D 1981 *Helv. Phys. Acta* **54** 552
Hughes V W and Kinoshita T 1977 *Muon Physics* vol. 1. ed. V W Hughes and C S Wu (New York: Academic)
Hughes V W, McColm D W, Ziock K and Prepost R 1960 *Phys. Rev. Lett.* **5** 63
Iakovleva I V 1958 *Zh. Eksp. Teor. Fiz.* **35** 970 (Engl. transl. 1959 *Sov. Phys.–JETP* **35** 676)
Ito Y, Miyake Y, Tabata Y, Nishiyama K and Nagamine K 1982 *Chem. Phys. Lett.* **93** 361
Ivanter I G 1968 *Zh. Eksp. Teor. Fiz.* **56** 1419 (Engl. transl. 1969 *Sov. Phys–JETP* **29** 761)
Ivanter I G, Minaichev E V, Myasishcheva G G, Obukhov Yu V, Roganov V S, Savelev G I, Smilga V P and Firsov V G 1972 *Zh. Eksp. Teor. Fiz.* **62** 14 (Engl. transl. 1972 *Sov. Phys.–JETP* **35** 9)
Ivanter I G, Nikolskii B A, Ponomarev A N, Selivanov V I and Suetin V A 1978 *Zh. Eksp. Teor. Fiz.* **75** 376 (Engl. transl. 1978 *Sov. Phys.–JETP* **48** 189)
Ivanter I G and Smilga V P 1967 *Zh. Eksp. Teor. Fiz.* **54** 559 (Engl. transl. 1968 *Sov. Phys.–JETP* **27** 301)
—— 1968 *Zh. Eksp. Teor. Fiz.* **55** 1521 (Engl. transl. 1969 *Sov. Phys.–JETP* **28** 796)
—— 1971 *Zh. Eksp. Teor. Fiz.* **60** 1985 (Engl. transl. 1971 *Sov. Phys.–JETP* **33** 1070)
—— 1972 *Zh. Eksp. Teor. Fiz.* **61** 2176 (Engl. transl. 1972 *Sov. Phys.–JETP* **34** 1167)
Katayama-Yoshida H and Shindo K 1983 *Phys. Rev. Lett.* **51** 207
Kiefl R F, Patterson B D, Holzschuh E, Odermatt W and Harshman D R 1984a *Hyperfine Interactions* **17–19** 563
Kiefl R F, Warren J B, Marshall G M, Oram C J, Brewer J H, Judd D J and Spires L D 1979 *Hyperfine Interactions* **6** 185
Kiefl R F, Warren J B, Oram C J, Marshall G M, Brewer J H, Harshman D R and Clawson C W 1982 *Phys. Rev.* B **26** 2432
Kudinov V I, Minaichev E V, Myasishcheva G G, Obukhov Yu V, Roganov V S, Savelev G I, Samoilov V M and Firsov V G 1975 *Zh. Eksp. Teor. Fiz. Pis. Red.* **21** 49 (Engl. transl. 1975 *Sov. Phys.–JETP Lett.* **21** 22)
de Launey J 1956 *The Theory of Specific Heats and Lattice Vibrations* vol. 2 *Solid State Physics* ed. F Seitz and D Turnball (New York: Academic)
Mainwood A and Stoneham A M 1983 *Physica* **116B** 101
Manninen M and Meier P F 1982 *Phys. Rev.* B **26** 6690
Marshall G M, Warren J B, Garner D M, Clark G S, Brewer J H and Fleming D G 1978 *Phys. Lett.* **65A** 351
Meier K 1984 *Hyperfine Interactions* **17–19** 3

Meier P F 1982 *Phys. Rev.* **A25** 1287
Messiah A 1965 *Quantum Mechanics* vol. 1 3rd reprint (Amsterdam: North-Holland)
Minaichev E V, Myasishcheva G G, Obukhov Yu V, Roganov V S, Savelev G I and Firsov V G 1970 *Zh. Eksp. Teor. Fiz.* **58** 1586 (Engl. transl. 1970 *Sov. Phys.–JETP* **31** 849)
Myasishcheva G G, Obukhov Yu V, Roganov V S and Firsov V G 1967 *Zh. Eksp. Teor. Fiz.* **53** 451 (Engl. transl. 1968 *Sov. Phys.–JETP* **26** 298)
—— 1969 *Zh. Eksp. Teor. Fiz.* **56** 1199 (Engl. transl. 1969 *Sov. Phys.–JETP* **29** 645)
Nosov V G and Yakovleva I V 1962 *Zh. Eksp. Teor. Fiz.* **43** 1750 (Engl. transl. 1963 *Sov. Phys.–JETP* **16** 1236)
Orear J, Harris G and Bierman E 1957 *Phys. Rev.* **107** 322
Pantelides S T 1979 *Hyperfine Interactions* **6** 145
Patterson B D 1984 *Hyperfine Interactions* **17–19** 517
Patterson B D, Hintermann A, Kündig W, Meier P F, Waldner F, Graf H, Recknagel E, Weidinger A and Wichert T 1978 *Phys. Rev. Lett.* **40** 1347
Patterson B D, Holzschuh E, Kiefl R, Blazey K W and Estle T L 1984a *Hyperfine Interactions* **17–19** 599
Patterson B D, Odermatt W, Holzschuh E, Kündig W, Meier P F, Sellschop J P F and Stemmet M C 1984b *Hyperfine Interactions* **17–19** 605
Percival P W 1981 *Hyperfine Interactions* **8** 315
Percival P W, Brodovitch J-C, Leung S-K and Newman K E 1984a *Hyperfine Interactions* **17–19** 543
Percival P W, Brodovitch J C and Newman K E 1984b *Hyperfine Interactions* **17–19** 721
Percival P W, Brodovitch J-C, Newman K E and Spencer D P 1982 *Chem. Phys. Lett.* **93** 366
Percival P W and Fischer H 1976 *Chem. Phys.* **16** 89
Percival P W, Fischer H, Camani M, Gygax F N, Rüegg W, Schenck A, Schilling H and Graf H 1976 *Chem. Phys. Lett.* **39** 333
Percival P W, Roduner E and Fischer H 1978 *Chem. Phys.* **32** 353
Perlson B D and Weil J A 1974 *J. Magn. Res.* **15** 594
Resta R 1977 *Phys. Rev.* **B 16** 2717
Roduner E 1981 *Chem. Phys. Lett.* **81** 191
Roduner E and Fischer H 1981 *Chem. Phys.* **54** 261
Sahoo N, Mishra K C and Das T P 1985 *Preprint* State University of New York, Albany
Sahoo N, Mishra S K, Mishra K C, Coker A, Das T P, Mitra C K, Snyder L C and Glodeann A 1983 *Phys. Rev. Lett.* **50** 913
Simanek E and Orbach R 1966 *Phys. Rev.* **145** 191
Slichter C P 1978 *Principles of Magnetic Resonance* 2nd edn (Berlin: Springer)
Spencer D P, Fleming D G and Brewer J H 1984 *Hyperfine Interactions* **17–19** 567
Stambaugh R D, Casperson D E, Crane T W, Hughes V W, Kaspar H F, Souder P, Thompson P A, Orth H, zu Putlitz G and Denison A B 1974 *Phys. Rev. Lett.* **33** 568
Swanson R A 1958 *Phys. Rev.* **112** 580
Tawara H 1978 *At. Data Nucl. Data Tables* **22** 491

Tawara H and Russek A 1973 *Rev. Mod. Phys.* **45** 178
Vinsome P K W and Richardson D 1971 *J. Phys. C: Solid State Phys.* **4** 2650
Walker D C 1981 *Hyperfine Interactions* **8** 329
—— 1983 *Muon and Muonium Chemistry* (Cambridge: Cambridge University Press)
Wang J S-Y and Kittel C 1973 *Phys. Rev. B* **7** 713
Wangsness R K and Bloch F 1953 *Phys. Rev.* **89** 728
Weidinger A, Albert E, Möslang A and Recknagel E 1983 *Yamada Conf. Muon Spin Rotation Associated Problems (Shimoda, Japan)* University of Tokyo, Meson Science Laboratory, abstract booklet
Weidinger A, Balzer G, Graf H, Recknagel E and Wichert Th 1981 *Phys. Rev. B* **24** 6185
Weil J A 1981 *Hyperfine Interactions* **8** 371

8

Muon Spin Rotation in Insulators

8.1 Introduction

The free 'diamagnetic' μ^+ precession in insulators has only been studied systematically for a very few selected systems, primarily the magnetic oxides. This lack of studies is caused, in part, by the fact that free μ^+ precession for many insulators appears with a reduced asymmetry only, pointing to the formation of muonium (see, for example, table 3 in Brewer *et al* 1975). Insulators also present more complex systems in other respects, as the fate and whereabouts of implanted μ^+ are more difficult to elucidate. Chemical effects may be particularly important in contrast to metals. Nevertheless, early studies in insulators have contributed significantly to the development of the μSR technique as an important spectroscopic tool. The first observation of a damped μSR signal was performed in boron carbide (B_4C) by Swanson (1958). Later the observation of several partially resolved frequencies in the μSR signal from a single crystal $CaSO_4 \cdot 2H_2O$ sample, which could be traced back to the proton–μ^+ dipole–dipole interaction, visibly displayed the close correspondence with the NMR technique (Schenck and Crowe 1971). In the latter study it was found that the observable fraction of 'diamagnetic' μ^+ (accounting for about 70% of all implanted μ^+) was associated with the water molecules through one of the protons being replaced by a μ^+. This was thought to be the result of a muonium epithermal hot atom reaction. Interestingly enough, the 'diamagnetic' fraction of μ^+ was only lowered slightly (to $\simeq 60\%$) in an anhydrous $CaSO_4$ sample. A similar behaviour was also observed in $CuSO_4 \cdot 5H_2O$ and in pure $CuSO_4$ (see table 3 in Brewer *et al* 1975). These examples perhaps best show that a 'diamagnetic' μ^+ signal may be the result of complicated processes, and they need to be clarified before the μ^+ can be used as a probe. It is also dangerous to expect a certain behaviour analogous to observations in a different system. For instance, it was found that no replacement of protons takes place in the system $CoCl_2 \cdot 2H_2O$, contrary to the expectations based on the gypsum results (see §8.4).

In this chapter we discuss some of the results obtained in magnetic insulators. It is worth noting that all substances, with the exception of MnF_2,

CoF_2 and $CoCl_2 \cdot 2H_2O$ are oxides (see table 8.1). Perhaps the ability to observe a diamagnetic signal in these compounds is related to the presence of oxygen and the subsequent formation of an oxygen–muon bond, analogous to a hydrogen bond, with muonium as a short-lived precursor state. The measurements do indeed lend evidence to such a picture. The μ^+ can obviously be used as a probe for studying the static and dynamic aspects of the magnetic properties of these substances once a 'diamagnetic' signal has been detected.

8.2 αFe_2O_3 and Cr_2O_3

These were the first magnetic oxides successfully investigated by μSR (Graf et al 1978, Rüegg et al 1979, 1981, Boekema et al 1981, Rüegg 1981). Since the results are also representative for the other magnetic oxides they will be discussed here in somewhat greater detail. Information on some of the properties of these substances are given in table 8.1. Figures 8.1 and 8.2 display the crystallographic and spin structures of αFe_2O_3 and Cr_2O_3 respectively. The magnetic moment per Fe or Cr ion is $5\mu_B$ in αFe_2O_3 and $2.76\mu_B$ in Cr_2O_3 (Corliss et al 1965, Jacobs et al 1971).

Figures 8.3(a) and (b) display the observed μSR frequencies as a function of temperature in zero external field. Quite a complex behaviour can be seen. Only one frequency component survives at higher temperatures while several show up at low temperatures. The Morin transition in αFe_2O_3 is reflected in a significant reduction of the μ^+ precession frequency.

The direction of the local field with respect to the crystalline axes was determined by applying an external field either along or perpendicular to the crystalline c axis (in the basal plane). The total field $\boldsymbol{B}_\mu$ at the μ^+ is then given by

$$\boldsymbol{B}_\mu = \boldsymbol{B}_{\text{loc}} + \boldsymbol{B}_{\text{ext}} \quad (8.1)$$

and

$$|\boldsymbol{B}_\mu| = (B_{\text{loc}}^2 + B_{\text{ext}}^2 + 2B_{\text{loc}}B_{\text{ext}}\cos\theta)^{1/2} \quad (8.2)$$

where θ is the angle between $\boldsymbol{B}_{\text{ext}}$ and $\boldsymbol{B}_{\text{loc}}$. There is no demagnetisation field as long as the antiferromagnetic structure is not destroyed by the applied field.

In the case of αFe_2O_3, the field dependence of B_μ was measured below and above the Morin temperature. Below the Morin temperature at 60 K, for $\boldsymbol{B}_{\text{ext}}$ parallel to the c axis all frequencies split into two such that

$$|\boldsymbol{B}_\mu| = B_{\text{ext}} \pm B_{\text{loc}}$$

showing $\boldsymbol{B}_{\text{loc}}$ is parallel or antiparallel to the c axis. At room temperature, well above the Morin transition, the field dependence of B_μ indicates that $\boldsymbol{B}_{\text{loc}}$ is now perpendicular to the c axis. In order to determine its polar angle in the basal plane the external field was also applied perpendicular to the c axis. From the

Table 8.1 List of magnetic insulators investigated by μSR (see also Landolt–Börnstein: New series II.2, III.4, 12).

Compounds	Crystal structure	Magnetic order	T_N, T_c (K)	Other properties	References
αFe_2O_3	Corundum	antiferro-	963	Morin transition at $T = 263$ K	Rüegg et al (1981)
Cr_2O_3	$R3c$ (D_{3d}^6)	antiferro-	308		Rüegg et al (1979)
Fe_3O_4		ferri-	858	Verwey (metal–insulator) transition at $T = 121$ K	Boekema (1984)
$SmFeO_3$	perovskite	antiferro-	674	spin reorientation Γ_2–Γ_4: 440–460 K	
$EuFeO_3$	orthorombic	and	662	spin configuration Γ_4	
$DyFeO_3$	$Pbmn$ (D_{2h}^{16})	weak ferro-	645	spin configuration Γ_4, $T \geq 36$ K	Holzschuh et al (1983)
$HoFeO_3$			639	spin configuration Γ_4, $T \geq 50$ K	
$YFeO_3$			644	spin configuration Γ_4	
$ErFeO_3$			620–636	spin reorientation Γ_2–Γ_4: 40–80 K	
MnO	simple cubic	antiferro-	122	Heisenberg antiferromagnet	Uemura et al (1984)
V_2O_3	corundum $R3c$ (D_{3d}^6)	antiferro-	160		
MnF_2	tetragonal rutile	antiferro-	67.3	Heisenberg antiferromagnet	Renzi et al (1984)
CoF_2	($P4_2/mnm$)	antiferro-	37.7	Ising antiferromagnetic	
$CoCl_2 \cdot 2H_2O$	monoclinic (C2/m)	antiferro-	17.5		Brewer et al (1981)

αFe_2O_3 and Cr_2O_3

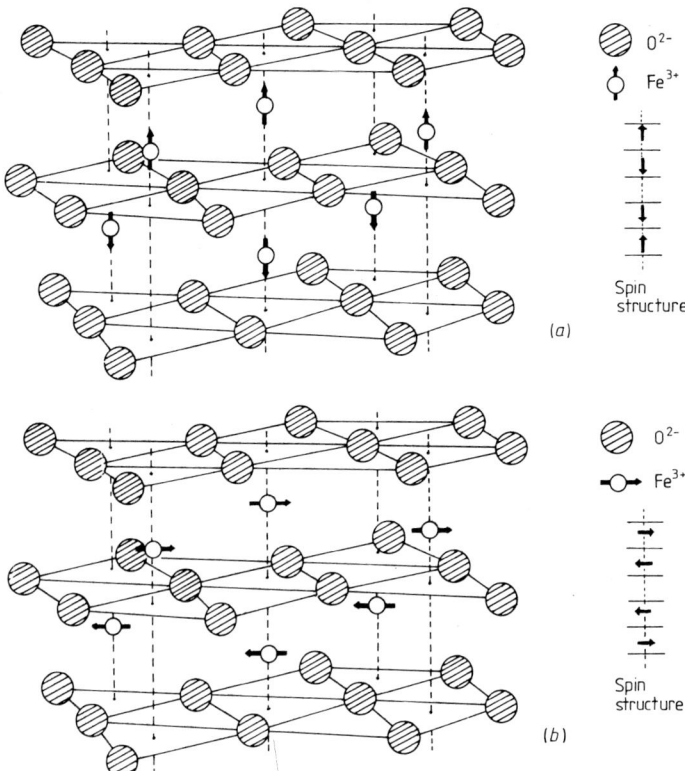

Figure 8.1 Crystallographic and spin structure of αFe_2O_3 for (a) configuration below the Morin temperature, T_M, and (b) configuration above T_M (Rüegg 1981).

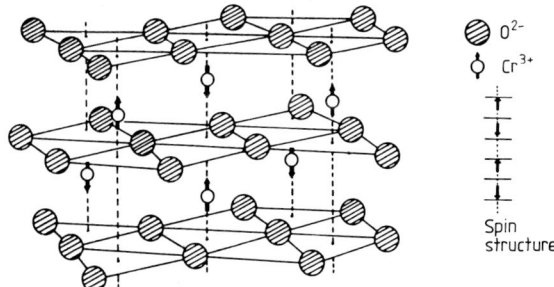

Figure 8.2 Crystallographic and spin structure of Cr_2O_3 (Rüegg 1981).

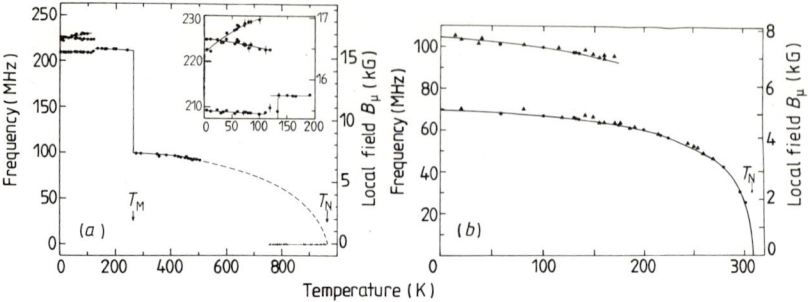

Figure 8.3 μ^+ Larmor frequencies and associated local fields in (a) $\alpha\text{Fe}_2\text{O}_3$ (Rüegg et al 1981) and (b) Cr_2O_3 (Rüegg et al 1979) as a function of temperature in zero applied field. ●, single crystal; ▲, powder.

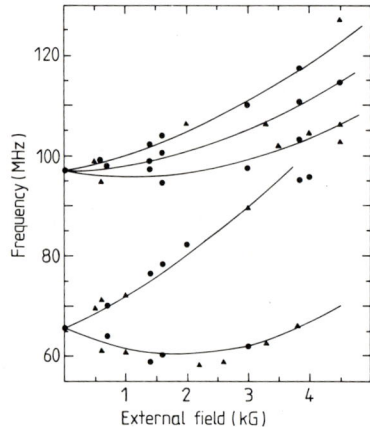

Figure 8.4 External field dependence of the μ^+ Larmor frequencies at 130 K in Cr_2O_3. The external field is applied along the crystallographic c axis (Rüegg 1981).

observed splitting it was deduced that $\boldsymbol{B}_{\text{loc}}$ is probably parallel to the crystalline a axis.

Similar measurements in Cr_2O_3 revealed that $\boldsymbol{B}_{\text{loc}}$ makes an angle of $\pm 29°$ with the basal plane in the case of the lower frequency line, and angles of $1 \pm 9°$ for the upper one, apparently independent of temperature. Figure 8.4 displays the field dependence of B_μ and its splitting for both the upper and lower line with $\boldsymbol{B}_{\text{ext}}$ parallel to the c axis.

Before we discuss these observations, it is important to mention some other features of the μSR data. In $\alpha\text{Fe}_2\text{O}_3$, the collapse of the three lines into one at about 110 K seems to reflect a transition into a common state, since the asymmetry of the new line is equal to the sum of the asymmetries of the three lines below 110 K. The nature of such a transition has, as yet, not found any substantial interpretation. The disappearance of the zero field signal around 500 K, and reappearance of a signal when in an applied field, with a frequency corresponding to this field, around 750 K has been consistently explained in terms of diffusion (Rüegg et al 1981). The disappearance of the upper line in Cr_2O_3 at about 170 K seems to be accompanied by an increased amplitude of the free μ^+ signal when observed with an applied field present. Compared to Fe_2O_3, the asymmetries of the μSR signal are generally reduced.

The local field at the μ^+, as in metals, may be composed of a contact hyperfine field, caused by non-zero local spin density, plus contributions from the local atomic magnetic dipole moments

$$B_{loc} = B_{hf} + B_{dip}. \tag{8.3}$$

The calculation of B_{dip} may proceed as described in §4.6. Since we are dealing with an antiferromagnetic system, the Lorentz field, B_L will equal zero, and $B_{dip} = B'_{dip}$. The calculation requires knowledge of the μ^+'s position, and of possible lattice relaxation around the μ^+.

The contact hyperfine field in an insulator may result from direct overlapping by the wavefunction tails of the magnetic electron and the μ^+, or may be caused by a super-transferred hyperfine field. The first possibility has not been seriously considered, since the magnetic electron is usually well localised at its atomic site. The second possibility, however, may be of some importance, particularly in the oxides investigated here where covalency effects have to be taken into account. The partially covalent character of the Fe–O and Cr–O bond leads to a spin polarisation of the p electrons at the oxygen site. Assuming a further bond between the oxygen and the μ^+, a 'muoxyl bridge', a non-zero spin density is also created at the μ^+ site. The situation is sketched in figure 8.5. The $3d^3$(Cr_2O_3) and the $3d^5$(αFe_2O_3)

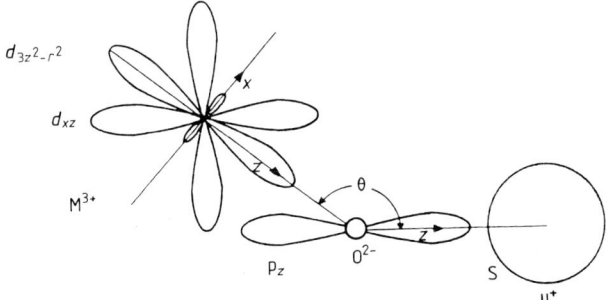

Figure 8.5 Muon–Oxygen–Metal linkage: schematic arrangement of μ^+ s, oxygen p, and metal d electron orbitals, illustrating the covalent bonding features and the origin of super-transferred hyperfine fields (Boekema 1984).

electron configurations are split by the crystal field into two orbitals with t_{2g} (lowest state) and e_g symmetry. The t_{2g} orbital is three-fold degenerate (d_{xy}, d_{xz}, d_{yz}) and the e_g orbital is two-fold degenerate ($d_{3z^2-r^2}$, $d_{x^2-y^2}$). In Cr_2O_3 with three 3d electrons, only the t_{2g} orbital is occupied, while in αFe_2O_3 both orbitals are fully occupied. The e_g orbitals are directed towards the oxygen ions and the t_{2g} orbitals are parallel and perpendicular to the c axis. The p_z orbital of the oxygen ion is linked up with the μ^+ and overlaps with an assumed 1s electronic state at its site. This model leads to a hyperfine field at the μ^+ which can be expressed in the following way (Sawatzky and Van der Woude 1974, see also Boekema 1984)

$$B_{hf} = C[(A_\sigma^2 - A_\pi^2)\cos^2\theta_i + A_\pi^2] \qquad (8.4)$$

where C is a constant, and A_σ and A_π are the spin polarisation magnitudes of the oxygen ion orbitals p_z and p_x with σ and π symmetry, respectively. θ_i is the angle between the ith metal ion–oxygen direction and the oxygen–μ^+ direction. The direction of B_{hf} is parallel to the 3d moment. In contrast to B_{dip}, B_{hf} will not change its value at the Morin transition. The total super-transferred hyperfine field at the μ^+ is then a sum over contributions from all the metal ions which form a direct linkage with the μ^+ of the form shown in figure 8.5.

In further analysis, various calculations of the potential surface of a μ^+ in αFe_2O_3 (see Rüegg 1981) were combined with calculations of the local field B_{loc}. Proper assumptions thus had to be made on the super-transferred hyperfine field B_{hf}. It was hoped that at one of the calculated potential minima, the estimated B_{loc} would reproduce the data. This procedure was only partially successful. No unique site assignment was possible, but it was found that all possible sites compatible with the measured local fields were at a distance to the oxygen ions that suggested a kind of 'hydrogen bridge' formation between at least two oxygen ions. The possible sites are shown in figure 8.6. μ^+ trapping at a defect site could also not be ruled out. For instance, the lower frequency line in Cr_2O_3 can be best reproduced if the μ^+ is trapped near an unmagnetic Al^{3+} ion which has substituted a Cr^{3+} ion (Rüegg 1981).

Recent µSR results in ferrimagnetic Fe_3O_4 are discussed in a similar fashion, including covalency effects, as described in Boekema (1984).

8.3 Orthoferrites

µSR signals can be observed in all the rare earth orthoferrites (Holzschuh et al 1980, 1981, 1983) and these are listed in table 8.1 together with information on their crystalline and magnetic properties. Figure 8.7 displays an idealised unit cell of the orthoferrites. The indicated spin structure corresponds to the so-called Γ_4 configuration. The indicated slight canting of the Fe spins shows that

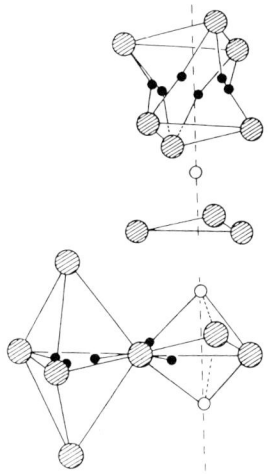

Figure 8.6 Candidate sites, ●, for the μ^+ in corundum structured oxides (Rüegg 1981).

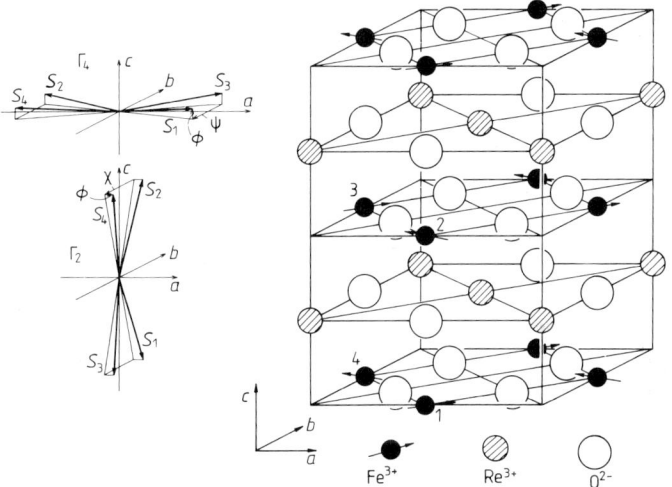

Figure 8.7 Idealised unit cell of the orthoferrites. The indicated Fe spin arrangement corresponds to the high temperature Γ_4 configuration. The Γ_2 and Γ_4 spin configurations are also indicated on the left side (Holzschuh et al 1983).

weak ferromagnetism is another property of these compounds. The rare earth spins remain unordered typically down to 4 K. Besides the Γ_4 configuration, there exists also a Γ_2 configuration where the Fe spins are nearly confined to the cb plane with a weak ferromagnetic component in the a direction (see table 8.1 for the transition temperatures).

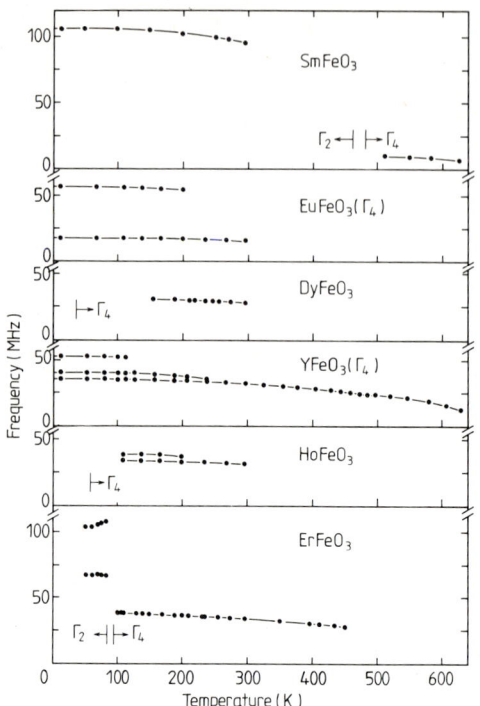

Figure 8.8 Temperature dependence of μSR frequencies in all rare earth orthoferrites investigated below the Néel temperature, T_N, in zero applied field (Holzschuh et al 1983).

Figure 8.8 compiles zero field frequencies as a function of temperature from all the single crystalline compounds. With the exception of SmFeO$_3$ and DyFeO$_3$, the substances display several lines at low temperature. In ErFeO$_3$ the two lines disappear at the onset of the Γ_2–Γ_4 transition, and a different single line is observed upon completion of this transition. Obviously the change in the spectra must be related to the Γ_2–Γ_4 transition. The disappearance of the high frequency components in the other compounds at certain temperatures is not related to any of their other known properties. Holzschuh et al (1983) speculate that the higher frequency lines could be associated with metastable μ^+ states as proposed by Browne and Stoneham (1982). These states decay at a certain temperature to a ground state configuration associated with the lower frequency line the latter being visible up to much higher temperatures. Such transitions are indeed indicated by the observation that the lowest frequency line's amplitude increases with temperature up to a maximum possible value. This suggests that all implanted μ^+ are finally found in this ground state configuration.

The analysis of the local fields was, therefore, restricted to the lowest frequency line associated with an internal field B_{loc}. The direction of the internal field was again determined by measuring the field dependence of the total field at the μ^+, $|B_\mu|$. An example is presented in figure 8.9 for YFeO$_3$. If the external field is applied along the c axis, which for the Γ_4 configuration is the easy axis of weak ferromagnetism, a linear rise of $|B_\mu|$ with the external field is observed above $B_{ext}=23$ G. Below this field B_μ is independent of B_{ext} (see insert in figure 8.9). The origin of this field independence is the same as previously discussed in the case of ferromagnetic metals. The external field is screened by the demagnetisation field as long as the external field is smaller than the saturation field. Above the saturation field B_{ext} penetrates the sample unhindered. The weak ferromagnetism of this compound leads to a very small saturation field. The linear rise of $|B_\mu|$ with B_{ext} indicates that the internal field has the same direction as the applied field, that is, it is oriented along the crystalline c axis. This at first sight, appears strange, since as with the case of αFe_2O_3 one would have expected the internal field to appear with both signs, because of differently oriented sublattice magnetisations at sites otherwise crystallographically equivalent. The explanation is found in the particular spin structure of the Γ_4 configuration for the ferromagnetically saturated case. Figure 8.9 also displays the field dependence of $|B_\mu|$ for B_{ext} parallel to the $\langle 110 \rangle$ direction, and hence perpendicular to the easy axis of weak ferromagnetism. We notice a small splitting and a nonlinear field dependence. The origin of this behaviour is found in the field induced canting of the Fe spins (Holzschuh et al 1983).

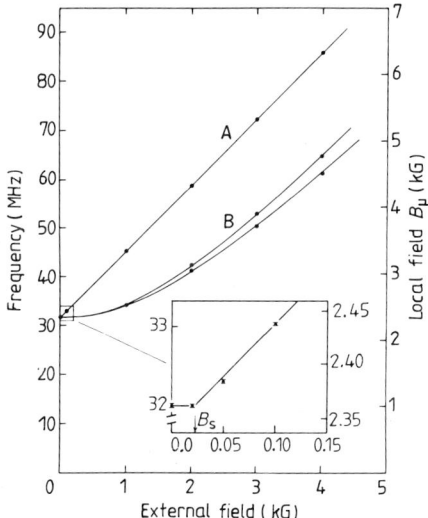

Figure 8.9 μ^+ Larmor frequency and corresponding local field in YFeO$_3$ at 296 K as a function of external field applied: A, parallel to the c axis; B, perpendicular to the c axis $\|\langle 110 \rangle$ (Holzschuh et al 1983).

310   *Muon spin rotation in insulators*

The temperature dependence of $|B_{loc}|$ largely follows the sublattice magnetisation as deduced from Mössbauer effect measurements.

A detailed analysis of B_{loc} in relation to the μ^+ lattice site allows some definite conclusions to be drawn. First, B_{loc} is only of dipolar field origin. No contact hyperfine field is indicated by the analysis. Secondly, only one site in the lattice is compatible with all aspects of the measured B_{loc}. This site is indicated in figure 8.10 by an asterisk. It is $\simeq 1$ Å from the centre of an oxygen ion. Furthermore, the position is almost located on the line connecting the two oxygen ions. The distance of 1 Å is exactly the bond length of an OH^- molecule, indicating that an analogous $(OMu)^-$ molecule has been formed. Its position between two oxygen ions suggest the establishment of a 'hydrogen bridge' (OMu)—O.

For details of the analysis of B_{loc} leading to the determination of the μ^+ site the reader is referred to Holzschuh *et al* (1983).

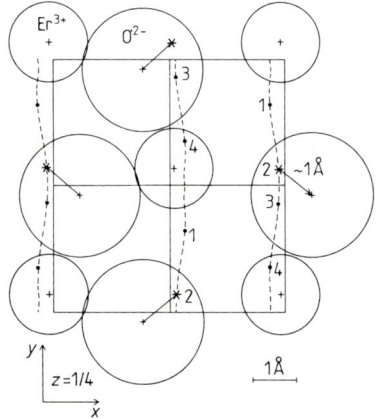

Figure 8.10 The most likely μ^+ sites in ErFeO$_3$ indicated by asterisks. The figure shows the $z = \frac{1}{4}$ crystalline plane. The Er and oxygen ions are drawn with their ionic radii (Holzschuh *et al* 1983).

8.4 MnO and V$_2$O$_3$

Zero field measurements in these crystals below the Néel temperature revealed a single μSR line which could be followed up to T_N (Uemura *et al* 1984). This suggests a unique μ^+ lattice site with high crystal symmetry and the absence of diffusion. The temperature dependence of the zero field μSR frequencies are shown in figure 8.11. The discontinuity of the frequency at T_N, and in particular the appearance of a hysteresis in V$_2$O$_3$ too, characterises the magnetic transition as a first order phase transition.

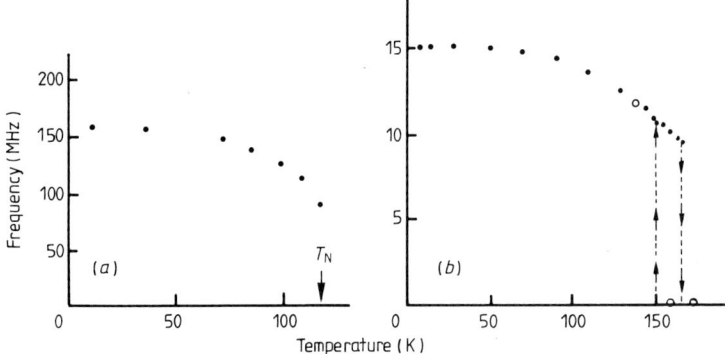

Figure 8.11 Temperature dependence in zero applied field of the μ^+ Larmor frequency in the antiferromagnets (a) MnO and (b) V_2O_3: ●, from lower temperature; ○, from higher temperature (Uemura et al 1984).

The internal field at the μ^+ in MnO, extrapolated to 0 K ($\simeq 11.4$ kG), can be reasonably well understood if the μ^+ is placed at the body centred interstitial site of the simple cubic lattice of MnO. The calculated dipolar field at this site amounts to 14.8 kG, while a contact hyperfine field of -4.8 kG is estimated from the Knight shift of -0.2% measured in the paramagnetic state of MnO (Uemura et al 1981).

8.5 MnF$_2$ and CoF$_2$

MnF$_2$ and CoF$_2$ are two of the classical antiferromagnetic insulators. The magnetic properties of MnF$_2$ are well described by a Heisenberg model with a small anisotropy, while those of CoF$_2$ are best represented by an Ising model and a much larger anisotropy. For other properties consult table 8.1.

μSR measurements were successfully performed, both in the paramagnetic and the antiferromagnetic phase (De Renzi et al 1982, 1984). In the paramagnetic phase just above the T_N whilst applying an external field B_{ext} of 3 kG, the μ^+ local field shift was measured as a function of crystal orientation with respect to B_{ext} (see figure 8.12). In fact, a line splitting was observed which depended measurably on the crystal orientation. Since a direct contact spin density was excluded in these insulators, the frequency shift and splitting must originate from local moments on the Co and Mn ions which have been polarised by the external field. The effective local moment can be calculated from the known susceptibility: $\langle \mu \rangle = \chi^{at} B_{ext}$. The rest is a matter of calculating the resulting dipole fields at various possible positrons of the μ^+ in the lattice, and comparing the results with the measured values. As it turns out, only the octahedral position midway between two nearest neighbour Co or Mn sites is compatible with the data.

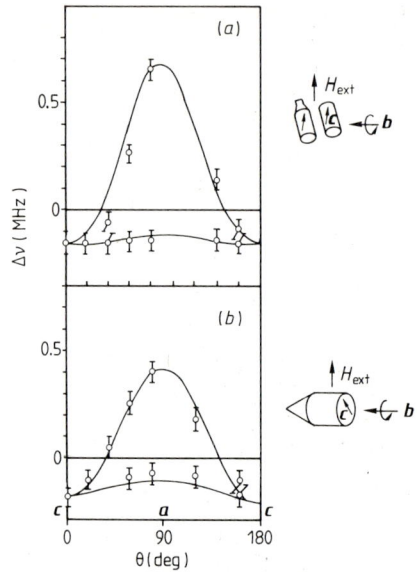

Figure 8.12 Orientation dependence and splitting of the paramagnetic frequency shift of the μ^+ Larmor frequency in the paramagnetic state of (a) CoF_2 at 40 K and (b) MnF_2 at 71 K. The crystal is rotated around the crystallographic b axis. θ is the angle between the c axis and the applied field $H_{ext} = 3$ kG (De Renzi et al 1984).

This position is also compatible with a unique spontaneous internal field at the μ^+ observed in CoF_2 below the Néel temperature ($B_{loc} = 2288$ G at 2 K). In contrast, at 15 K in MnF_2 two lines were observed representing less than 5% of the total asymmetry. These lines were not observed at higher temperatures.

In CoF_2 the temperature dependence of the spontaneous field at the μ^+ could be measured near to the T_N, and it is displayed in figure 8.13. The relaxation of the μSR signal above and below T_N is very interesting. The relaxation rate, λ, is shown in figure 8.13 too. On approaching T_N either from above or below, a divergent behaviour of λ is observed obeying a power law of the form $\lambda \simeq |T - T_N|^{-\beta}$. This critical damping is due to the dynamics of the Co spins and reflects the critical slowing down of the antiferromagnetic spin fluctuations near T_N. A similar behaviour has already been discussed in §5.1 concerning measurements in the weak itinerant ferromagnetic MnSi.

8.6 $CoCl_2 \cdot 2H_2O$

This compound has been intensively studied by proton NMR in the antiferromagnetic phase (Narath 1969). A unique internal magnetic field at the proton

$CoCl \cdot 2H_2O$ 313

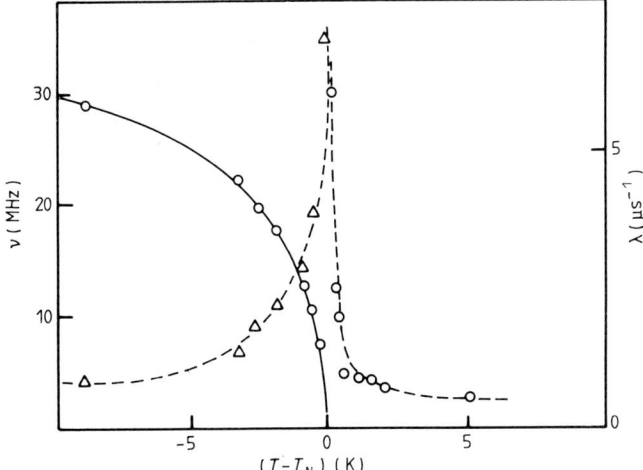

Figure 8.13 Temperature dependence of the spontaneous hyperfine field (full curve) at the μ^+ and the transverse μ^+ spin relaxation rate (broken curve) in CoF_2 with the applied field parallel to the c axis, and $T_N = 37.7$ K (De Renzi et al 1984).

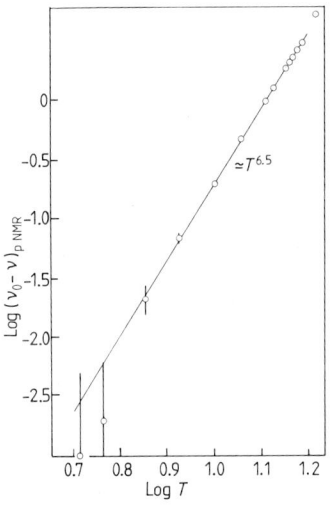

Figure 8.14 Double logarithmic plot of the temperature dependence of the proton NMR frequency $[\nu_p(T=0) - \nu_p(T)]$ in antiferromagnetic $CoCl_2 \cdot 2H_2O$ (Narath, private communic.; see also Narath 1969).

site in the water molecules has been found, indicating that all proton sites are magnetically equivalent. The internal field at the proton site amounts to 4.2 kG in the zero degree Kelvin limit. The temperature dependence of the internal field over a large temperature range follows a power law of the form (see figure 8.14)

$$[B_p(0\text{ K}) - B_p(T)] \propto T^{6.5}.$$

The μSR measurements were started with the expectation of finding the μ^+ at one of the proton sites, as suggested by the μSR results in gypsum ($CaSO_4 \cdot 2H_2O$) (Schenck and Crowe 1971). Instead, two frequencies were found, corresponding to internal fields of 2.73 kG and 2.83 kG in the zero degree Kelvin limit (Brewer et al 1981). The most remarkable features of the data are their temperature dependence, which, in contrast to the proton NMR data, displays a power law with a quite different exponent

$$[B_\mu(0\text{ K}) - B_\mu(T)] \propto T^4.$$

The data are plotted in figure 8.15 as a function of temperature, and figure 8.16 in a double logarithmic fashion so as to display the power law dependence. This different temperature dependence is quite surprising since one would expect the μ^+, as well as the proton, to sample the same sublattice

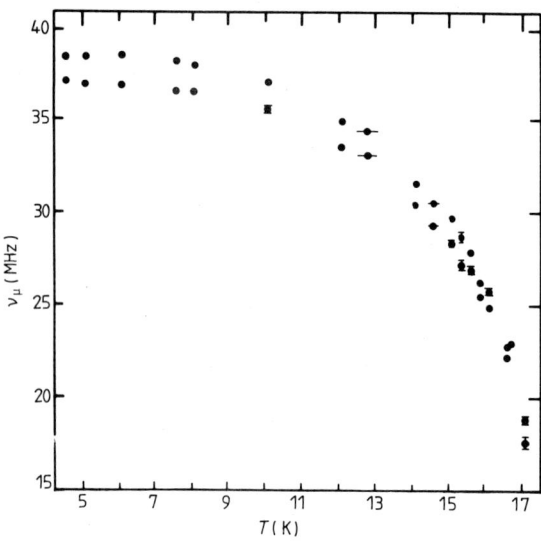

Figure 8.15 Temperature dependence of the μ^+ Larmor frequency doublet in antiferromagnetic $CoCl_2 \cdot 2H_2O$ in zero applied field (Brewer et al 1981).

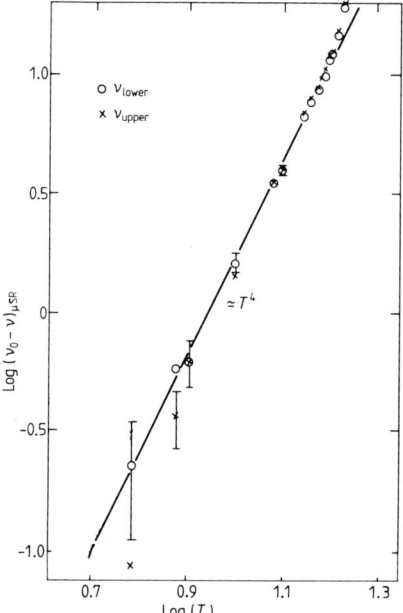

Figure 8.16 Double logarithmic plot of the temperature dependence of $[v_\mu(T=0) - v_\mu(T)]$ in antiferromagnetic $CoCl_2 \cdot 2H_2O$ in zero applied field (Brewer *et al* 1981).

magnetisation. This behaviour is totally different from the observations in the other antiferromagnetic insulator, where the temperature dependence of B_{loc} generally follows the sublattice magnetisation. No explanation of the results in $CoCl_2 \cdot 2H_2O$ has been put forward so far.

References

Boekema C 1984 *Hyperfine Interactions* **17–19** 305
Boekema C, Rüegg K and Hofmann W P 1981 *Hyperfine Interactions* **8** 609
Brewer J H, Crowe K M, Gygax F N and Schenck A 1975 *Muon Physics* vol. III ed. V W Hughes and C S Wu (New York: Academic)
Brewer J H, Koster E, Schenck A, Schilling H and Williams D Ll 1981 *Hyperfine Interactions* **8** 619
Browne A M and Stoneham A M 1982 *J. Phys. C: Solid State Phys.* **15** 2709
Corliss L M, Hastings J M, Nathans R and Shirane G 1965 *J. Appl. Phys.* **36** 1099
De Renzi R, Guidi G, Bucci C, Podini P, Tedeschi R and Cox S F J 1984 *Hyperfine Interactions* **17–19** 479
De Renzi R, Tedeschi R, Guidi G and Bucci C 1982 *Solid State Commun.* **43** 683

Graf H, Hofmann W, Kündig W, Meier P F, Patterson B D, Reichart W and Rodriguez A 1978 *Hyperfine Interactions* **4** 452

Holzschuh E, Boekema C, Denison A B, Kündig W, Meier P F and Rüegg K 1980 *Hyperfine Interactions* **8** 77

Holzschuh E, Boekema C, Kündig W, Rüegg K and Patterson B D 1981 *Hyperfine Interactions* **8** 615

Holzschuh E, Denison A B, Kündig W, Meier P F and Patterson B D 1983 *Phys. Rev.* B **27** 5294

Jacobs I S, Beyerlein R A, Foner S and Remeika J P 1971 *Int. J. Magn.* **1** 193

Landolt-Börnstein (Springer) New series II.2; III.4, 12

Narath A 1969 *Phys. Rev.* **136A** 766

Rüegg K 1981 *Doctoral thesis* University of Zürich

Rüegg, K, Boekema C, Hofmann W, Kündig W and Meier P F 1979 *Hyperfine Interactions* **6** 99

Rüegg K, Boekema C, Kündig W, Meier P F and Patterson B D 1981 *Hyperfine Interactions* **8** 547

Sawatzky G A and Van der Woude F 1974 *J. Phys. C: Solid State Phys.* **6** 47

Schenck A and Crowe K M 1971 *Phys. Rev. Lett* **26** 57

Swanson R A 1958 *Phys. Rev.* **112** 580

Uemura Y J, Imazato J, Nishida N, Hayano R S, Takigawa M and Yamazaki T 1981 *Hyperfine Interactions* **8** 725

Uemura Y J, Yamazaki T, Kitaoka Y, Takigawa M and Yasuoka H 1984 *Hyperfine Interactions* **17–19** 339

Index

Accelerator laboratories with μSR facilities, 11
Accidentals, 23
AC susceptibility cusp, 184
Alkaline earth metals, 94
Aluminium (Al), 63, 64, 66, 69, 105, 107, 112–6, 142, 146
Aluminum oxide (Al_2O_3), 251, 252
  powder, 291, 292
Ambiguous events, 22
  second μ, e, 22, 23
Ammonia (NH_3), 247, 248
Analogue detection technique, 26
Anisotropy energy, 170
Annealing temperature, 116
Anomalous muonium, 266, 267, 279, 286
  diffusion in diamond, 287
Antimony (Sb), 106, 107, 142, 161
Argon (Ar), 245, 249
Arsenic (As), 161
Arrhenius law, 75, 113, 118, 122, 285
Arrhenius plot, 260, 261
Atomic impurities, 87
Auto-correlation function, 89
  see also self-diffusion function

Band propagation, 72–3, 77, 82, 112
  diffusion constant of, 98, 100
  in periodic magnetic field array, 97, 99
  see also coherent propagation

Band structure calculations, 138, 139, 175
  of hyperfine fields, 174
Barium (Ba), 142, 143, 146
Beryllium (Be), 106, 107, 142, 143, 144, 153
Beryllium oxide (BeO), 251
Bethe–Bloch theory, 246
Bismuth (Bi), 106, 107, 142, 161
Bloch electrons, 153
Bloch equations, 41
Bloch–Siegert shift, 32
Bonding states, 131
Boron carbide (B_4C), 300
Bound states, 136
Breit–Rabi diagram, 218, 219, 222
Breit–Rabi equations, 218
Brillouin function, 161, 162
Brillouin zone, 65
  K symmetry point in hexagonal, 149, 150, 152, 153

Cadmium (Cd), 142–4, 148–54
  CdHg alloy, 148–53
  $\underline{Cd}$Mg alloy, 148–53
Calcium (Ca), 142, 143
Calcium oxide (CaO), 251
$CaSO_4 \cdot 2H_2O$, 300, 314
Capture probability by traps, 87, 88
Carbon dioxide (CO_2), 251
Cavity model, 287, 288
Čerenkov counter, 26

Cerium (Ce) (α, γ phase), 155–60
  CeSn$_3$, 155, 156, 159, 160
  Ce$_{1-x}$Th$_x$, 155, 156
Caesium (Cs), 142, 143, 146
Channelling of μ^+, e^+, 286
Charge density enhancement factor
  $\eta_F(r_\mu)$, 136, 141
Charge density functional (Hohenberg–Kohn–Sham, HKS)
  formalism, 136, 141
Charge exchange processes, 246–8
  involving protons, 247
Chemical exchange reaction, 90
Chemical radical, 240, 287
  muonic, 251
Chemical reactions of Mu, 230, 234, 251
Chemical shift, 135
  Lamb expression of, 135
Chodorov potential, 128
Classical jumping over the barrier, 75
Clogston–Jaccarino plot, 157
Cluster calculations, 138, 139, 175
  of hyperfine fields, 174
Cobalt (Co), 101, 166, 167, 168–76
  structural phase transition in, 167
CoCl$_2$·2H$_2$O, 300, 312–5
  sublattice magnetisation in, 315
CoF$_2$, 301, 311–3
  critical slowing down of spin fluctuations, 312, 313
  Ising model of, 311
Coherence length $\xi(T)$, 207, 209
Coherent diffusion (propagation), 97, 99, 114
  dynamical destruction of, 73
  see also band propagation
  spin relaxation due to, 97–100
  transition to incoherent, 76, 77
Coincidence configuration, 74, 75
Colour centres, 5
Condon approximation, 77
Conduction electrons, 128–30, 136
Copper (Cu), 23, 49, 63–5, 69, 77, 83, 105–11, 142, 146, 159
Cu$_3$Au structure, 159
CuSO$_4$·5H$_2$O, 300
Core polarisation, 131, 136, 143

Correlation function, 79, 80, 82, 83
Correlation time, 84, 87, 88, 91
Counter telescope, 26
Critical exponents, 182, 312
Critical fields in superconductors, 203
Critical phenomena, 4, 182
Critical points (band structure), 149, 150, 152
Critical slowing down, 199
Chromium (Cr), 106, 107
Cr$_2$O$_3$, 301–6

Dead time, 24
Debye–Waller approximation, 177
Debye–Waller factor, 72
Demagnetisation field, 57, 133, 163
Density of states, 130, 131, 144, 152, 153
  atomic, 145, 146
Depolarisation by paramagnetic impurities, 94–7
Desorption of Mu from surfaces, 293
Detection efficiency, 16
Detectors, 20
  light pipes, 20
  plastic scintillators, 20
Diamagnetic screening, 135, 136
Diamond, 264–90
  structure of, 269
  transition Mu→Mu* in, 283–5, 287
Dielectric function, 287
Dielectric function approach, 288
Diffusion constant, 70, 81, 88, 122
  for band propagation, 98, 100
Diffusion rate, 69, 72, 73, 82
  incoherent, 74
Dipolar relaxation, 5, 43
  in transverse fields, 43–8
  in zero and longitudinal fields, 48–56
Dipole–dipole interaction, 53
  Hamiltonian of, 43, 44, 51, 53, 217
Dipole fields, 162, 172
Dipole field tensor, 163–5, 171
  in BCC metals, 97
Dipole force tensor, 67
Domain (saturation) magnetisation, 132, 166, 170, 176, 177

Index   319

Domain (saturation) magnetisation (*contd.*)
  pressure dependence in Ni, Fe, 175, 178
Double Born–Mayer potential, 67
Double electron muon magnetic resonance (DEMUR), 43, 261–4
Dynamical destruction of coherent propagation, 73
Dysprosium (Dy), 167, 168, 172–5
  helical antiferromagnet, 172

Easy axis of magnetisation, 98, 166, 169, 171
Edwards–Anderson model, 189, 196
  order parameter, 189
Effective electron mass, 135
Effective mass theory (EMT), 287
Einstein's relation, 70
Elasticity theory, 128
Electric field gradient (EFG), 47, 52, 53, 69, 111
  non radial, 69
Electron capture by μ^+, 246–8
Electron charge density
  distribution, 128, 129, 133, 136, 137
Electron gas
  free, 136, 138, 139, 141, 146, 175
  homogeneous, 131, 139, 141, 147
  inhomogeneous, 141, 175
  interacting, 147
Electronic detection scheme, 21, 22
  time resolution of, 24
Electronic specific heat, 144
Electronic structure
  hydrogen in metals, 136
  μ^+ in metals, 128, 129, 135
Electron irradiation, 106, 115, 116
Electron Landé (g) factor, 275, 276
  anisotropic, 135, 154
  tensor, 224
Electron nuclear double resonance (ENDOR), 43, 261
Electron photomultiplier, 20
Electrons (s, d, f), 130–2
Electron spin relaxation in Mu, 230, 237, 251
Electron spin resonance (ESR), 5, 224, 286

Electrostatic potential
  2nd derivative of, 134
Energy momentum dispersion, 149
$ErRh_4B_4$, 214
Ewald method, 163, 172
Exchange field, 162
Exchange interaction, 131, 132
Exchange integral, 133
Exchange narrowing, 96, 278
Exchange polarisation of conduction electrons, 159
Exchange scattering of conduction electrons, 140, 281
Exchange splitting in ferromagnetic Ni, 176
Extended tunnelling state, 78

αFe_2O_3, 301–6, 309
  crystal field splitting in, 306
  crystallographic structure of, 303
  Morin transition in, 301
  spin structure of, 303
  supertransferred hf fields in, 305, 306
Fe_3O_4, 306
Fermi energy, 144, 145, 149, 150
Fermi momentum, 130
Fermi surface, 130, 131, 135, 151
Field correlation function, 80, 81, 87, 101
Field gradient tensor, 134
Flux quantum ϕ, 203
Flux vortices, 203, 207
Fourier analysis, 27
Fourier transform, 28
Frequency space, 28

Gadolinium (Gd), 101, 167, 168, 169, 172, 173, 174
$GdFe_2$, 167, 169, 171, 173, 174, 175, 176
Gallium (Ga), 106, 107, 142
GaAs, 266, 286
GaP, 266, 286
Gaussian field distribution, 46, 186
  second moment of, 46, 47
Gaussian–Markovian process, 79, 84, 86, 87, 186, 189
Gaussian relaxation function, 79, 82
Germanium (Ge), 245–90

Germanium oxide (GeO), 251
Ginzburg–Landau equations, 207
Gold (Au), 94, 105, 107, 117, 118, 142, 159
Golden rule, 8
Gorkov equations, 209, 211
Grafoil, 39
Green's function, 61, 290

Hanle effect, 32
Harmonic approximation, 61
Heat of solution, 138
Heisenberg representation, 226
Helium (He), 245, 247–9
Histogram, rate versus time, 22, 23, 27, 56
$Ho_{0.7}Lu_{0.3}Rh_4B_4$, 214
Homogeneous line broadening, 193
Hot atom reactions, 249, 300
Hybridisation, 131, 175
Hydrogen bridge, 306, 310
Hydrogen in metals, 155
   diffusion in Cu, 109
   diffusion in Ag, Au, 118
   electronic structure of, 60, 136–41, 173
Hydrogen molecule (H_2), 247, 248
Hydroxyl (OH^-) ion, 293
Hyperfine (hf) field (interaction), 1, 4, 128, 129
   band structure calculations of, 138, 140
   cluster calculations of, 138
   dipole field contributions, 133
   Fermi contact, 94, 118, 129, 131, 133, 162, 166, 173, 174, 215
   induced, 144
   jellium calculations of, 136–8
   normalised (per unpaired electron per atom), 132
Hyperfine interaction Hamiltonian of Mu
   anisotropic, 216
   axially symmetric, 217
   isotropic, 215
Hyperfine tensor of Mu*, 216

Ice, 245, 251

Incoherent diffusion, 73–8, 99
   see also tunnelling
Indium (In), 106, 107
Inhomogeneous line broadening, 193
Insulators, μ^+ precession in, 300–15
Interaction potential (μ^+ metal), 60, 128
Intermediate phase in superconductors, 204, 206
Intermediate valence, 159
Interstitial sites in FCC, BCC and HCP crystals, 62, 63
Intrasite scattering, 72
Iron (Fe), 101, 102, 105, 107, 163, 166, 167, 168, 170, 171, 174, 175, 176
   compressibility of, 175, 178
   FeCr, 167
Isotope effect
   in diffusion, 70, 110
   in Knight shift, 141, 156
Ivanter–Smilga equations, 236, 237

Jellium model, 136, 138
   calculations, 138, 140, 146, 147
   pseudo, 141

Kanzaki formalism (force), 61, 65, 110
Knight shift (K), 4
   isotropic, 129–33
   axial, 139, 153
   definition of, 129
   effect of μ^+ trapping on, 142
   isotope effects on, 141, 156
   nuclear K, 136
   in Cd, 149, 153
   in Li, 144
   in Ni, 157
   proton K, 155
   anisotropic, 133–5
Kohn–Korringa–Rostoker (KKR) band structure calculations, 140
Korringa peak, 190
Korringa relaxation, 96, 278
   in PdH, PdD, 141
K symmetry point in hexagonal Brillouin zone, 149, 150, 152, 153
Krypton (Kr), 247–9

Kubo–Toyabe function (signal), 50, 51, 54, 56, 83, 87, 93, 111, 185, 187, 194, 206

Landau diamagnetism, 135
$LaAg_xI_{1-x}$, 155, 156
$LaNi_5H_x$, 106
Laplace transformation, 85, 90, 187, 235
Large polaron, 61
Larmor precession (frequency), 17, 18, 27, 28, 56
  amplitude, 27
  nuclear, 79
Lattice imperfections, 78
Lattice relaxation, 61–3
Lattice relaxation energy, 75, 110
Lattice strain, 87
Lattice vibrations, 74
Lead (Pb), 142, 144, 205
PbIn, 209
Lepton, 7
  number, 7
Lifetime of band state, 114
Lifshitz phase transition, 149
Limitations of μSR technique, 24, 25, 33
Linear combination of atomic orbitals (LCAO), 290
Linear elasticity theory, 65
Linear response theory, 289
Linear Zeeman regime, 253, 256
Lithium (Li), 142, 146
Local moment formation, 161
Localised impurity state, 288
Localised state, 61
Logic signature
  stopped μ, 20
  decay electrons, 20
Longitudinal relaxation in magnetically ordered systems, 100–4
Lorentz field, 57, 133, 166
Lorentz sphere, 56–8, 133, 163

Macroscopic detection technique, 26
Magic field, 275, 276
Magnesium (Mg), 142, 144
MgO, 251
Magnetic energy, 170
Magnetic flux pinning in V, 210, 211

Magnetic insulators, compilation of, 302
Magnetic moment density (magnetisation), 133
Magnetic ordering, 129
Magnetic superconductors, 214
Magnetic susceptibility, 58
  Curie type, 130, 158, 159
  d electrons, 157
  diamagnetic, 157
  local moments, 130, 132
  orbital, 152
  Pauli spin, 130–1, 135, 147
  transition metals, 131
  Van Vleck para-, 157
Magnetisation, bulk, 98
Markovian process, 84, 89
Mean free path, 72, 97
Meissner effect, 203, 204, 210
Mercury (Hg), 142, 143
Meson factories, 3
  list of, 11
Metal hydrides, 106
  see also, PdH, VH_x, NbH_x, $LaNi_5H_6$, ZrH_x
Metastable μ^+ states, 72
Methane (CH_4), 247, 248
Missing fraction, 245, 248
  diamond, 285, 286
Mixed valence system, 157
MnF_2, 300, 311, 312
  Heisenberg model of, 311
MnO, 310, 311
  μ^+ Knight shift in, 311
MnSi, 182, 183, 312
Moments of field distribution, 46
Morin transition
  αFe_2O_3, 301
Mössbauer effect, 2, 3
  orthoferrites, 310
Motional narrowing, 78–87, 91, 97, 98, 113, 119, 120, 185, 186
  extreme limit of, 79, 83
  relaxation function, 91, 113
Mott transition, 158
Multi-domain structure, 163
Multistop option, 26
Muons (μ^+, μ^-)
  properties of, 8

Muons (μ^+, μ^-) (*contd.*)
   range in matter, 14
   range width, 14
Muon beam
   DC beam, 17
   decay beam line, 13
   kinematic depolarisation of, 14
   microscopic time structure, 28
   momentum resolution of, 14
   pulsed structure, 17, 25–7
   range in matter, 14
   spin polarisation of, 11
Muon decay, 7
   angular distribution of e^+, e^-, 9, 10
   energy distribution of e^+, e^-, 9
Muon diffusion studies
   survey of, 107
Muon magnetic resonance (MMR), 27, 35–43
   first application of, 41
   water, 41
Muon nearest neighbour force constant, g, 65, 110, 128
Muonic radicals, 251
Muonium ($\mu^+ e^-$, Mn), 1, 5, 19, 24, 161
   cavity model, 287, 288
   chemical reaction of, 1, 230, 234, 251
   DEMUR, 261–4
   desorption from surface, 293
   diffusion in SiO_2, 260
   eigenfunctions, 216, 219–23
   electron reduced mass, 215
   electron spin relaxation, 230, 237, 251
   energy eigen values, 216–21, 229
   formation, 245–51
   formation probabilities, 281–3
   hyperfine interaction of, 215–24
   magic field, 275, 276
   on surfaces (SiO_2, Al_2O_3, MgO), 291–4
   quenching of depolarisation, 228, 251
   relaxation in Si, Ge, 276–81
   residual polarisation
      in longitudinal fields, 227, 228, 237
      in transverse fields, 236
   singlet state of, 216
   spur model of formation, 249–51

Muonium ($\mu^+ e^-$, Mn) (*contd.*)
   superhyperfine (shf) interaction, 240–5, 252, 292
   transitions among Mu states, 237–40, 282–6, 287
   triplet state of, 216
   two-dimensional diffusion on surface, 293
   two-frequency precession of, 254, 266
   Zeeman splitting of, 215, 217, 218, 229
   Zero-point motion of, 289, 290

Nearest neighbour displacement, 65, 67, 68, 128
Negative muons, 2
   atomic capture of, 4
   atomic cascade of, 4
      spin–orbit coupling during cascade, 4
   depolarisation of, 3
   nuclear capture of, 4
   Primakoff (Z^4) law, 4
Neon (Ne), 245, 247–9
Neopentane, 251
Neutrino, 7
   helicity of, 7, 11
Neutron irradiation, 106, 112, 115
Neutron scattering, 2, 3, 174
   inelastic, 67
Neutron spin echo technique, 188
Nickel (Ni), 101, 106, 107, 140, 155–9, 166–8, 170, 171, 174–6, 182
   compressibility of, 175, 178
   exchange splitting of, 176
   permeability of, 170
<u>Ni</u>Co, <u>Ni</u>Cr, <u>Ni</u>Cu, 167
Niobium (Nb), 49, 63, 64, 66–9, 105, 107, 118–20, 155, 156, 205, 209, 210
NbH_x, 106
Nitrogen molecule (N_2), 247, 248
Nuclear magnetic resonance (NMR), 2, 25, 35, 43, 50, 100
   free induction decay in, 20
   line shape of, 91
   of protons in $CoCl_2 \cdot 2H_2O$, 312, 313
   spin echo technique in, 20

Nuclear quadrupole interaction, 47, 48, 63, 80, 83
  Hamiltonian of, 48

One-phonon process, 74, 78
Orthoferrites, 306–10
  compilation of, 302
  spin structure of, 307
  unit cell of, 307
  weak ferromagnetism of, 307, 309
Ortho-positronium, 291

Palladium (Pd), 94, 155, 156, 157, 158, 159
Palladium hydride, (PdH, PdD), 141, 155, 156
Paramagnetic hydrogen centres in group IV semiconductors, 286
Parity violation, 1, 7
Particle occupation number, 80
Paschen–Back effect, 228, 232, 243, 244, 251, 264, 268, 275
Pauli spin matrices, 225
Pauli spin paramagnetism, 129, 130
Pauli spin susceptibility, 130, 131, 133, 135, 147
Penetration depth $\lambda(T)$, 207, 209, 212
Percolation, 200
Perturbed angular correlations (PAC), 2, 3, 20, 29, 35
Petzinger zero field function, 93, 94
Phonon overlap factor, 72
Pions (π^+, π^-), 10
  decay of, 10, 11
  production of, 10
Platinum (Pt), 94, 155, 156, 157, 158, 159
Polarised neutron capture, 35
Positron annihilation, 2, 3, 291
Positron trapping, 87
Potassium (K), 142, 146
Potassuum chloride (KCl), 251
Pseudo-jellium model, 141
Pseudo-potential, 128, 288, 289
Presolvated electron, 250
Primakoff (Z^4) law, 4
Proton stopping power, 246

Quantum diffusion, 4
  theory of, 72–8, 98
  see also band (coherent) propagation
    incoherent diffusion
    tunnelling
Quarks, 7
Quartz (SiO_2), 245, 252, 253–64
  diffusion of Mu in, 260
  ESR in, 258, 259
  hydrogen in, 258, 259
Quasi-particle, 61
Quenching of depolarisation, 228, 251

Raman relaxation, 279
Radiolysis products, 249
Random hopping, 98
Random walk, 81, 89
Range straggling, 14
Redfield theory, 101
Relaxation function, 19
  caused by coherent motion, 98
  longitudinal field, 19
  presence of trapping, 87–92
  transverse field, 19, 82
  zero field, 83–7, 93–4
Residence time (mean time of stay), 69, 82, 83, 87, 88, 91, 113
Residual polarisation, 235
  longitudinal field, 227, 228, 237
  transverse field, 236
Rotating coordinate system, 37
Rubidium (Rb), 142, 143, 146
Ruderman–Kittel–Kasuya–Yosida (RKKY) exchange mechanism, 94, 132, 140, 159, 175, 185, 193, 194

Saddle point field, 207
Scaling laws in spin glasses, 194, 198
Scattering rate between bands, 98
Schrödinger equation, 287, 289
Schrödinger representation, 226
Screening calculations in non-linear jellium, 128, 136, 138
Screening charge, 48, 128, 135
Second moment, 46, 79, 89, 91, 99, 116
  effective, 83
  field dependence of, 49, 69

Second moment (*contd.*)
  with quadrupole interaction, 47–9, 52–5
  of tunnelling configuration, 67
  Van Vleck formula of, 46, 52, 62
  Van Vleck limit, 48, 63, 69
  in zero field, 51–5
Self-diffusion function, 80, 81, 89
Self-trapping, 60, 62
  site of, 128
  time delay to, 71, 72
Semi-metals, 161
Shallow acceptors, donors, 287
Shubnikov (mixed) phase in type II superconductor, 203, 207–10
Silicon (Si), 245, 264–90
  bulk modules of, 274
  Mu diffusion in 273, 282
  transition Mu→μ^+ in, 282
Silver (Ag), 94, 107, 117–18, 142, 159
Simple (non-transition) metals, 130
Single particle (Stoner) excitation, 177
Site representation, 80
Small polaron, 60–2, 71, 110
  effective bandwidth of, 72
  effective mass of, 72
  group velocity of, 72
$SmRh_4B_4$, 219
Sodium (Na), 142, 146
Solvated electrons, 250
Spectral density of field distribution in mixed phase of type II superconductor, 207, 208
Spherical harmonics addition theorem, 134
Spherical solid model, 138, 140, 144, 147, 148, 289
Spin auto-correlation function, 188
Spin density, 129, 131, 136, 137, 173, 174, 175
Spin density enhancement factor, $\rho_s(r_\mu)$, 130, 136, 138, 144, 146, 147
Spin density functional (SDF) formalism, 136, 138, 146
Spin density matrix (operator), 17, 45, 225, 230, 238, 240, 241, 242
Spin flip transition, 79, 83, 100

Spin fluctuations
  self-consistent renormalisation (SCR) theory of, 182
  in spin glasses, 185–9
Spin glasses, 4, 5
  AC susceptibility cusp in, 184, 190
  critical slowing down in, 194
  definition of, 184
  dipole fields in, 185
  Edwards–Anderson model of, 189, 196
  Edwards–Anderson order parameter, 189
  field induced static line broadening, 193
  Gaussian field distribution in, 186
  inhomogeneous spin dynamics, 200
  Korringa peak in, 190
  Lorentzian field distribution in, 185
  percolation in, 200
  range of correlation times accessible, 190
  RKKY mechanism in, 185, 193, 194
  scaling laws in, 194, 198
  single correlation time model, 188, 198, 199
  spin auto-correlation function, 188
  spin fluctuations, 185–9
  $\underline{Ag}$Mn, 191–200
  $\underline{Au}$Fe, 191–200
  $\underline{Cu}$Mn, 191–200

$(CoO)_{40} \cdot (Al_2O_3) \cdot (SiO_2)_{50}$, 191
$(FeTiO_3)_{88} \cdot (Fe_2O_3)_{12}$, 191
$MnO \cdot Al_2O_3 \cdot SiO_2$, 191
Spin–lattice relaxation, 79, 97, 101, 182, 183, 184, 188, 278, 279
Spin–orbit coupling, 135, 154
Spin polarisation, 131, 132
  inhomogeneous, 141, 175
  conduction electrons, 162
Spur model, 249–51
Static phonon Green's function, 61
Strain field, 72
  long range, 61
Stroboscopic μSR, 27–35
  gate function, 29, 31
  precision of, 34

Stroboscopic μSR (*contd.*)
  range of detectable precession frequencies, 33
  structure function, 30
Strong collision approximation (model), 83–5, 87–90, 187
Strontium (Sr), 142, 143
Substitutional impurities, 117
Superhyperfine (shf) interaction, 240–5, 252, 292
Supertransferred hf fields, 305, 306

Tantalum (Ta), 49, 63, 64, 66, 69, 105, 107, 155, 156, 205
T_1 process (spin–lattice relaxation), 278, 279
T_2 process (inhomogeneous line broadening), 278
Terminal spur, 249
Thermalisation of μ^+, 245–8
Thermal volume expansion effect on hf field, 176, 177
Thomas–Fermi equation, 288
Time differential μSR signal, 27, 28
Total volume expansion induced by protons, 65
Transient digitiser, 26
Transition metals, 130, 131, 139
Transverse relaxation time, 78, 83, 87, 96
Trapping, 87, 112, 114, 115, 116, 119, 121, 122
  capture probability, 87, 88
  centres, 78, 87, 118
  see also self-trapping
Tunnelling, 72, 110
  incoherent, 73–8, 115
  matrix element, 72–5, 77
  one-phonon assisted, 74
  phonon assisted, 73–8
  two-phonon assisted, 74, 75
Tunnelling configuration, 66
  second moment of, 67
Two-dimensional diffusion of Mu on surfaces, 293
Two-frequency precession, 254, 266
Two-state model, 113, 114
Type I, II superconductors, properties of, 203

Unrestricted Hartree–Fock cluster calculations, 290

Vacancies, 87
  cluster of, 87
  clustering of, 115
  on He surface film, 292
  migration of, 123
Valence fluctuations, 4, 161
Valence transition, 4, 157
Vanadium (V), 49, 63, 64, 66, 69, 105, 107, 155, 156, 210–4
VH_x, 106
V_2O_3, 310, 311
  first order magnetic phase transition in, 310
V_3Ga, 205
Van Hove singularities, 149, 150
Van Vleck paramagnetism, 130
Vibrational excitation energies of protons in Nb, 67
Vibrational motion, 78
Vortex lattice, 203, 208
Vortex lattice constant, 209

Wangsness–Bloch equations, 231, 281
Water (H_2O), 245, 249
Weak collision model (WCM), 188
Weak interaction, 1, 7
  Hamiltonian of, 8
  parity violation in, 7
  (V–A) theory of, 7
Wigner angle, 14
Wigner–Eckhart theorem, 243

X-rays (muonic, pionic), 2
Xenon (Xe), 245, 247–9

YFe_2, 167, 173, 174, 175

Zeeman interaction, 17, 36, 217, 218, 229
Zero-point motion
  of μ^+, 65, 141, 165, 176, 177
  of muonium, 289, 290
Zimmerman–Brittin theory, 90, 92
Zinc (Zn), 105, 107, 142, 143, 144, 153
ZrH_x, 106